MODELING LANGUAGES IN MATHEMATICAL OPTIMIZATION

Applied Optimization

Volume 88

MODELING LANGUAGES IN MATHEMATICAL OPTIMIZATION

Edited by

JOSEF KALLRATH

BASF AG, GVC/S (Scientific Computing), D-67056 Ludwigshafen, Germany
Dept. of Astronomy, Univ. of Florida, Gainesville, FL 32611

Kluwer Academic Publishers

Boston/Dordrecht/London

Distributors for North, Central and South America:
Kluwer Academic Publishers
101 Philip Drive
Assinippi Park
Norwell, Massachusetts 02061 USA
Telephone (781) 871-6600
Fax (781) 871-6528
E-Mail <kluwer@wkap.com>

Distributors for all other countries:
Kluwer Academic Publishers Group
Post Office Box 322
3300 AH Dordrecht, THE NETHERLANDS
Telephone 31 78 6576 000
Fax 31 78 6576 474
E-Mail <orderdept@wkap.nl>

Electronic Services <http://www.wkap.nl>

Library of Congress Cataloging-in-Publication

Kallrath, Josef
Modeling Languages in Mathematical Optimization
ISBN 1-4020-7547-2

Printed on acid-free paper.

Printed in the United States of America

Contents

List of Figures xiii

List of Tables xv

Preface xvii

Contributing Authors xxiii

Introduction xxix
Josef Kallrath

Part I Theoretical and Practical Concepts of Modeling Languages

1
Mathematical Optimization and the Role of Modeling Languages 3
Josef Kallrath
1.1 Mathematical Optimization 3
1.2 Classes of Problems in Mathematical Optimization 6
1.2.1 A Deterministic Standard MINLP Problem 7
1.2.2 Constraint Satisfaction Problems 9
1.2.3 Multi-Objective Optimization 9
1.2.4 Multi-Level Optimization 10
1.2.5 Semi-Infinite Programming 11
1.2.6 Optimization Involving Differential Equations 12
1.2.7 Safety Programming 12
1.2.8 Optimization Under Uncertainty 13
1.2.8.1 Approaches to Optimization Under Uncertainty 13
1.2.8.2 Stochastic Optimization 15
1.2.8.3 Beyond Stochastic Programming 16
1.3 The History of Modeling Languages in Optimization 17
1.4 Conventions and Abbreviations 24

2
Models and the History of Modeling 25
Hermann Schichl
2.1 The History of Modeling 25
2.2 Models 28
2.3 Mathematical Models 29
2.4 The Modeling Process 31
2.4.1 The Importance of Good Modeling Practice 32
2.4.2 Making Mathematical Models Accessible for Computers 35

3
Mathematical Model Building 37
Arnold Neumaier
3.1 Why Mathematical Modeling? 37
3.2 A List of Applications 38
3.3 Basic numerical tasks 39
3.4 The Modeling Diagram 40
3.5 General Rules 41
3.6 Conflicts 42
3.7 Attitudes 43

4
Theoretical Concepts and Design of Modeling Languages 45
Hermann Schichl
4.1 Modeling Languages 45
4.1.1 Algebraic Modeling Languages 46
4.1.2 Non-algebraic Modeling Languages 51
4.1.3 Integrated Modeling Environments 54
4.1.4 Model-Programming Languages 55
4.1.5 Other Modeling Tools 55
4.2 Global Optimization 56
4.2.1 Problem Description 57
4.2.2 Algebraic Modeling Languages and Global Optimization 58
4.3 A Vision — What the Future Needs to Bring 59
4.3.1 Data Handling 59
4.3.2 Solver Views 60
4.3.3 GUI 60
4.3.4 Object Oriented Modeling — Derived Models 61
4.3.5 Hierarchical Modeling 61
4.3.6 Building Blocks 62
4.3.7 Open Model Exchange Format 62

5
The Importance of Modeling Languages for Solving Real-World Problems 63
Josef Kallrath and *Josef Liesenfeld*
5.1 Modeling Languages and Real World Problems 64
5.2 Requirements from Practitioners towards Modeling Languages and Modeling Systems 65

Part II The Modeling Languages in Detail

6
The Modeling Language AIMMS 71
Johannes Bisschop and *Marcel Roelofs*
6.1 AIMMS Design Philosophy, Features and Benefits 72
6.2 AIMMS Outer Approximation (AOA) Algorithm 78
6.2.1 Problem Statement 79
6.2.2 Basic Algorithm 79
6.2.3 Open Solver Approach 81

6.2.4 Alternative Uses of the Open Approach 83
6.3 Units of Measurement 84
6.3.1 Unit Analysis 86
6.3.2 Unit-Based Scaling 87
6.3.3 Unit Conventions 87
6.4 Time-Based Modeling 88
6.4.1 Calendars 90
6.4.2 Horizons 90
6.4.3 Data Conversion of Time-Dependent Identifiers 92
6.5 The AIMMS Excel Interface 93
6.5.1 Excel as the Main Application 94
6.5.2 AIMMS as the Main Application 94
6.6 Multi-Agent Support 96
6.6.1 Basic Agent Concepts 96
6.6.2 Examples of Motivation 97
6.6.3 Agent-Related Concepts in AIMMS 99
6.6.4 Agent Construction Support 101
6.7 Future Developments 101
Appendix 103
A AIMMS Features Overview 103
A.1 Language Features 103
A.2 Mathematical Programming Features 103
A.3 End-User Interface Features 103
A.4 Connectivity and Deployment Features 104
B Application Examples 104

7
Design Principles and New Developments in the AMPL Modeling Language 105
Robert Fourer, David M. Gay and *Brian W. Kernighan*
7.1 Background and Early History 106
7.2 The McDonald's Diet Problem 107
7.3 The Airline Fleet Assignment Problem 111
7.4 Iterative Schemes 114
7.4.1 Flow of Control 114
7.4.2 Named Subproblems 115
7.4.3 Debugging 118
7.5 Other Types of Models 118
7.5.1 Piecewise-Linear Terms 119
7.5.2 Complementarity Problems 120
7.5.3 Combinatorial Optimization 123
7.5.4 Stochastic Programming 126
7.6 Communicating with Other Systems 128
7.6.1 Relational Database Access 128
7.6.2 Internet Optimization Services 129
7.6.3 Communication with Solvers via Suffixes 133
7.7 Updated AMPL Book 135
7.8 Concluding Remarks 135

8
General Algebraic Modeling System (GAMS) 137
Michael R. Bussieck & Alex Meeraus
8.1 Background and Motivation 138
8.2 Design Goals and Changing Focus 139
8.3 A User's View of Modeling Languages 141
8.3.1 Academic Research Models 141
8.3.2 Domain Expert Models 145
8.3.3 Black Box Models 147
8.4 Summary and Conclusion 149
Appendix 150
A Selected Language Features 150
B GAMS External Functions 151
C Secure Work Files 151
D GAMS versus FORTRAN Matrix Generators 152
E Sample GAMS Problem 153

9
The LINGO Algebraic Modeling Language 159
Kevin Cunningham and *Linus Schrage*
9.1 History 159
9.2 Design Philosophy 160
9.2.1 Simplified Syntax for Small Models 160
9.2.2 Close Coupled Solvers 162
9.2.3 Interface to Excel 163
9.2.4 Model Class Identification 166
9.2.5 Automatic Linearization and Global Optimization 167
9.2.6 Debugging Models 168
9.2.7 Programming Interface 170
9.3 Future Directions 171

10
The LPL Modeling Language 173
Tony Hürlimann
10.1 History 173
10.2 Some Basic Ideas 175
10.3 Highlights 176
10.4 The Cutting Stock Problem 178
10.5 Liquid Container 181
10.6 Model Documentation 182
10.7 Conclusion 183

11
The MINOPT Modeling Language 185
Carl A. Schweiger and *Christodoulos A. Floudas*
11.1 Introduction 186
11.1.1 Motivation 186
11.1.2 MINOPT Overview 188

11.2 Model Types and Solution Algorithms 189
11.2.1 Mixed-Integer Nonlinear Program (MINLP) 190
11.2.1.1 Generalized Benders Decomposition (*GBD*) 190
11.2.1.2 Outer Approximation/Equality Relaxation/ Augmented Penalty (*OA/ER/AP*) 192
11.2.2 Nonlinear Program with Differential and Algebraic Constraints (NLP/DAE) 195
11.2.3 Mixed-Integer Nonlinear Program with Differential and Algebraic Constraints (MINLP/DAE) 197
11.2.4 Optimal Control Problem (OCP) and Mixed Integer Optimal Control 200
11.2.5 External Solvers 204
11.3 Example Problems 204
11.3.1 Language Overview 205
11.3.2 MINLP Problem–Nonconvex Portfolio Optimization Problem 205
11.3.3 Optimal Control Problem–Dow Batch Reactor 207
11.4 Summary 209

12

Mosel: A Modular Environment for Modeling and Solving Optimization Problems 211

Yves Colombani, Bob Daniel & Susanne Heipcke

12.1 Introduction 211
12.1.1 Solver Modules 212
12.1.2 Other Modules 212
12.1.3 User Modules 213
12.1.4 Contents of this Chapter 213
12.2 The Mosel Language 213
12.2.1 Example Problem 214
12.2.2 Types and Data Structures 215
12.2.3 Initialization of Data/Data File Access 217
12.2.4 Language Constructs 218
12.2.4.1 Selections 218
12.2.4.2 Loops 219
12.2.5 Set Operations 219
12.2.6 Subroutines 220
12.3 Mosel Libraries 220
12.4 Mosel Modules 222
12.4.1 Available Modules 223
12.4.2 QP Example with Graphical Output 224
12.4.3 Example of a Solution Algorithm 225
12.5 Writing User Modules 230
12.5.1 Defining a New Subroutine 231
12.5.2 Creating a New Type 233
12.5.2.1 Module Context 234
12.5.2.2 Type Creation and Deletion 235
12.5.2.3 Type Transformation to and from String 235
12.5.2.4 Overloading of Arithmetic Operators 236
12.6 Summary 238

13
The MPL Modeling System 239
Bjarni Kristjansson & Denise Lee
13.1 Maximal Software and Its History 239
13.2 Algebraic Modeling Languages 241
13.2.1 Comparison of Modeling Languages 242
13.2.1.1 Modeling Language 243
13.2.1.2 Multiple Platforms 243
13.2.1.3 Open Design 243
13.2.1.4 Indexing 244
13.2.1.5 Scalability 244
13.2.1.6 Memory Management 244
13.2.1.7 Speed 245
13.2.1.8 Robustness 245
13.2.1.9 Deployment 245
13.2.1.10 Pricing 245
13.3 MPL Modeling System 245
13.3.1 MPL Integrated Model Development Environment 246
13.3.1.1 Solve the Model 247
13.3.1.2 View the Solution Results 248
13.3.1.3 Display Graph of the Matrix 249
13.3.1.4 Change Option Settings 250
13.4 MPL Modeling Language 250
13.4.1 Sparse Index and Data Handling 251
13.4.2 Scalability and Speed 252
13.4.3 Structure of the MPL Model File 252
13.4.3.1 Sample Model in MPL: A Production Planning Model 253
13.4.3.2 Going Through the Model File 254
13.4.4 Connecting to Databases 257
13.4.5 Reading Data from Text Files 259
13.4.6 Connecting to Excel Spreadsheets 259
13.4.7 Optimization Solvers Supported by MPL 260
13.5 Deployment into Applications 262
13.5.1 Deployment Phase: Creating End-User Applications 263
13.5.2 OptiMax 2000 Component Library Application Building Features 263

14
The Optimization Systems MPSX and OSL 267
Kurt Spielberg
14.1 Introduction 267
14.2 MPSX from its Origins to the Present 268
14.2.1 Initial Stages Leading to MPSX/370 268
14.2.2 The Role of the IBM Scientific Centers 270
14.2.3 An Important Product: Airline Crew Scheduling 271
14.2.4 MPSX Management in White Plains and Transition to Paris 272
14.2.5 A Major Growth Period in LP and MIP: MPSX/370; 1972-1985 272
14.2.6 MPSX as an Engine in Research and Applications 273

14.2.6.1 CASE A: Algorithmic Tools for Solving Difficult Models 273
14.2.6.2 CASE B: New Solver Programs with ECL 274
14.2.6.3 CASE C: Application Packages – Precursors to Modeling 275
14.2.7 Business Cases for MPSX 276
14.2.8 Changes in Computing, Development and Marketing Groups 277
14.2.9 Transition to OSL 277

15
The NOP-2 Modeling Language 279
Hermann Schichl and *Arnold Neumaier*
15.1 Introduction 279
15.2 Concepts 280
15.3 Specialties of NOP-2 282
15.3.1 Specifying Structure — The Element Concept 283
15.3.2 Data and Numbers 286
15.3.3 Sets and Lists 287
15.3.4 Matrices and Tensors 288
15.3.5 Stochastic and Multistage Programming 290
15.3.6 Recursive Modeling and Other Components 291
15.4 Conclusion 291

16
The OMNI Modeling System 293
Larry Haverly
16.1 OMNI Features as they Developed Historically 293
16.1.1 Early History 293
16.1.2 Activities Versus Equations 296
16.1.3 Recent and Current Trends 297
16.2 Omni Features to Meet Applications Needs 299
16.3 OMNI Example 300
16.4 Omni Features 305
16.5 Summary 306

17
The OPL Studio Modeling System 307
Pascal Van Hentenryck and *Laurent Michel* and *Frederic Paulin and Jean-Francois Puget*
17.1 Introduction 308
17.2 Overview of OPL 310
17.3 Overview of OPL Studio 311
17.4 Mathematical Programming 313
17.5 Frequency Allocation 314
17.6 Sport Scheduling 319
17.7 Job-Shop Scheduling 322
17.8 Scene Allocation 325
17.9 The Trolley Application 331
17.10 Visualization 338
17.11 Conclusion 343

Appendix: Advanced Models 344
A A Round-Robin Model for Sport-Scheduling 344
B The Complete Trolley Model 347

18
PCOMP: A Modeling Language for Nonlinear Programs with Automatic Differentiation 351
Klaus Schittkowski
18.1 Introduction 351
18.2 Automatic Differentiation 353
18.3 The PCOMP Language 356
18.4 Program Organization 360
18.5 Case Study: Interactive Data Fitting with EASY-FIT 361
18.6 Summary 367

19
THE TOMLAB OPTIMIZATION ENVIRONMENT 369
Kenneth Holmström and *Marcus M. Edvall*
19.1 Introduction 369
19.2 MATLAB as a Modeling Language 370
19.3 The TOMLAB Development 371
19.4 The Design of TOMLAB 372
19.4.1 Structure Input and Output 372
19.4.2 Description of the Input Problem Structure 372
19.4.3 Defining an Optimization Problem 374
19.4.4 Solving Optimization Problems 375
19.5 A Nonlinear Programming Example 376

Part III The Future of Modeling Systems

20
The Future of Modeling Languages and Modeling Systems 379
Josef Kallrath

References 383

Index 403

List of Figures

2.4.1	Modeling cycle	31
2.4.2	A typical model for resource optimization	34
2.4.3	Detailed modeling cycle	35
3.4.1	Modeling Diagram	40
4.3.1	Optimization model displayed in a GUI front-end	61
6.1.1	Graphical overview of AIMMS components	73
6.1.2	Example of a model tree	74
6.1.3	Declaration attribute window	75
6.1.4	Key benefits of AIMMS from different perspectives	78
6.2.5	A non-convex linearization cutting off part of the feasible region	81
6.2.6	Solver communication illustrated	82
6.4.7	Timetable relating calendar and horizon	89
6.5.8	The AIMMS Interface Setup dialog box	95
6.6.9	Communication between agents	97
6.6.10	Messages with only input arguments	100
8.3.1	The orginial GAMS model for trim loss optimization	143
8.3.2	The Scalar GAMS and AMPL versions of the trim loss model	144
8.3.3	Scenario Management Tool: VEDA	147
8.3.3	Geographical Information with MapInfo	147
8.3.3	Visualization with MATLAB	147
9.2.1	OLE in LINGO	165
11.2.1	Schematic flowsheet for the MINLP/DAE algorithm	201
12.2.1	IVE display after model execution	216
12.4.2	Graph of the efficient frontier	226
12.4.3	Solution of the Fiveleaper problem	230
12.5.4	Module definition with IVE	231
13.3.1	The Status Window	247
13.3.2	The MPL Error Message Window	248

13.3.3	Viewing the Solution File in a Window	248
13.3.4	Model Definition Window	249
13.3.5	The Graph of Matrix Window	250
13.3.6	CPLEX Parameter Options Dialog	251
13.4.7	MPL Database Connection	258
13.5.8	MPL OptiMax Design Overview	264
13.5.9	OptiMax Object Model	266
17.4.1	A MIP Model for Capacitated Warehouse Location	314
17.5.2	The Frequency-Allocation Problem (`alloc.mod`)	315
17.5.3	Instance Data for the Frequency-Allocation Problem	317
17.6.4	A Solution to the Sport-Scheduling Application with 8 Teams	319
17.6.5	A Simple Model for the Sport-Scheduling Model	320
17.7.6	A Job-Shop Scheduling Model (`jobshop.mod`)	323
17.8.7	The Scene Allocation Model	326
17.8.8	Instance Data for the Scene-Allocation Problem	327
17.8.9	The Scene-Allocation Model with Symmetry Removal	329
17.8.10	The Scene-Allocation Model with Redundant Modeling	330
17.9.11	The Trolley Problem: Part I	333
17.9.12	The Trolley Problem: Part II	334
17.9.13	The Trolley Problem: the Instance Data	334
17.9.14	A Script for the Trolley Problem (`trolley.osc`)	336
17.9.15	C++ Code for the Trolley Problem	338
17.10.16	Gantt chart visualization for a scheduling problem	339
17.10.17	Search Tree Visualization with Five Collapsed Subtrees	340
17.10.18	The Choice Stack and the Corresponding n-ary Search Tree	341
17.10.19	The Propagation Spy for the Pheasants and Rabbits Problem	342
17.10.20	Concept Mapping between a Traditional and a CP Debugger	343
17.10.21	Christmas Tree	344
17.A.1	A Round-Robin Model for Sport-Scheduling: Part I	345
17.A.2	A Round-Robin Model for Sport-Scheduling: Part II	346
17.A.3	Experimental Results for the Sport-Scheduling Model	346
17.B.4	The Final Trolley Model: Part I	349
17.B.5	The Final Trolley Model: Part II	350
18.5.1	Experimental Data	366
18.5.2	Model Structure and Discretization	367
18.5.3	Dynamical Equations	368
18.5.4	Prepare Data Fitting	368
18.5.5	Interpretation of Results	368
19.4.1	Chain call for least squares problem in TOMLAB.	375

List of Tables

6.1	Basic SI quantities and their base units.	85
6.2	Selected derived SI quantities and their base units.	85
6.3	Conversion rules for "during" quantities.	92
6.4	Conversion rules for interpolated data.	93
7.1	McDonald's Diet Problem foods and nutrients.	107
7.2	McDonald's Diet Problem data.	108
7.3	Concrete McDonald's Diet Problem.	108
7.4	Abstract model for the McDonald's Diet Problem.	109
7.5	Diet Model in AMPL (`mcdiet.mod`).	109
7.6	AMPL data statements for McDonald's Diet Problem (`mcdiet1.dat`).	110
7.7	Continuous-variable solution of McDonald's Diet Problem.	110
7.8	Integer solution of McDonald's Diet Problem.	111
7.9	McDonald's Diet for 63 foods, 12 nutrients.	111
7.10	Airline Fleet Assignment Model, part 1 — column-wise specification of network flow costs and balance constraints.	113
7.11	Airline Fleet Assignment Model, part 2 — row-wise specification of flight coverage and fleet size limitations.	114
7.12	Simple script for sensitivity analysis.	115
7.13	Output from the sensitivity analysis script.	116
7.14	Model for cutting optimization with given patterns.	116
7.15	Model for new pattern generation.	117
7.16	Script for cutting-stock problem declarations and initialization.	117
7.17	Script switching problems for cutting-stock optimization.	117
7.18	Using single-step mode with the cutting-stock script.	118
7.19	Expanding the cutting-stock constraints.	119
7.20	Nonlinear objective in Kallrath's problem.	120
7.21	Piecewise-linear objective in Kallrath's problem.	121
7.22	Production cost minimization as an equilibrium model.	122
7.23	Bounded-variable cost minimization as an equilibrium model.	122

7.24 Economic equilibrium with price-dependent demands. 123

7.25 Assigning Distributions to Random Variables. 127

7.26 Reading and writing database tables for the diet problem. 129

7.27 Problem types accepted by the NEOS Server. 130

7.28 Kestrel example, part 1: sending a problem instance to a remote solver. 130

7.29 Kestrel example, part 2: receiving the solver listing and displaying results from the remote solver. 131

7.30 Kestrel parallel processing. 132

7.31 Retrieval of auxiliary solver information via solver-defined suffixes. 134

11.1 Notation for variable and constraint used in the algorithm descriptions. 189

11.2 External solvers used by MINOPT. 204

15.1 Predefined elements with one dimensional result. 285

15.2 Elements having higher dimensional results or involving matrices and tensors. 286

18.1 Work Ratios for Forward, Reverse, and Numerical Differentiation. 356

19.1 Model information in the TOMLAB *Prob* structure. 373

19.2 Fields defining sub-structures in *Prob*. 373

19.3 Problems and assignments in TOMLAB. 374

Preface

This book deals with the aspects of modeling and solving real-world optimization problems in a unique combination. It treats systematically the major modeling languages and modeling systems used to solve mathematical optimization problems. The book is an offspring of the 71st Meeting of the GOR (Gesellschaft für Operations Research) Working Group *Mathematical Optimization in Real Life* which was held under the title *Modeling Languages in Mathematical Optimization* during April 23–25, 2003 in the German Physics Society Conference Building in Bad Honnef, Germany. The modeling language providers

AIMMS	Johannes Bisschop, Paragon Decision Technology B.V, Haarlem, The Netherlands,
AMPL	Bob Fourer, Northwestern Univ.; David M. Gay, AMPL Optimization LLC., NJ,
GAMS	Alexander Meeraus, GAMS Development Corporation, Washington D.C.,
Mosel	Bob Daniel, Dash Optimization, Blisworth, UK,
MPL	Bjarni Kristjansson, Maximal Software, Arlington, VA,
NOP-2	Hermann Schichl, Vienna University, Austria,
PCOMP	Klaus Schittkowski, Bayreuth University, Germany, and
OPL	Sofiane Oussedik, ILOG Inc., Paris, France

gave deep insight into their motivations and conceptual design features of their software, highlighted their advantages but also critically discussed their limits. The participants benefited greatly from this symposium which gave a useful overview and orientation on today's modeling languages in optimization.

Roughly speaking, a modeling language serves the need to pass data and a mathematical model description to a solver in the same way that people, especially mathematicians describe those problems to each other. Of course, in reality this is not done in exactly the same way, but the resemblance has to be close enough to spare the user any significant translation. As in this book we focus on modeling languages used in mathematical optimization let us give an example from that discipline. When practitioners or researchers describe large-scale linear programs to each other, they use summations and subscripting. To give a negative definition first: Probably one would not consider a language lacking these features to be a modeling language for large-scale *Linear Programming*. This can be turned into a positive definition: A modeling language in mathematical optimization needs to support the expressions and symbols used in the mathematical optimization community. Therefore, it is natural that algebraic modeling languages support the concepts of data, variables, constraints and objective functions. Those entities are not only connected by algebraic operations (+,-,$\cdot$) but also by nonlinear functions. Algebraic models are embedded in a larger class of differential-algebraic models including ordinary or partial differential equations. They may also be extended towards relationships appearing in constraint programming. Examples are membership relations or all-different constraints.

The earliest algebraic modeling languages appeared in the late 1970's and early 1980's. They were already very useful, supporting analysts in the input of their problems to solvers. In the middle of the 1980's, when, for instance, AMPL, GAMS, MPL and mp-model appeared, the software developers were already trying to improve on previous designs, by taking advantage of faster computers and better computing environments. Some modeling languages supporting differential equations in addition to algebraic terms have their roots in the process industry. Examples are MINOPT discussed in Chapter 11, PCOMP described in detail in Chapter 18, and gPROMS.

While a more precise definition of modeling languages is given in Chapter 4, it is appropriate at this place to focus for a moment on the terms *modeling language* and *modeling systems*. Some people use these expressions synonymously, and this is also reflected in the acronyms AMPL and GAMS, where the former translates into *A Mathematical Programming Language* while the latter stands for *General Algebraic Modeling System*. In this book we rather keep the following meaning. In its purest sense a modeling language in mathematical optimization is a means to give a declarative representation[1] of an optimization problem; AMPL, GAMS and mp-model are good examples of this. A modeling system is rather a complete support system to solve real-world problems. It usually contains a modeling language but offers many other features supporting the solution process, *e.g.*, passing commands to the solver, analyzing infeasibilities, incorporating the solver's output back into the model or its next steps, visualizing the branch&bound tree or the structure of the matrix; many more features could be listed. AIMMS, Mosel, MPL and OPL Studio are good examples of modeling systems. Of course, one might argue that the existing modeling languages are usually somewhat in between those extreme definitions, *e.g.*, the early versions of AMPL was almost completely declarative, while GAMS from the beginning had also procedural features supporting, for instance, to input the output of one model into subsequent models.

The book is aimed at researchers of mathematical programming, scientists in various disciplines who use optimization methods to model and solve problems, operations research practitioners, supply chain management consultants, and decision makers in the area of tool selection for optimization tasks as well as students and graduates in mathematics, physics, operations research, and businesses with interest in modeling and solving real optimization problems. Often application software has implemented an optimization model without an

[1] As the book, in the sense of this definition, focuses on algebraic declarative modeling languages, the reader should not be surprised to find not too much on mathematical modeling systems such as Mathematica, MathCad or MATLAB. These systems are conceptionally rather procedural tools.

algebraic modeling language. The people responsible for maintaining or further developing such applications might be looking for improvements to put their software on a safer footing. They will definitely benefit from this book. Assuming some background in mathematics and optimization or at least a certain willingness to acquire the skills necessary to understand the described algorithms and models, this book provides a sound overview on model formulation and solving as well as implementation of solvers in various modeling language packages. It demonstrates the strengths and characteristic features of such languages. May it provide a bridge for researchers, practitioners and students into a new world of excitement: solving real-world optimization problems with the most advanced modeling systems.

Structure of this Book

This book benefits from contributions of experienced practitioners and developers of modeling languages. The languages presented during the symposium as well as a few others such as `LINGO` and `MINOPT` are described in great detail in chapters of their own. Readers with some inclination towards the history of the field will also find the chapters on MPSX and `OMNI` interesting.

The first part of the book contains introductory material. Chapter 1 starts with an introduction into mathematical optimization including a brief overview on classes of optimization problems. The main entities in optimization — variables, constraints, and the objective function — are explained. The chapter continues with a description of the various classes of problems typically appearing in applications, and concludes with a brief overview of the history of modeling languages. Chapter 2 deals with models, mathematical models, and the history of modeling, and Chapter 3 describes the modeling process itself. Chapter 4 discusses theoretical aspects and conceptual design features of modeling languages, while Chapter 5 focuses on the importance algebraic modeling languages have for solving real-world problems.

The second and main part of the book contains 14 chapters, 6 to 18, covering the modeling languages, `AIMMS`, `AMPL`, `GAMS`, `LINGO`, `LPL`, `MINOPT`, `Mosel`, `MPL`, `NOP-2`, `OMNI`, `OPL Studio`, `PCOMP`, and `TOMLAB` in great detail. Those chapters reflect the personal views and focus of the authors. It is left to the reader to compare the languages and see which one serves his personal needs best. The `AIMMS` modeling language is due to its object-oriented design quite different from all others and provides agent-based simulation techniques. `AMPL` has a dominant position in universities, is often used for prototyping and has many innovative ideas in its pipeline. The `GAMS` contribution focuses on the purposes of modeling languages with respect to different group of models or customers, resp., and describes a few special features of this modeling language.

The LINGO developers have introduced general global optimization algorithms which nonconvex models into a number of smaller convex models and then use a B&B manager to find the global optimal point over the entire variable domain. NOP-2 is entirely focused on global optimization. Tony Hürlimann's chapter on LPL provides additional conceptual insights and some features not found in any other language. The chapter on MPL represents its strength on computer science and covers topics such as speed, scalabilty, data management and deployment aspects of optimization. More on language elements related to constraint programming is found in the OPL Studio chapter. A major novelty in a modeling language is Mosel's concept of *modularity*, which makes it possible to extend this language according to one's needs and to provide new functionality, in particular to deal with other types of problems and solvers. The MINOPT and PCOMP chapters demonstrate how to use models involving differential equations in modeling languages. How to connect optimization models to a mathematical modeling systems such as MATLAB is shown in the TOMLAB chapter. The chapters on MPSX and OMNI bridge to the time before real modeling languages appeared.

A summary of the findings and some thoughts about the future of modeling systems are given in Chapter 20, which establishes the third part of the book.

By showing the strengths and characteristic features we hope to give novices and practitioners in mathematical optimization, supply chain management, financial industry, and other areas of industry a useful overview. It may help decision makers to select the best modeling system satisfying their needs.

Acknowledgements

Thanks is addressed at first to the contributing authors. It was a pleasure to work with such a group of distinguished individuals who formed the field and manifested their ideas into the modeling languages. The book benefited greatly from the fact that most of them contributed in time and used a clear LaTeX so that most of the editorial time was free to be spent on content and consistency. The book has been proof-read by Dr. Reuben Brown (BASF-AG, Ludwigshafen), Dr. Prof. Dr. Jürgen Garloff and Andrew P. Smith (FH Konstanz). Johannes Bisschop, Bjarni Kristjansson, Susanne Heipcke and Herman Schichl gave valuable feedback and checked selected chapters. This book has been produced with *Scientific Workplace* kindly provided by Additive GmbH (Wiesbaden, www.additive-net.de). The publisher and his competent personnel (Angela Qulici, Deborah Doherty, and Amy Hendrickson) were very helpful while producing this book. Last but not least I would like to thank Julia for all her patience while I was working on the book.

Ludwigshafen, July 2003 Josef Kallrath

to my wife, Julia, and
daughter, Diana

Contributing Authors

Johannes Bisschop is president of Paragon Decision Technology in Haarlem, The Netherlands, and has a part-time position as full professor in the Applied Mathematics Department at the Technical University of Twente. His main interests are in the areas of computational optimization and modeling. His main contribution has been towards the development of the AIMMS modeling system.

Michael Bussieck is a senior research analyst at GAMS Development Corporation which he joined in 1998 after receiving his PhD from Technical University Braunschweig, Germany. He has lectured at George Washington University, U.S.A. teaching students the art of modeling and theoretic concepts of mathematical programming. His focus is on optimization applications to problems in science and engineering.

Yves Colombani is a computer scientist from Marseille (France). For his PhD, awarded in 1997, he worked on scheduling problems and implemented a constraint solver on intervals of integers. He then joined Dash Optimization. Yves has entirely designed and implemented Mosel and continues the development of its core functionality.

Kevin Cunningham implemented the first version of the `LINGO` modeling language and the first version of the What'sBest spreadsheet solver (the first version was on Visicalc.) He was co-developer of the LINDO LP/IP solver. He developed the first version of the algebraic formula management system that is central to the recently released Global Optimization capability in `LINGO`.

Bob Daniel is a principal in Dash Optimization which he co-founded in 1983. He has a long experience in modeling and solving real-world problems. He has lectured in Operations Research and Management Science at the Universities of Edinburgh and Buckingham (UK), and he finally retired from Buckingham as Professor of Management Science in 1998, where he remains an Honorary Professor.

Marcus M. Edvall is a Business Development Manager at Tomlab Optimization Inc., the leading provider of optimization software for Matlab. His primary experience is in model predictive control (MPC) applications and software solutions for the pulp and paper industry.

Chris A. Floudas is Professor of Chemical Engineering at Princeton University, Associated Faculty in the Program of Computational and Applied Mathematics at Princeton University, and Associated Faculty in the Department of Operations Research and Financial Engineering at Princeton University. He earned his Ph.D. in December 1985 at Carnegie Mellon University. His research interests lie at the interface of chemical engineering, applied mathematics, and operations research, with principal areas of focus including chemical process synthesis and design, process control and operations, discrete-continuous nonlinear optimization, local and global optimization, and computational chemistry and molecular biology. Among over 160 refereed publications are two graduate text books: Nonlinear Mixed-Integer Optimization (Oxford University Press, 1995), and Deterministic Global Optimization (Kluwer Academic Publishers, 2000). Besides other numerous awards for teaching and research, in 2001 he received that the 2001 AIChE Professional Progress Award for Outstanding Progress in Chemical Engineering.

Larry (C.A.) Haverly graduated from Ohio State University with a Bachelors of Chemical Engineering and Masters of Business Administration and has completed advanced courses in Operations Research. He has lectured, given many talks and written some papers but mostly has been active in Linear Programming, and related applications. He worked at Merck and then at Esso in the early days of the application of LP to petroleum refineries. He formed Haverly Systems Inc. (HSI) in 1962 and has been president ever since. Refineries have continued to be their major area of applications. His son, Victor, is a Vice President of HSI and has a Bachelors Degree in Engineering from Stanford and a Masters Degree in Computer Science from the University of California and expects to carry on HSI. Larry was the inventor of `MaGen` (the first modeling language) and later `OMNI`, has developed optimizer codes and been involved in many applications. At 79 Larry still works more than full time and says it is more fun than being retired and the optimization field remains challenging and exciting. His primary efforts these days are in the area of refinery scheduling and scheduling optimization (as contrasted to planning optimization).

Susanne Heipcke worked for BASF-AG (Germany) before joining Dash Optimization in 1998. Her PhD research (1999) focussed on the solution of large-scale industrial problems by the combination of constraint programming and mixed integer programming. More recently she has been involved in the development of teaching material for Mosel, culminating in the book *Applications of Optimization with Xpress-MP* published in September 2002.

Pascal Van Hentenryck is professor of computer science at Brown University. His book, Constraint Satisfaction in Logic Programming (MIT Press), pioneered constraint programming over finite domains, the core of all modern constraint programming systems. He is recipient of an NSF NYI Award and the 2002 INFORMS ICS Award, the author of 3 books, and a main designer/implementor of the software systems CHIP, Newton, Numerica, OPL, and Comet.

Kenneth Holmström is professor of optimization at Mälardalen University, Sweden, and founder and president of Tomlab Optimization in Sweden (1986) and United States (2001). He formed the Center for Mathematical Modeling at Mälardalen University in 1996, and has been doing algorithm development in nonlinear parameter estimation and global optimization, as well as modeling and optimization research in many application areas like exponential fitting, computational finance, energy optimization, system identification and industrial plant control. He has written about 70 papers in applied optimization and numerical analysis, developed several software packages, the most well-known is Tomlab, and been working as a consultant in more than 100 projects. For 3.5 years he worked with distributed computer control systems for ABB Automation in Västerås, Sweden. He is currently optimization expert in a European Committee project for the pulp and paper industry.

Tony Hürlimann is a researcher and a lecturer at the Department of Informatics at the University of Fribourg, Switzerland. He teaches algorithms and theoretical concepts of programming languages, as well as mathematics for economists. The main focus of his research is to conceptualize and implement a mathematical modeling language, called LPL, which is in productive use in several companies in Switzerland. He also works in various ongoing research projects (Campus Virtual – electronic online textbooks in modeling; Assumption Based Reasoning – Uncertainty in modeling; Food Shortage Planning in Switzerland – Large mathematical models).

Josef Kallrath obtained his PhD in astrophysics from Bonn University in 1989, is professor at the University of Florida (Gainesville, FL), and solves real-world problems in industry using the broad spectrum of methods in scientific computing reaching from modeling physical systems to supporting decisions processes by mathematical optimization. He holds a lectureship at Heidelberg University teaching students mathematical modeling and solving real-world problems. He has written several review articles on this subject, about 60 research papers in astronomy and applied mathematics and several books on mixed integer optimizing and eclipsing binary stars.

Bjarni Kristjansson is the president and founder of Maximal Software, Inc. He attended the University of Iceland where he obtained a BS degree in Computer Science. In 1987, he formed Maximal Software, which specializes in making software for optimization modeling, and is the developer of the MPL Modeling System and the OptiMax 2000 Component Library. His research and development interests have always centered around optimization modeling. In the early years this included language design for the MPL language, implementing Graphical User Interfaces for effective model development and how to connect relational databases to optimization models. In the recent years, implementation of component libraries for optimization have become more important, as well as creating distributed applications over the Internet with optimization services embedded in the background.

Josef Liesenfeld is the managing directory of MaBOS GmbH, a company providing consultancy to industry in the area of mathematical modeling and optimization in particular.

Alexander Meeraus is a principal at GAMS Development Corporation which he founded in 1987 after having worked for 16 years at the World Bank's research complex in Washington, DC. He has wide interests and experience in the application of optimization techniques to problems in science, engineering and government.

Laurent Michel received his Ph.D.degree in Computer Science from Brown University in 1999. He currently is an Assistant Professor in the Computer Science and Engineering Department at the University of Connecticut. His research interests span programming languages, constraint programming, artificial intelligence and non-linear programming. He especially focuses on the design and implementation of languages for combinatorial optimization with contributions to systems like Newton, Numerica, Localizer, OPL and more recently Comet.

Arnold Neumaier is Professor for Computational Mathematics at the University of Vienna (Austria). He has published three books and over 100 research articles in pure and applied mathematics, optimization, statistics, and physics. He maintains the 'Global (and Local) Optimization' website, the most comprehensive collection of links to online information, software and test problems on global optimization; in addition, extensive web pages on public domain software for numerical analysis, optimization, and statistics. His research group is active in developing state of the art software for global optimization by integrating methods from nonlinear programming, interval analysis, constraint programming and combinatorial optimization, and an associated modeling language.

Frédéric Paulin received his engineer diploma from ENSEEIHT in 1989 with a specialization in artificial intelligence. He is project manager of ILOG OPL Studio since September 2000.

Jean-Francois Puget is a former student of Ecole Normale Superieure at Paris, and received his PhD from Université Paris XI in 1990. Puget joined ILOG in 1989 where he developed ILOG Solver, one of the prominent Constraint Programming software. He is now Vice President of Optimization R&D and manages the development of all ILOG optimization products, including ILOG CPLEX, ILOG Solver and ILOG OPL Studio among others.

Marcel Roelofs received his Ph.D. on the application of Computer Algebra in Mathematical Physics in 1993, and subsequently worked as a post-doc at the Center for Mathematics and Computer Science in Amsterdam. In 1995 he became technical director of Paragon Decision Technology in Haarlem, The Netherlands. His main responsibilities have been the design and documentation of the `AIMMS` 3.x language and interface.

Hermann Schichl is Assistant Professor for Computational Mathematics at the University of Vienna (Austria). His PhD thesis (1998) is on infinite dimensional differential geometry, and he has since moved his interest towards deterministic global optimization and rigorous computing. He is the leading designer of the algorithmic system developed in the COCONUT (IST-2000-26063) EU-project, which combines solving methods from nonlinear programming, interval analysis, constraint programming and combinatorial optimization, and he has co-developed the `NOP-2` modeling language.

Klaus Schittkowski is Professor for Applied Computer Science at the Department of Mathematics of the University of Bayreuth, Germany. His main research areas are nonlinear programming, especially SQP algorithms, data fitting in dynamical systems, automatic differentiation, and related topics. His special interest is the implementation, testing and performance evaluation of numerical software, and especially the practical usage and application of these codes. Among the 100 papers published, there are many based on long-term industrial cooperations. He is author and co-author of several books, editor of the journal *Optimization Methods and Software*, and organizer of about 15 research seminars, workshops, and tutorials on nonlinear optimization, mechanical structural optimization, topology optimization, and modeling of pharmaceutical systems.

Linus Schrage has been involved in the development of optimization software since the late 1970's. He was one of the co-developers of the LINDO LP/IP solver, the What'sBest spreadsheet optimizer, and the `LINGO` modeling system. He has taught in Operations Research and Operations Management at the Graduate School of Business, University of Chicago during the same period of time.

Carl Schweiger is a Senior Research Engineer at Pavilion Technologies in Austin, TX. His work focuses on the application of modeling and optimization technology to solve problems arising in the areas of planning, scheduling, and process control. His research is in the area of decomposition and hybrid optimization techniques for solving large-scale mixed-integer linear and mixed-integer nonlinear programming problems. The scope of his work ranges from the development of optimization algorithms to the design of the software to implement the algorithms to the application of the algorithms to solve client problems.

Kurt Spielberg is currently under part-time contract at IBM; B.S. City College 1951; Ward Medal in Physics, Varsity Letter Soccer; Awarded Higgins Fellowship at Columbia 1951; Ph. D. Physics Vienna 1956; Asst. Professor (Tenure) of Physics at CCNY. Joined IBM in 1958: initial projects: Chebyshev and Rational Approximations of Functions (publications in Numerische Mathematik): in charge of the first FORTRAN Function Package.

Consultant at the IBM New York Scientific Center: First "Global Weather Forecasting System on STRETCH", with US Weather Bureau. Integer Programming. One year temporary Fellow IBM Research, 1971. Manager of Management Science at Philadelphia Scientific Center: (details in Chapter 14). Nominated IBM FELLOW by two Scientific Center Managers. Scientific Marketing White Plains 1973-1987 (MPSX). Associate Editor MP in first 5-7 years (ed. Balinsky). 1987-90 Philadelphia Scientific Marketing. Retired 1990. IBM contract (MPSX,OSL) 1990 to present. Still enjoying new research in (0,1) MIP.

Introduction

Josef Kallrath

For many, let me call them the functionalists, a modeling language is simply a tool for passing the problem and its mathematical representation to the computer and for making it available to a solver. Their view is that there is not too much difference in using this or another modeling language. However, can this view survive a deeper analysis?

There are people who think similarly that the human languages are just for communicating with each other. Probably, there is great consensus that a language should support communication. But is this really all? Is there not much more? For many, the spoken languages, English, French, Chinese, Russian, etc. but also classical languages like Greek and Latin reflect a cultural and philosophical approach to the world. Certain concepts, especially in philosophy, cannot be easily, if at all, mapped from one language to another one. Some native inhabitants of North America used languages which did not support the tenses past, present, and future.

It is agreed upon by many people that modeling real-world problems is rather an art than a science. Of course, there are scientific elements involved and certain rules should be observed but there still is some consensus that some models are elegant and beautiful while others are considered aesthetically lacking, even if they are not considered nasty or cumbersome.

There are model formulations which perfectly reflect the structure of a problem, and there are modeling languages which are more appropriate to implement these model formulations than others. So, in the ideal case a problem formulated in a modeling language reflects the inner structure of that problem. But this in turn means that a modeling language has an inner dimension which may be more or less appropriate to a problem. A modeling language and its conceptual design is probably more akin a composer's method of composing to produce a symphony. Some modeling languages are so different from each other such that mapping a model written in one of them to another one is as difficult or impossible as mapping a Beethoven Symphony to Stravinsky's style. The reader may or may not agree with this comparison but there might be at least consensus that a modeling language can be appealing or repulsive to one's own personal taste and style.

It has been said that modeling is rather an art than a science, but the history of both, art and science, shows that personal taste and style is definitely relevant for success. Models, their formulation and implementation in a modeling language are subject to aesthetics. History also shows that the tools available decide which problems can be attacked and solved successfully. Most companies have only one or two modeling languages available. This might be appropriate or not. But maybe the words in this introduction will initiate some process to reflect on the connection of problem solvability and available tools.

Thus, as a contributing author to this book and also as the editor, I hope that a broad class of readers with different personal tastes and styles may find in this book many aspects and areas appealing to them, inspiring new ideas, helping to solve challenging real-world problems whose solution might put decisions on a safe rational fundament and contribute to a somewhat better world.

I

THEORETICAL AND PRACTICAL CONCEPTS OF MODELING LANGUAGES

Chapter 1

MATHEMATICAL OPTIMIZATION AND THE ROLE OF MODELING LANGUAGES

Josef Kallrath

BASF-AG, GVC/S-B009 (Scientific Computing) & Dept. of Astronomy, University of Florida

D-67056 Ludwigshafen, Germany *Gainesville, FL 32611*

josef.kallrath@t-online.de kallrath@astro.ufl.edu

Abstract We give an introduction into mathematical optimization including a brief overview on classes of optimization problems. The main entities in optimization — variables, constraints, and the objective function — are explained. The chapter continues with a description of the various classes of problems typically appearing in applications, and concludes with a brief overview on the history of modeling languages.

Keywords: Models, mathematical optimization, variables, constraints, objective function, classes of optimization problems, Stochastic Optimization, history of modeling languages

1.1 Mathematical Optimization

This book is about modeling languages in mathematical optimization. Thus, it seems appropriate to explain briefly what mathematical optimization is and what optimization problems are. In an optimization problem, one tries to minimize or maximize a quantity associated with a decision process such as elapsed time or cost, by exploiting available degrees of freedom under a set of restrictions (constraints). Optimization problems arise in almost all branches of industry, *e.g.*, in product and process design, production, logistics, and even strategic planning (see the list below). While the word *optimization*, in nontechnical language, is often used in the sense of *improving*, the mathematical optimization community sticks to the original meaning of the word related to finding the *best value* either globally or at least in a local neighborhood. Except for very simple cases, optimization problems cannot be solved by *simulation* [also called *parameter studies*], *i.e.*, by simulating the processes under investigation, evaluating the objective function and comparing the results. This is because the inspection of a finite number of points or scenarios can never generate a complete knowledge which proves a solution to be optimal. Nevertheless, since experts of simulation techniques have developed intuition and heuristics to select appropriate scenarios to be evaluated, and simulation software exists to perform their evaluation. Simulation may lead to reasonable results, but there is no guarantee that the optimal solution or even a solution close to the optimum will be found. This is especially troublesome for complex problems, or those which require decisions of large financial impact.

What do we need when we want to solve a real-world problem by mathematical optimization? The first thing we need to do is to represent our problem by a *mathematical model*, that is, a set of mathematical relationships (*e.g.*, equalities, inequalities, logical conditions) which represent an abstraction of our real-world problem. This translation is part of the model building phase (which is part of the whole modeling process outlined in more detail in Section 2.4 and Chapter 3), and is not trivial at all. Usually, a model in mathematical optimization consists of four key objects:

- data [also called the *constants* of a model],
- variables (continuous, semi-continuous, binary, integer) [also called decision variables or *parameters*],
- constraints (equalities, inequalities) [sometimes also called *restrictions*], and
- the objective function.

The data may represent cost or demands, fixed operation conditions of a reactor, capacities of plants and so on. The variables represent the degrees of freedom, *i.e.*, what we want to decide: How much of a certain product is to be produced, whether a depot is closed or not, or how much material will we store in the inventory for later use. Classical optimization (calculus, variational calculus, optimal control) treats those cases in which the variables represent continuous degrees of freedom, *e.g.*, the temperature in a chemical reactor or the amount of a product to be produced. Mixed integer, combinatorial, or discrete optimization involve variables restricted to integer values, for example counts (numbers of containers, ships), decisions (yes-no), or logical relations (if product A is produced then product B also needs to be produced). The constraints can be a wide range of mathematical relationships: algebraic, analytic, differential or integral. They may represent mass balances, quality relations, capacity limits, and so on. The objective function expresses our goal: minimize costs, maximize utilization rate, minimize waste, and so on. Mathematical models for optimization usually lead to structured problems such as:

- *linear programming* (LP) problems,
- *mixed integer linear programming* (MILP) problems,
- *nonlinear programming* (NLP) problems, and
- *mixed integer nonlinear programming* (MINLP) problems.

In addition, although not strictly belonging to the mathematical programming community, *constraint programming* (CP) shows increasing popularity and can be used to solve certain problems, such as scheduling problems, more efficiently.

Below we list some areas in which applications of (linear and nonlinear) mixed integer optimization are found. This list is by no means complete but it illustrates the wide range of possible applications:

application area / problem type	problem class
production planning (production, logistics, marketing)	MILP, MINLP
sequencing problems (putting activities into order)	MILP
scheduling (planning activities subject to resource limits)	CP
allocation problems (resources to orders, people to tasks)	MILP, CP
distribution and logistics (supply chain optimization)	MILP
blending problems (production and logistics)	LP, MILP, NLP, MINLP
refinery planning and scheduling	NLP, MINLP
process design (chemical and food industry, refineries)	MINLP
engineering design (all areas of engineering)	NLP, MINLP
market clearing problems (energy markets)	LP, MILP, NLP
portfolio optimization (production, energy industry, finance)	MILP, MINLP
selection and depot location problems (strategic planning)	MILP
investment and de-investment problems (strategic planning)	MILP
network design (planning, strategic planning)	MILP, MINLP
financial problems (strategic planning)	MILP, MINLP

In the model building phase, real-world optimization problems are structured according to their basic objects, variables, objective function and constraints. Then we are faced with the problem of how to put such a model into the computer. In the literature we find several approaches to achieve that goal:

- *Unstructured approaches* in which a programming language such as FORTRAN, C, or C++ is used to combine the model and the data and either produce a matrix or other input file for a commercial solver, or in addition include the solution algorithms in one big program. This approach suffers from several disadvantages: the difficulty to understand and to be read by others, the difficulty to maintain the code, the dependence on operating systems and compilers, to mention a few. Nevertheless, computer scientists often prefer this approach.

- *Table or spreadsheet approaches* in which the mathematical structure is extracted from table templates to be filled with data. In the past this approach was often found in software packages used in the process industry and especially refineries. The models were usually confined to linear problems or nonlinear ones with a very special structure. Those approaches are usually subject to data management problems, they are restricted to two dimensions, and scale-up can be a tedious work.

- *Algebraic modeling languages*, the subject of this book, are by far the best approach. They allow one to implement an optimization problem close to its mathematical formulation, they are flexible and open to fast reformulations. Most of these languages have been developed by individuals in mathematical optimization.

Although there exist big commercial software packages developed for treating some of the applications listed above, which contain models coded in programming languages such as FORTRAN, C, or C++, a significant portion of industrial problems and models have been developed or implemented in one of the modeling languages discussed in this book.

Once the model has been coded we need a *solver*, *i.e.*, a piece of software which has an algorithm implemented capable of solving the problem listed above. In the ideal case, the optimal solution is returned. In reality, when trying to solve real-world problems, we often experience that a problem is returned with the statement 'problem is infeasible'. Thus, a modeling system should also support the identification of infeasibilities.

What is the difference between *simulation* and *mathematical optimization*? In contrast to simulation (parameter values are fixed by the user, feasibility has to be checked separately, no proof on optimality), mathematical optimization methods compute directly an optimal solution and guarantee that the solution satisfies all constraints. While in optimization feasible solutions are specified *a priori* and implicitly by the constraints, in simulation somebody has to ensure that only those combinations of parameter values are evaluated or considered which represent 'appropriate scenarios'. Regarding the subject of this book, modeling languages, there is another fundamental difference. Modeling languages for mathematical optimization, at least in the past, are declarative representations of problems while model formulations using simulation packages appear much more procedural.

1.2 Classes of Problems in Mathematical Optimization

In this section we will briefly review some classes of problems often occurring in mathematical optimization. Some of these problems are already supported by the modeling languages discussed in this book. Maybe some others will find the interest of modeling language providers, so that they will be supported in the future. Besides those described below in greater detail, more details about *complementarity problems* are found in Subsection 7.5.2 while *combinatorial problems* are discussed in Subsection 7.5.3.

1.2.1 A Deterministic Standard MINLP Problem

If we restrict ourselves to algebraic optimization[1], *i.e.*, we do not allow differential or integral relationships or black-box functions to appear in our models, the largest class of problems are mixed integer nonlinear optimization problems which may formally be described as follows. For vectors $\mathbf{x}^{\mathrm{T}} = (x_1, ..., x_{n_c})$ and $\mathbf{y}^{\mathrm{T}} = (y_1, ..., y_{n_d})$ of n_c continuous and n_d discrete variables, the augmented vector $\mathbf{x}_\oplus^{\mathrm{T}} = \mathbf{x}^{\mathrm{T}} \oplus \mathbf{y}^{\mathrm{T}}$, an objective function $f(\mathbf{x}, \mathbf{y})$, n_e equality constraints $\mathbf{h}(\mathbf{x}, \mathbf{y})$ and n_i inequalities constraints $\mathbf{g}(\mathbf{x}, \mathbf{y})$, an optimization problem

$$\min \left\{ f(\mathbf{x},\mathbf{y}) \;\middle|\; \begin{array}{lll} \mathbf{h}(\mathbf{x},\mathbf{y}) = 0 & \mathbf{h}: X \times U \to \mathrm{I\!R}^{n_e} & \mathbf{x} \in X \subseteq \mathrm{I\!R}^{n_c} \\ \mathbf{g}(\mathbf{x},\mathbf{y}) \geq 0, & \mathbf{g}: X \times U \to \mathrm{I\!R}^{n_i}, & \mathbf{y} \in U \subseteq \mathbb{Z}^{n_d} \end{array} \right\} \tag{1.2.1}$$

is called *Mixed Integer Nonlinear Programming* (MINLP) problem, if at least one of the functions $f(\mathbf{x}, \mathbf{y})$, $\mathbf{g}(\mathbf{x}, \mathbf{y})$ or $\mathbf{h}(\mathbf{x}, \mathbf{y})$ is nonlinear. The vector inequality, $\mathbf{g}(\mathbf{x}, \mathbf{y}) \geq 0$, is to be read component-wise. Any vector $\mathbf{x}_\oplus^{\mathrm{T}}$ satisfying the constraints of (1.2.1) is called a *feasible point* of (1.2.1). Any feasible point, whose objective function value is less or equal than that of all other feasible points is called an *optimal solution*. From this definition it follows that the problem might not have a unique optimal solution.

The continuous variables in (1.2.1) could for instance describe the states (temperature, pressure, etc.), flow rates or design parameters of plants or chemical reactors. The discrete variables, often binary variables, may be used to describe the topology of a process network or to represent the existence or non-existence of plants. Let us consider the following pure integer nonlinear problem with two integer variables y_1 and y_2

$$\min_{y_1, y_2} \left\{ 3y_1 + 2y_2^2 \;\middle|\; \begin{array}{l} y_1^4 - y_2 - 15 = 0 \\ y_1 + y_2 - 3 \geq 0, \end{array} \quad y_1, y_2 \in U = \mathrm{I\!N}_0 = \{0, 1, 2, ...\} \right\}$$

A feasible point is $y_1 = 3$ and $y_2 = 66$. The optimal solution $\mathbf{y}^* = (y_1, y_2)^* = (2, 1)$ and $f(\mathbf{y}^*) = 8$ is unique.

Depending on the functions $f(\mathbf{x}, \mathbf{y})$, $\mathbf{g}(\mathbf{x}, \mathbf{y})$, and $\mathbf{h}(\mathbf{x}, \mathbf{y})$ in (1.2.1) we get the following structured problems known as

acronym	*type of optimization*	$f(\mathbf{x},\mathbf{y})$	$\mathbf{h}(\mathbf{x},\mathbf{y})$	$\mathbf{g}(\mathbf{x},\mathbf{y})$	n_d
LP	Linear Programming	$\mathbf{c}^{\mathrm{T}}\mathbf{x}$	$\mathsf{A}\mathbf{x} - \mathbf{b}$	$\mathbf{x}$	0
MILP	Mixed Integer LP	$\mathbf{c}^{\mathrm{T}}\mathbf{x}_\oplus$	$\mathsf{A}\mathbf{x}_\oplus - \mathbf{b}$	$\mathbf{x}_\oplus$	≥ 1
MINLP	Mixed Integer NLP				≥ 1
MIQP	Mixed Integer QP	$\mathbf{x}_\oplus^{\mathrm{T}}\mathsf{Q}\mathbf{x}_\oplus + \mathbf{c}^{\mathrm{T}}\mathbf{x}_\oplus$	$\mathsf{A}\mathbf{x}_\oplus - \mathbf{b}$	$\mathbf{x}_\oplus$	≥ 1
NLP	Nonlinear Programming				0
GLOBAL	Global Optimization				≥ 0

[1]This book also covers some modeling languages supporting the formulation of optimization problems including differential equations. Such problems are usually solved by appropriate discretization which transforms the problem into an algebraic optimization problem, in the worst case a MINLP problem.

with a matrix A of m rows and n columns, *i.e.*, $\mathsf{A} \in \mathcal{M}(m \times n, \mathbb{R})$, $\mathbf{b} \in \mathbb{R}^m$, $\mathbf{c} \in \mathbb{R}^n$, and $n = n_c + n_d$. Real-world problems lead much more frequently to LP and MILP than to NLP or MINLP problems. QP refers to quadratic programming problems. They have a quadratic objective function but only linear constraints. QP and MIQP problems often occur in applications of the financial service industries. Since some problems occur as subproblems of others it is very important that the algorithms to solve the subproblems are well understood and exploited efficiently. While LP problems can be solved relatively easy (the number of iterations, and thus the effort to solve real-world LP problems with m constraints grows approximately linearly in m), the computational complexity of MILP and MINLP grows exponentially with n_d. Numerical methods to solve NLP problems work iteratively and the computational problems are related to questions of convergence, getting stuck in 'bad' local optima and availability of good initial solutions. Global optimization applies to both NLP and MINLP problems and its complexity increases exponentially in the number of all variables entering nonlinearly into the model.

Modern modeling systems usually offer a variety of solvers which are based on various algorithms. Thus, those who are active in problem solving need to know a lot about good modeling practice and how to connect models and algorithms. For that reason, we strongly recommend that practitioners and consultants obtain a deep understanding of the algorithms involved. This also puts some stress on the developers of modeling systems to provide a reasonable amount of technical documentation. The following table contains a series of books which we consider as valuable regarding the background of algorithms or in the context of solving real-world optimization problems:

LP problems	[184]
MILP problems	[181], [250]
NLP problems	[180], [85], [223]
MINLP problems	[97], [83], [61], [247]
global optimization	[68], [116], [142], [115], [230]
real-world problems	[249], [140], [45], [138]

Besides the exact mathematical programming algorithms in this table, there also exists constraint programming (CP) useful for solving scheduling problems, and heuristic methods, better called *metaheuristics*, which can find feasible points of optimization problems, but can only prove optimality or evaluate the quality of these feasible points when used in combination with deterministic approaches. Such methods include *Simulated Annealing*, *Tabu Search*, *Genetic Algorithms*, *Evolution Strategy*, *Ant Colony Optimization*, and *Neural Networks*. These methods are not optimization methods in the strict sense as they cannot prove optimality or generate safe bounds. However, for very hard problems, they can be used to find feasible points of optimization problems, and recently, some modeling language packages such as `GAMS` or `Mosel` provide connections, for instance, to Glover's Tabu Search method ([86]).

1.2.2 Constraint Satisfaction Problems

Constraint satisfaction problems are not strictly optimization problems. They lack any objective function. However, there are various goals associated with these problems:

- Find one solution (*e.g.*, map coloring problem).
- Find all solutions (in case of discrete solution sets).
- Find outer and inner coverings of the solution set to provide, *e.g.*, a visualization of the solution space.
- Find N sufficiently distinct solutions, which can be presented to the end-user (*e.g.*, an engineer) who picks one of the computed solutions by personal judgement.
- Prove that there is no solution.

All of these goals can either be fulfilled approximately or rigorously also in the presence of rounding errors.

1.2.3 Multi-Objective Optimization

Multi-objective optimization, also called multi-criteria optimization deals with problems containing more than one objective function. A simple approach to solving multi-criteria problems, *i.e.*, *vector minimization problems*, is to express all objectives in terms of a common measure of goodness, *e.g.*, money, but that is very often not possible. The problem really is how to compare different objectives on a common scale. Basically, one can distinguish two cases. Either the search is for *Pareto-optimal* solutions, or the problem has to be *solved for every objective function separately*.

If the problem has to be solved independently for all objective functions, this should not be handled simply by passing each instance independently to the solver, because that is a highly inefficient method.

When minimizing several objective functions simultaneously the concept of *Pareto optimal solutions* turns out to be useful. In a multi-criteria optimization problem a solution is said to be *Pareto optimal* if there exists no other solution that is at least as good according to every objective, and is strictly better according to at least one objective. When searching for Pareto-optimal solutions, the task might be to *find one*, *find all*, or *cover the extremal set*.

A special solution approach to multiple objective problems is to require that all the objectives should come close to some targets, measured each in its own scale. The targets we set for the objectives are called *goals*. Our overall objective can then be regarded as to minimize the overall deviation of our goals from their target levels. The solutions derived are Pareto optimal.

Goal programming can be considered as an extension of standard optimization problems in which targets are specified for a set of constraints. There are two basic approaches for goal programming: the preemptive *(lexicographic)* approach and the *Archimedian* approach.

In preemptive goal programming goals are ordered according to importance and priorities. The goal at priority level, i, is considered to be infinitely[2] more important than the goal at the next lower level, $i + 1$, but they are relaxed by a certain absolute or relative amount when optimizing for the level $i + 1$. Our reactor example above might have the ranking: reactor size ($i = 1$), safety issues ($i = 2$), and eventually production output-rate ($i = 3$). Preemptive goal programming is recommended if there is a ranking between incommensurate objectives available. In the Archimedian approach weights or penalties are applied for not achieving targets. A linear combination of the violated targets weighted by some penalty factor is added, or establishes the objective function.

Goal programming offers an alternative approach but should not be regarded as without defects. The specific goal levels selected greatly determine the answer. Therefore, care is need when selecting the targets. It is also important in which units the targets are measured. Therefore, some modelers still prefer to work with one objective function and then re-run the model with variations in the objective function coefficients.

Detailed treatment of goal programming appears in such books as Ignizio (1976,[126]) and Romero (1991,[198]) who introduce many variations on the basic idea, as well as in Schniederjans (1995,[212]). It would be very desirable if multiple objective functions would be supported in a modeling system.

1.2.4 Multi-Level Optimization

To explain the idea of multi-level optimization, let us consider the following situation. A company wants to check whether a given material is suitable for building a certain structure. The material can be accepted only, if the maximum stress in the structure does not exceed a critical value specified. For given

[2]It would also be possible to define weights which express how much the ith objective is more important than the $(i+1)^{th}$ objective.

material design parameters[3] x we have to solve the optimization problem

$$\begin{aligned} \max\ & g(x,y) \\ \text{s.t.}\ & G(x,y) \in \mathbf{G} \quad , \quad y \in Y. \end{aligned}$$

If the maximum stress g is below a given bound s_{max}, the material is suitable for building the structure.

A company building the structure is interested in keeping the fabrication costs minimal. The cost c depends on the material used, and hence on the material parameters. The real problem, which is relevant for our company is therefore

$$\begin{aligned} \min\ & c(x) \\ \text{s.t.}\ & F(x) \in \mathbf{F} \quad , \quad x \in X \\ & \max\ g(x,y) \leq g_{\text{max}} \\ & \text{s.t.}\ \ G(x,y) \in \mathbf{G} \quad , \quad y \in Y. \end{aligned}$$

This problem is a *bi-level optimization problem* [see, for instance, Dempe (2002, [53])]. Problems of this type are called *multi-stage optimization problems*. The inner or stage-1-problem has the variables y, objective function g, and constraints G. Furthermore, the material parameters x are *free variables* in the inner problem. The outer or stage-2-problem depends on the variables x, has objective function c and constraints F.

In general, multi-stage problems are *hierarchical structures* of two or more optimization and constraint satisfaction problems, the simplest ones are the *minimax problems*.

For modeling these kinds of problems, it would be useful if models could be used as functions in the formulation of other models by regarding some of the parameters as free variables, *i.e.*, the function arguments.

1.2.5 Semi-Infinite Programming

Semi-Infinite Programming deals with optimization problems of the form

$$\min_{\mathbf{x}\in X} \left\{ f(\mathbf{x}) \;\middle|\; \begin{array}{ll} \mathbf{h}(\mathbf{x}) = 0 & \mathbf{h} : X \to \mathrm{I\!R}^{n_e} \\ \mathbf{g}(\mathbf{x}) \geq 0, & \mathbf{g} : X \to \mathrm{I\!R}^{n_i}, \\ \mathbf{s}(\mathbf{x},\mathbf{t}) \geq 0, & \mathbf{s} : X \times T \to \mathrm{I\!R}^{n_s}, \end{array} \quad \begin{array}{l} \mathbf{x} \in X \subseteq \mathrm{I\!R}^{n} \\ \forall \mathbf{t} \in T \subseteq \mathrm{I\!R}^{n_t} \end{array} \right\}$$

While $\mathbf{h}(\mathbf{x})$ and $\mathbf{g}(\mathbf{x})$ represent the usual constraints involving a finite set of unknown variables represented by the vector $\mathbf{x}$, $\mathbf{s}(\mathbf{x},\mathbf{t})$ is an infinite constraint which has to hold for all value of the parameter $\mathbf{t}$.

[3] Engineers usually do not talk about variables but rather use the term *parameter*.

Semi-Infinite programming can be called *standard* or *generalized.* While in standard semi-infinite programming the set T does not depend on the unknown variables represented by the vector $\mathbf{x}$, it does in the generalized case. The problem of approximating t^2 over the interval $[0, 2]$ by a combination of t and e^t leads to the following standard semi-infinite optimization problem with $n = 3$, $n_e = n_i = 0$ and $n_s = n_t = 1$:

$$\min_{\mathbf{p},d \in \mathrm{I\!R}^{2+1}} \left\{ d \;\middle|\; d - \left|t^2 - (p_1 t + p_2 e^t)\right| \geq 0,\ \forall t \in [0,2] \right\}.$$

For a good survey of standard Semi-infinite Programming see [103] and [228] for the generalized case. Standard Semi-infinite Programming is supported by an `AMPL` extension called `SIPAMPL` [246], which is publicly available for free.

1.2.6 Optimization Involving Differential Equations

In design optimization (*e.g.*, aircraft design) or optimization problems in the process industry often constraints are defined involving *ordinary* or *partial differential equations*. Sometimes also *differential algebraic systems* or *integral equations* have to be used in the modeling process (*e.g.*, parameter optimization for ovens used in silicon wafer production).

There exist only very few modeling systems such as `MINOPT` (Chapter 11) and `PCOMP` (Chapter 18) which provide the possibility to specify differential equations.

1.2.7 Safety Programming

An important problem in engineering is the *verification of safety constraints for uncertain designs.*

We again consider the company of Section 1.2.4 and the structure it wants to build. The problem with material parameters is that usually real materials do not have guaranteed parameters. For instance, the elasticity module in steel bars might vary by more than 15% from one bar to another. When building larger composite structures, these parameter uncertainties play a considerable role.

In *finite element* structural analysis for buildings, *e.g.*, one has to verify that the stresses in all steel bars do not exceed the maximal acceptable amount. A typical problem formulation is: Prove or find a counterexample to

$$\begin{aligned} &C(x,u) \leq 0 \\ &F(x,u) = 0 \quad \forall x, u \\ &\|x - x_0\| \leq \Delta x. \end{aligned}$$

In the finite element example the function F is linear in u.

Such problems have to be solved by rigorous computing, and the number of variables reaches several thousand in industry-relevant cases. Such problems can only be solved if *all structural knowledge* is exploited. So matrix operations must not be flattened, as is done in most algebraic modeling languages.

1.2.8 Optimization Under Uncertainty

1.2.8.1 Approaches to Optimization Under Uncertainty. Usually when people talk about optimization they use data treated as without any errors or uncertainty leading to deterministic optimization problems. In contrast to this class, there exists optimization under uncertainty, *i.e.*, optimization problems in which at least some of the input data are subject to errors or uncertainties, or in which even some constraints hold only with some probability or are just soft. Those uncertainties can arise for many reasons:

- Physical or technical parameters which are only known to a certain degree of accuracy. Usually, for such input parameters safe intervals can be specified.

- In strategic design optimization problems such as described and solved, for instance, by Kallrath (2002, [137]), demand and price forecast are used over a long planning horizon of 10 to 20 years. These predictions are subject to all sorts of uncertainties and can be at best quantified with some probability distribution or scenario approach.

Despite possible uncertainties in some input data, there are optimization problems which can be solved under the assumption of deterministic data without serious problems. Either the data show only small errors or uncertainties, or the problem and its structure is not sensitive with respect to such variations.

However, especially in long term or strategic planning the deterministic assumption is clearly not reasonable but the results obtained at least provide enough intuition into the problem to justify the model. This reflects the fact that a clear decision on what approach is the best for solving a problem under uncertainty cannot be made *a priori* but rather after the problem is solved. Unfortunately, it may also happen, that assuming deterministic data does not provide accurate enough results to justify the model. Sometimes making that assumption can even be the reason that misleading results are obtained.

Regarding the conceptual problems involved in optimization under uncertainty it is not a surprise that dealing with uncertainty in the context of an algebraic modeling language is cumbersome at best. The current approaches

- *sensitivity analysis* which is conceptually a difficult problem in the context of MIP,

- *robust optimization* [see, for instance, Ben-Tal *et al.* (2000, [12])],

- *multi-stage stochastic optimization* [see, for instance, Kall & Wallace (1994,[136]), Schultz (1995, [215]) or Carøe and Schultz (1999, [37])], a detailed discussion on various algorithms for stochastic integer optimization can be found in the survey of Klein-Haneveld & van der Vlerk (1999, [145]) and, more recently, in Schultz (2003,[216]),

- *fuzzy set modeling* supporting uncertainties which fall into the class of vague information and which are expressible as linguistic expressions [fuzzy set theory in the context of LP problems has been used, for instance, by Zimmermann (1987,[253]; 1991,[252] and Rommelfanger (1993,[199])],

- *global optimization* and the techniques, such as interval arithmetic, used in this field [see Section 4.2 in Chapter 4, or [68], [116], [142] and [115]]

to mention a few, show that the field of optimization under uncertainty is not yet a settled one but shows a variety of activities. Some of the current modeling systems such as `AMPL`, `MPL`, and `Mosel` have made a start and support, for instance, distribution and scenario based recourse problems, or chance constrained problems. However, despite the evolution in solution techniques and computer processing power which has made some of the problems more tractable, the situation in modeling optimization under uncertainty still needs more support in the modeling languages. What would be really desirable is to see language and syntax elements which support modeling or abstracting modeling issues, which are essential aspects of stochastic optimization problems.

There are a wide spectrum of different methods and schools for specifying and treating uncertainties. It is useful to distinguish three different categories of uncertainties found in Hürlimann (1999,[120]).:

1 partial or incomplete knowledge of the world (the knowledge is too expensive to get, or is hidden in the future);

2 vague or fuzzy knowledge (sometimes exact quantitative measures are absent, and the more fuzzy natural languages are used);

3 inconsistent, *i.e.*, contradictory, input data resulting from physical observations subject to measurements.

If the probability distributions of the random data are discrete, the enumeration of all possible realizations defines a set of scenarios. Whether discrete representations of data points in the universe of possible events are meaningful for the specific problem needs to be determined during the modeling process and is an important consideration. If discrete scenarios are not sufficient some distribution or probability function is required. Determining a meaningful sample size in this case is essential to validate the quality of the results obtained. It may be possible to treat some of the data before using it in the model. In addition, it may be necessary to obtain some of the said data from the output of simulation experiments, from databases, or from the result of some other computation.

Expected value (EV) models are constructed by replacing the random data by their expected values. Such an EV model is thus a deterministic optimization problem again as the uncertainty is dealt with before it is introduced into the underlying optimization problem. One might call the optimal objective function of the EV problem Z^{EV} and the optimal solution X^{EV}. By fixing the decision variables of the original optimization to X^{EV} it is possible to check the performance of such solutions in the presence of different realizations of the random data. The expected value of these objective values is called *expectation* of the expected value solution, indicated by Z^{EEV}.

1.2.8.2 Stochastic Optimization. Most stochastic optimization problems (see also Subsection 7.5.4) are almost invariably built as recourse problems, *i.e.*, multi-stage stochastic optimization problems [see, for instance, Schultz (1995, [215]) or Carøe and Schultz (1999, [37])]. Recourse problems are dynamic optimization problems characterized by uncertain future outcomes of some input data. There is a set of first-stage decisions to be made with (assumed) deterministic data and some unknown data, followed by a set of recourse decisions taken for each realization of the data using entirely known data and the results of the previous decisions. The objective of the problem is to come up with first-stage decisions that are robust enough in the sense that uncertain events can be addressed by taking different recourse actions, *i.e.*, the optimal solution hedges against all possible events that may occur in the future. This implies the existence of time periods, or stages. These may be natural to the problem, or artificially introduced by the modeler. There could be one or many of them, the latter case being a recursion of the first. Depending on the freedom of the recourse actions, the structure of the problem, and other variables, it could be the case that examining the problem as a stochastic optimization problem is not necessary. Interpreting and measuring the value of the stochastic solutions is an important part of modeling optimization under uncertainty.

Many important real-world problems like *portfolio optimization* depend on *stochastic processes* present in the stock market.

Building models for these problems requires the introduction of *random variables* in addition to the variable classes further discussed in Section 4.2.1; let us call them *stochastic variables*. They can be continuous or discrete. Usually *distribution information* such as *distribution functions* and *covariance matrices* is at least partially available. Subsection 7.5.4 contains an example of how distributions can be assigned to random variables in `AMPL`.

The new class of variables also yields new types of constraints involving *expectation values*, *variances*, *covariances*, and *probabilities*, and because complex stochastic processes need to be modeled, very often multi-stage models are the result of the modeling process. Hence, a typical stochastic optimization problem has the form

$$\begin{aligned} \min\ & f^{(k)}(x^{(k)}, \xi^{(k)}) \\ \text{s.t.}\ & F^{(k)}(x^{(k)}, x^{(<k)}, \xi^{(<k)}) \in \mathbf{F} \\ & x^{(k)} \in \mathbf{x}_0^{(k)} \\ & \xi^{(k)} \sim g^{(k)}(\theta^{(k)}, t), \end{aligned}$$

where the $\xi^{(k)}$ stand for the stochastic variables in stage k, and g_k describes their distribution information. The $x^{(k)}$ are the deterministic variables of stage k, and $\mathbf{x}_0^{(k)}$ denotes their initial bounds.

Today's algorithms use *scenarios* to transform stochastic problems to a sequence of deterministic problems, which in turn are solved using an algorithm for standard nonlinear problems.

Therefore, some modeling systems provide the possibility to include scenarios in the model. In the author's opinion, the scenarios are not part of the model but rather part of the model-solver interface. In some sense, the scenarios represent algorithmic knowledge and not declarative knowledge. So there should be a clean separation between the stochastic models and the scenarios needed nowadays for solving them.

1.2.8.3 Beyond Stochastic Programming. Stochastic models are still not the most general optimization problems which arise from real-world applications. The stochastic variables are sometimes simplifications of the real situation.

However, usual modeling languages do not provide modelers with the possibility to record all their knowledge about a real-world problem in their mathematical model. Stochastic models are then generated by trying to translate the knowledge into stochastic information. But this sometimes hides the real problem structure.

Typical knowledge which has to be kept in the models is, *e.g.*,

- the informed opinion of one or more experts,
- the fact that some unknown stays within certain bounds,
- knowledge about the mean value and/or the variance of some unknown,
- distribution functions, probably with missing parameters (*e.g.*, the unknown is normally distributed with mean 0 but unknown variance),
- time series information,

- sets of discrete values (with or without probabilities),
- data and parameter uncertainty,
- data characterization: exact values, approximations, estimates & guesses.

The relations between the variables and the data lead to constraints, which can be hard or soft, and the degree of softness plays a role as well. There might also be cases where constraints are only valid with some probability, or where they are based on expert knowledge.

From the point of view of model structure, for global optimization and other complex algorithms it would be useful to combine *mathematically equivalent formulations* of one model in order to get more flexibility in the solution process.

Although this may not be an easy task, it would be interesting to see whether modeling systems would permit the modeler to specify all his knowledge without the need to translate his intuitive knowledge into a certain predefined form. This might be one of the visionary enhancements listed in Chapter 20 which would increase the flexibility of the model formulations and would help the modeler to formulate as closely as possible to the real problem.

1.3 The History of Modeling Languages in Optimization

Writing a section on the history of modeling languages is not an easy task. The author may be unaware of certain aspects. Others might be focussed on too strongly. At best, it seems possible to give a review and to look back on the last 40 years of modeling languages using all sources available, among them the article by Ciriani *et al.* (2002,[43]), and ask the reader's forgiveness if something is missing or not correct. Unlike the usual situation of a historian there is still the opportunity to get in direct contact with the individuals who formed the field and many of them are still active. But even then in some sense the history is partially constructed because the presentation of history is based on selected historical sources, write-ups or telephone conversations with people in the field when they relate their memories, and there is always a choice and a viewing angle present. And memories can be deceptive.

The review given here tries to put activities in some relationship with each other and to provide a coherent picture. One essence of this picture is that modeling languages play an important role in mathematical optimization. The history of modeling languages is tightly connected to the history of industry-related mathematical optimization. The availability of modeling languages increased the sharing of test problems among scientists which in turn lead to significant improvements of the algorithms. The structure and features in problem classes supported by modeling languages gave rise to treating these problems from an algorithmic point of view. Another important element is that very distinguished individuals formed the field and manifested their ideas into their modeling systems. It is interesting to see how the focus and technical constraints of such systems have changed in the last 30 years. As is outlined in Section 8.2, modeling systems underwent three major phases that shift the emphasis from the computational issues (the past: computational limits were the dominant constraint), to modeling issues (now: languages and system enabling to bring the modeling skills) and finally the application or the real problems (now, and entering into the future: the optimization models become a small piece, which to the end-user is a black-box module in a larger decision support system).

People new to the field but with some interest in history might ask questions such as "Who invented MPS?" or "Which was the first modeling language?". Deeper questions are "What was the motivation to design the language so-and-so in this way?" or "What type of person is the language designer anyway?". Unlike the creation of Earth and all life on Earth it took more than seven days to create the first industrial standard LP solvers, the first general, integrated matrix and report generator, the MPS standard input format, but fortunately enough, the origins are not yet lost in the dark of the past.

The first computer LP solver was in the late 1950's when William Orchard-Hays, working at the RAND Corporation in consultation with George Dantzig, developed a system with the computer code on punched cards. It was programmed in Assembler, largely by the wife of Orchard-Hays, to run on IBM's 7090 computer. At that time one loaded in the cards to start an iteration and when you wanted to run the next iteration you reloaded the cards. It was a beginning. But Bill Orchard-Hays was very creative and saw a better way. He was now, around 1958-1960, at CEIR, an English company, and conceived of a system where the code was all on tape and loaded in automatically as required. Esso, Shell, Union Carbide and several other oil companies provided some funding and LP90 (later LP94 on the IBM 7094) was created using the product form of the inverse as the algorithm. The LP90/94 code was programmed entirely in assembler, almost single-handedly, by Bill Orchard-Hays, according to Michael Grigoriadis who worked on optimization at IBM in the sixties. A book on the mathematics of the algorithms by Orchard-Hays was the bible for early LP implementors. Eli Hellerman was instrumental in getting the commercial version working and in helping new users. The input format was called "SHARE standard". The maximum size of models was 1000 rows, although this was increased later by the use of decomposition techniques.

In the beginning there was, besides pure Assembler programming, only FORTRAN II, and the world was good; except for the cold war, various other wars and disasters. This was around 1960; at that time FORTRAN was the best and modeling was as flexible as is possible in a procedural language. Of course, it was not flexible in the sense how modeling can be done using algebraic modeling languages and it was definitely not convenient. Life became somewhat easier in 1964, when Larry Haverly invented the first general, integrated matrix and report generator called MaGen which was widely used on the IBM 1401 computer.

When IBM developed the revolutionary 360 series of computers in the mid 1960's, the oil companies in particular asked for a efficient LP system. Various oil companies sent specialists to IBM to help insure the code was practical. The recording effort mounted for the 360 including the latest algorithms, ultimately resulted in the world best (at that time!) generally available LP solver known as MPS/360 (later MPSX), where MPS stands for *Mathematical Programming System*; the X for *extended* [see, for instance, Tomlin & Welch (1990,[233])].

According to Milt Gutterman, who was an intensive user of LP at Standard Oil of Indiana (now part of BP), the input formats for the MPS/360 were identical to its predecessors. Eight character names were used instead of six and bounds and ranges were added. These input format became known as the MPS input format. The following example

$$\max \quad 3x_1 + 2x_2$$
$$s.t. \quad x_1 + x_2 \leq 3.5 \quad ; \quad 0 \leq x_1, x_2 \leq 2 \quad ; \quad x_1, x_2 \quad \text{integer}$$

is represented in an MPS input file as

```
NAME          FARMER
ROWS
 L  BUY
 N  PROFIT
COLUMNS
    x1        BUY           1.000000   PROFIT      3.000000
    x2        BUY           1.000000   PROFIT      2.000000
RHS
    RHS00001  BUY           3.500000
BOUNDS
 UI BOUND001  x1            2.000000
 UI BOUND001  x2            2.000000
ENDATA
```

The correspondence is obvious. More details on MPS and MPSX are given in Chapter 14 by Kurt Spielberg; here we just summarize:

MPSX established a standard input format supported by all current commercial mathematical modeling languages and a unique representation of LP or MILP models, and many modeling systems transformed the model to this standard structure which basically kept the matrix and other data of LP problems. That is why this process was called matrix generation and many software packages (matrix generators, *e.g.*, Haverly's MaGen) were developed whose task was to allow users to enter their data in some user-friendly way into the computer and to generate MPS files. As is outlined by Ellison & Mitra (1982,[62]) the matrix generators were enhanced by report writers and lead to matrix-generator report-writer (MGRW) systems. DATAFORM (Ketron), GAMMA (Bonner&Moore), MGRW (IBM), MGG/RWG (Scicon) – a matrix generator generator which produced a FORTRAN matrix generator, which could then use FORTRAN to do lots of things not thought of by the matrix generator producers. A very clever idea. – MGVEN as well as MaGen and OMNI by Haverly Systems and UIMP (Unicom Consultants) belong to this class. There was also APEX. Most of those MGRW systems had been written in FORTRAN, ALGOL, or PL1 except MaGen and OMNI which were in assembly language.

In Europe the acceptance of declarative modeling languages was achieved already early in the late 1960's due to Martin Beale's influence and MGG and the wide use of MaGen. While MaGen was a specialized programming language for matrix generation, which one would rather not view as one of the modeling languages, MGG was already much more like a declarative modeling language.

The new, post-MPS, era of modeling languages was mostly initiated by individuals who were active in mathematical optimization and modeling real world problems. In general, the creators of established matrix generation software did not lead the switch to modeling languages; instead the first modeling languages tended to come from new developers, a number of whom were working independently toward the same goals. Robert Fourer [71] published a paper bringing together the arguments in favor of modeling languages and summarizing the progress that had already been made by the early 1980s.

In 1976 Alexander Meeraus, in collaboration with Johannes Bisschop, gave a talk at the ISMP conference in Budapest and started working on a *General Algebraic Modeling System* (now called GAMS) [23] whilst being employed at the Development Research Center of the World Bank. The World Bank at that time provided excellent working conditions for creative and intellectual people, especially for mathematical programmers. Being founded in 1946 as an international organization to rebuild the destroyed countries of Europe after World War II, the focus shifted after the Marshall Plan in 1948 to more general questions. In the mid 1970's, the senior vice president Hollis B. Chenery wanted to support economies with the best economists and techniques in the world. This included, of course, mathematical optimization, and he was famous for requesting that people put their arguments in terms of mathematical language and rigor. What a refreshing atmosphere that must have been compared to nowadays where responsible managers usually avoid making any clear statement in public. So it is not a surprise that together, Meeraus and Bisschop were able to get the necessary financial support through an internal research project. From the beginning, GAMS was designed to be a system for modeling nonlinear programming problems. While at the start GAMS was planned as a tool to research dissemination inside the World Bank, it fitted the organisation's attitude that good quality software should be able to survive the external market's request. One of the early companies interested in GAMS were EXXON Research and Shell Oil. Especially, Tom Baker who later with his product MIMI founded his own company Chesapeake Inc., was interested in GAMS and thus a fruitful collaboration between Washington and Florham Park resulted. The PROFIT project lead by Emilio Nuñez (Houston) was very crucial in the early phase of GAMS. Its success showed that an equation-based system such as GAMS really worked. It has to be said that in these early days of modeling languages there existed a positive climate for such systems. Larry Haverly and Martin Beale were leaders in the field which seemed to be concentrated somehow in Washington and London. In 1980 Bisschop left the World Bank, and Meeraus developed GAMS into an industrial-strength modeling system with extensive use inside and outside the World Bank. There then occurred that which always ends good times. The *Zeitgeist* changed, management was replaced by weaker people trying to follow the main trend. Mathematical optimization became less popular in the World Bank, the new buzz words were market service and market drive. People had only two choices: *stay and adjust*, or *leave*. Following a major reorganization of the World Bank in 1987, Meeraus founded the *GAMS Development Corporation* located in Washington; Tony Brooke and Arne Drud were also deeply involved. Nowadays, GAMS is much more advanced, *e.g.*, it can call itself recursively.

Having some background using MGG at Scicon, in 1983, Robert Daniel and Robert Ashford from Dash Optimization – the name *Dash* is composed from the authors' names, DAniel and ASHford – launched `mp-model`, a declarative modeling language for LP and MILP problems which in the year of writing, 2003, would have celebrated its 20th anniversary, were it not replaced by Dash's new modeling environment `Mosel` in 2001. `Mosel` has a more open architecture and provides more flexibility. Unlike `mp-model`, `Mosel` is no longer a purely declarative language but offers many procedural or algorithmic components. This modeling system follows the trend.

Another system that appeared in 1989 is LINGO. The predecessor of LINGO, LINDO was an interactive, equation oriented LP modeling system that became available on the DEC-20 computer in 1977. When the IBM PC became available in 1981, LINDO was immediately ported to it. The PC version was distributed worldwide by Scientific Press. A nonlinear, but still scalar version, called GINO was introduced in 1985. The Intel 80386 chip became widely available in 1987. This chip's 32-bit architecture, coupled with the DOS extenders of the time, made it possible to build programs that far exceeded the previous 640kB DOS memory limit. In addition, software designed for the new 32-bit architecture executed an order of magnitude faster than on their 16-bit brethren. Suddenly, the PC had become a viable machine for solving large, real-world optimization problems. At that point in time, the company's primary modeling environment was the LINDO platform, a scalar based modeling language without any of the "subscripted variable" capabilities that are the cornerstone of modern modeling languages. Although LINDO was simple to use for entering small models, a more powerful tool was needed to build the larger models users would soon want to run on the new generation hardware. Thus, the company began the development of the LINGO set-based algebraic modeling system in 1987, and shipped the first versions in 1988. The design of the LINGO language owes much to the structured modeling concepts that Art Geoffrion was beginning to develop around that time.

In the 1980's there was considerable interest at Bell Laboratories in "little languages" [14]. When Fourer spent a sabbatical leave at Bell Labs in 1985-86, this interest led to a collaboration with David Gay and Brian Kernighan on the initial design and implementation of AMPL. Advances in computing power by this time suggested that AMPL could be more ambitious than earlier languages in the power and naturalness of its model expressions. By the early 1990's AMPL had been developed to the point where it was no longer so little; a book-length introduction [75] included a variety of extensions and generalizations.

In the early nineties, Johannes Bisschop started Paragon Decision Technology B.V. in the Netherlands and developed the modeling system AIMMS. At that time, AIMMS distinguished itself from other modeling systems by offering integrated point-and-click facilities to build graphical end-user interfaces. In the years to follow, AIMMS became the first system with a model construction environment based on graphical tools instead of ASCII files. Additional features of interest are its facilities for capturing and checking units of measure, an open interface for solving mixed-integer nonlinear programs with the *outer approximation* algorithm, and a development environment for constructing distributed optimization applications involving multiple agents.

Nowadays, the modeling languages have turned themselves more into modeling systems. But it seems to make computer scientists more happy than the mathematical programmers who grew up with FORTRAN and purely declarative languages. We see many of the older systems which were command-line driven and based on ASCII files providing complete integrated modeling and development environments, where the user or his controller can measure the activity and progress over a day with a number of mouse clicks.

A small specialized algebraic language appeared in 1987 – called (Linear Programming Language) as the result of Tony Hürlimann's (1987,[117]) dissertation. The language had some interesting features still not found in present languages: (1) It was possible to define hierarchical indexed sets, (2) One could automatically generate databases in dBASE format, the algebraic structure guided the generation, and (3) it had some limited way of generating the bipartite activity-constraint graph, an extremely valuable tool for certain models to browse and understand the mathematical structure of a model instance. Hierarchical sets were especially appealing since many constraints were organized in a way which were best viewed as trees: Constraints were generated recursively from the top down the tree. LPL was also the first language to integrate unit of measurements (see [118], for example).

However, already in 1989 there appeared the modeling system MPL developed by Icelandic computer scientist Bjarni Kristjansson and his company Maximal Software. This system is remarkable because already in 1991 it had its own extended memory management and in 1992 a Windows-based environment. A professional database connection and open data access followed in 1993. In 1994 support for connecting to solvers directly in memory using dynamic link libraries (DLLs) was added and a full 32-bit release came in 1995. In 1997, the MPL modeling language reached full scalability which allows it to handle large-scale problems with millions of variables and constraints. This makes this system very useful to solve real-world problems, for instance, supply-chain optimization and manpower planning. In 1999, Maximal Software introduced the OptiMax component library for embedding or deploying optimization models into applications programs.

OPL Studio Studio, distributed by ILOG, a France and US based company, was from the beginning a modeling system in the sense of the definition given on page xviii, and it is based on the OPL Studio programming language developed by Pascal Van Hentenryck (1999, [235]), and Van Hentenryck *et al.* (2000, [237]) enhanced by OPLScript (Van Hentenryck & Michel, 2000, [235]). OPL Studio was the first attempt to combine the concept of algebraic modeling languages with the constraint programming framework and the ability to specify search procedures at a high level. OPL Studio as well as Mosel mentioned above and MPL/OptiMax described in the previous paragraph, are seen to be used in deployment.

There have been special purpose modeling languages around, such as NOP developed in 1997 by Arnold Neumaier from University of Vienna to pass models to their own solver GLOPT. While the original version had been written completely in FORTRAN, the author felt that this was not flexible enough and in 1998 the language had been rewritten and been transformed to C by Stefan Dallwig, Hermann Schichl, and Arnold Neumaier, leading to NOP-2. This language did not have many of the features found in other modeling languages, but it had a list of attractive exceptional features such as supporting the formulation that a matrix has to be positive definite. Although it is not supported any more, this language is quite instructive and may motivate others to implement such language features into their own modeling languages. For that reason this book also contains a detailed description of NOP-2 in Chapter 15.

Despite the availability of advanced modeling systems, many application software providers still code their models in C or C++. Speed is often a major issue when choosing a modeling language or system compared to programming languages. Over time, with improved technology and hardware this has become less of an issue and now, some of the modeling systems have been optimized for speed and are very fast. The reason for using C or C++ may lie in the fact that those companies hired mostly people educated in computer science but not mathematical programming. Some, but not all, modeling languages seem to pay more attention to this fact and enhance their declarative language syntax by procedural features.

So, today, we can observe that from the very few roots of modeling languages there have arisen many languages with different designs and focus. As we have seen during the symposium in Bad Honnef, the field is still active, some of the founders are still around and come up with new ideas which make the world of mathematical optimization more interesting.

1.4 Conventions and Abbreviations

The following table contains in alphabetic order the abbreviations used in this book.

Abbreviation	*Meaning*
AML	Algebraic Modeling Language
B&B	Branch and Bound
CP	Constraint Programming
ERP	Enterprise Resource Planning
GUI	Graphical User Interface
IP	Integer Programming
LP	Linear Programming
MCP	Mixed Complementarity Problem
MILP	Mixed Integer Linear Programming
MIP	Mixed Integer Programming
MIQP	Mixed Integer Quadratic Programming
MINLP	Mixed Integer Nonlinear Programming
MPEC	Mathematical Programming with Equilibrium Constraints
NLP	Nonlinear Programming
ODBC	Open DataBase Connectivity
OR	Operations Research
QP	Quadratic Programming
SCM	Supply-Chain Management
s.t.	subject to
w.r.t.	with respect to

Acknowledgments

I want to thank Hermann Schichl for his engaged discussions regarding part I of this book and his contributions to the Sections 1.2.2, 1.2.7, 1.2.8.2, and 1.2.8.3. Ismael Vaz (Minho University, Portugal) provided valuable feedback and literature references to Subsection 1.2.5 on Semi-Infinite Programming. Many details regarding the history of modeling languages have been provided by Johannes Bisschop, Bob Daniel, Bjarni Kristjansson, Alexander Meeraus, Larry Haverly, Linus Schrage and Kurt Spielberg in numerous phone calls and e-mails.

Chapter 2

MODELS AND THE HISTORY OF MODELING

Hermann Schichl*
Institut für Mathematik der Universität Wien
Strudlhofgasse 4, A-1090 Wien, Austria
Hermann.Schichl@esi.ac.at

Abstract After a very fast tour through 30,000 years of modeling history, we describe the basic ingredients to models in general, and to mathematical models in particular.

Keywords: Modeling, history of modeling, model, mathematical model

2.1 The History of Modeling

The word "modeling" comes from the Latin word *modellus*. It describes a typical human way of coping with the reality. Anthropologists think that the ability to build abstract models is the most important feature which gave homo sapiens a competitive edge over less developed human races like homo neandertalensis.

Although abstract representations of *real-world objects* have been in use since the stone age, a fact backed up by cavemen paintings, the real breakthrough of modeling came with the cultures of the Ancient Near East and with the Ancient Greek.

The first recognizable models were *numbers*; counting and "writing" numbers (*e.g.*, as marks on bones) is documented since about 30,000 BC. *Astronomy* and *Architecture* were the next areas where models played a role, already by about 4,000 BC.

It is well known that by 2,000 BC at least three cultures (Babylon, Egypt, India) had a decent knowledge of mathematics and used mathematical models

*funded by EU project COCONUT IST-2000-26063

to improve their every-day life. Most mathematics was used in an algorithmic way, designed for solving specific problems.

The development of philosophy in the Hellenic Age and its connection to mathematics lead to the deductive method, which gave rise to the first pieces of *mathematical theory*. Starting with *Thales of Miletus* at about 600 BC, *geometry* became a useful tool in analyzing reality, and analyzing geometry itself sparked the development of mathematics independently of its application. It is said that Thales brought his knowledge from Egypt, that he predicted the solar eclipse of 585 BC, and that he devised a method for measuring heights by measuring the lengths of shadows. Five theorems from elementary geometry are credited to him:

1. A circle is bisected by any diameter.
2. The base angles of an isosceles triangle are equal.
3. The angles between two intersecting straight lines are equal.
4. Two triangles are congruent if they have two angles and one side equal.
5. An angle in a semicircle is a right angle.

After Thales set the base, Pythagoras of Samos is said to have been the first pure mathematician, developing among other things the theory of numbers, and most important to initiate the use of *proofs* to gain new results from already known theorems.

Important philosophers like Aristotle, Eudoxos, and many more added lots of pieces, and in the 300 years following Thales, geometry and the rest of mathematics were developed further. The summit was reached by Euclid of Alexandria at about 300 BC when he wrote *The Elements*, a collection of books containing most of the mathematical knowledge available at that time. The Elements held among other the first concise axiomatic description of geometry and a treatise on number theory.

Euclid's books became the means of teaching mathematics for hundreds of years, and around 250 BC Eratosthenes of Cyrene, one of the first "applied mathematicians", used this knowledge to calculate the distances Earth-Sun and Earth-Moon and, best known, the *circumference of the Earth* by a mathematical/geometric model.

A further important step in the development of modern models was taken by Diophantus of Alexandria about 250 AD in his books *Arithmetica*, where he developed the beginnings of *algebra* based on symbolism and the notion of a *variable*.

For *astronomy*, Ptolemy, inspired by Pythagoras' idea to describe the celestial mechanics by circles, developed by 150 AD a mathematical model of the solar system with circles and epicircles to predict the movement of sun, moon, and the planets. The model was so accurate that it was used until the time of Johannes Kepler in 1619, when he finally found a superior, simpler model for planetary motions, that with refinements due to Newton and Einstein is still valid today.

Building models for real-world problems, especially mathematical models, is so important for human development that similar methods were developed independently in China, India, and the Islamic countries like Persia.

One of the most famous Arabian mathematicians is Abu Abd-Allah ibn Musa Al-Ḫwārizmī (late 8th century). His name, still preserved in the modern word *algorithm*, and his famous books *de numero Indorum* (about the Indian numbers — today called arabic numbers) and *Al-kitab al-muḫtaṣar fi ḥisāb al-ğabr wa'l-muqābala* (a concise book about the procedures of calculation by adding and balancing) contain many mathematical models and problem solving algorithms (actually the two were treated as the same) for real-life applications in the areas commerce, legacy, surveying, and irrigation. The term algebra, by the way, was taken from the title of his second book.

In the Occident it took until the 11th century to develop mathematics and mathematical models, in the beginning especially for surveying.

The probably first great western mathematician after the decline of Greek mathematics was *Fibonacci, Leonardo da Pisa* (ca. 1170–ca. 1240). As a son of a merchant, Fibonacci undertook many commercial trips to the the Orient, and in that time he became familiar with the Oriental knowledge about mathematics. He used the algebraic methods recorded in Al-Ḫwārizmī's books to improve his success as a merchant, because he realized the gigantic practical advantage of the Indian numbers over the Roman numbers which were still in use in western and central Europe at that time. His highly influential book *Liber Abaci*, first issued in 1202, began with a presentation of the ten "Indian figures" (0, 1, 2, ..., 9), as he called them. This date was especially important because it finally brought the number *zero* to Europe, an abstract model of *nothing*. The book itself was written to be an algebra manual for commercial use, and explained in detail the arithmetical rules using numerical examples which were derived, *e.g.*, from measure and currency conversion.

Artists like the painter Giotto (1267–1336) and the Renaissance architect and sculptor Filippo Brunelleschi (1377–1446) started a new development of geometric principles, *e.g.* perspective. In that time, *visual models* were used as well as mathematical ones (*e.g.*, for Anatomy).

In the later centuries more and more mathematical principles were detected, and the complexity of the models increased. It is important to note that despite the achievements of Diophant and Al-Ḫwārizmī the systematic use of variables was really invented by Vietá (1540–1603). In spite of that it took another 300 years until Cantor and Russell that the true role of variables in the formulation of mathematical theory was fully understood. *Physics* and the description of Nature's principles became the major driving force in modeling and the development of the mathematical theory. Later *economics* joined in, and now an ever increasing number of applications demand models and their analysis.

In the next section we will take a closer look at models, their function and their most prominent characteristics. Further information on mathematical history can, *e.g.*, be found in [84].

2.2 Models

As a basic principle we may say:

A *model* is a *simplified version* of something that is real.

The traits of the model can vary according to its use. They can vary in their *level of formality*, *explicitness*, *richness in detail*, and *relevance*. The characteristics depend on the basic function of the model and the *modeling goal*.

In building models, everything starts with the *real-world object* we are considering. In a *model* real-world objects are replaced by other, simpler objects, usually carrying the same names. The *knowledge* we possess about the real-world is *structured* by the model, and everything is *reduced* to those phenomena and aspects which are considered important. Of course, a model can only describe a part of the real-world phenomenon, and hence its usefulness is restricted to its *scope of application*.

Models can have many different functions.

Explain Phenomena. Most of the theories developed in physics belong to this category: Newton's mechanics, thermodynamics, Einstein's theory of relativity, quantum mechanics, the Standard Model of particle physics, and many more.

There is not only physics, however. The aggregate demand-price adjustment (AD-PA) model, the aggregate demand-inflation adjustment (AD-IA) model, or the Hicks-Hansen IS/LM Model are three examples of economic models describing macroeconomical equilibria.

Avalanche researchers build models based on statistical and phenomenological data to describe the state of snow on alpine slopes.

Biologists use predator-prey models or epidemiological models to investigate the relationship between various life-forms.

Make Predictions. After the models are built which explain the phenomena, these models can be used as a further step to make predictions about the future development of a real-world phenomenon.

The avalanche researchers, for example, take their state data and the topographical information of the slopes to make predictions on the probability that avalanches are triggered, on their likely strenghts and their presumed places.

Aerodynamical models make *e.g.*, predictions about the maneuverability of a constructed airplane.

Climatic models are used to forecast the effects of the increased amount of greenhouse gases in the atmosphere.

Decision Making. A car driver uses a model of his surroundings and the typical traffic on the streets to decide which route to take. Of course, this model of the real-world, reduced to streets and average traffic, is in no way a formal mathematical one. It is based on experience and rather vague, if it can be expressed in words at all.

A more formal model for decision making is the design problem for a chemical plant with hundreds of decision variables and thousands of additional variables and constraints, written as mathematical equations and inequalities. They represent space, capacity, cost limitations, and chemical principles like conservation of mass.

Communication. Another important aspect of models is that they can be used to communicate knowledge. If a person A wants to visit person B, he might ask for the way to drive. B will sketch the correct route on a sheet of paper with a few lines and some additional marks and text, like "here at the corner is a yellow house with small garden". This sheet of paper is a visual model for the surroundings of B's house; its purpose is to communicate a subset of B's knowledge about his city to A.

Others. The detailed check-lists for airplane maintenance are (extremely detailed and explicit ones) models as well, as are the collections of formal norms (ANSI, DIN, EU Norm,...) used to regulate public life in modern countries.

2.3 Mathematical Models

The models of interest for us are mathematical models. Here, the real-world object is represented by mathematical objects in a *formalized mathematical language*.

The advantage of mathematical models is that they can be analyzed in a precise way by means of *mathematical theory* and *algorithms*.

As we have seen in Section 2.1, mathematical models have been used long ago, and many problems have been formulated mathematically since hundreds of years. However, the sheer amount of computational work needed for solving the models restricted their use to qualitative analysis and to very small and simple instances.

The development of algorithms like the Runge-Kutta method or the Fast Fourier Transform made complex models accessible to computers. In the beginning of the twentieth century human workers (most of the time low paid women) were used as "computers", and the problem size was still very limited.

As ENIAC was started in 1945, models of previously unknown size became tractable. It was possible for the first time to use mathematical modeling for *solving* practical problems of significant size.

The improvements in computer technology in the years since and the enormous gain in storage capacity and speed have made mathematical modeling increasingly attractive for military and industry, and a special class of problems, *optimization problems*, became very important.

The success in solving real-world problems increased the demand for better and more complex models, and the modeling process itself was investigated. There are now several books on modeling, *e.g.*, [249] and [138], and branches of mathematics completely devoted to solving practical problems and developing theory and algorithms for working on models, like operations research. See also [119] for a short treatise of this theme.

The structure of the models has changed throughout history, as the mathematical community gained increasing insight into the foundations of mathematics and formal logic. The introduction of variables, function spaces, and of all the mathematical structural theory has made mathematical models increasingly formal.

To date a mathematical model consists of *concepts* like

variables: These represent *unknown* or changing parts of the model, *e.g.*, whether to take a decision or not (decision variable), how much of a given product is being produced, the thickness of a beam in the design of a ceiling, an unknown function in a partial differential equation, an unknown operator in some equation in infinite dimensional spaces as they are used in the formulation of quantum field theory, etc.

relations: Different parts of the model are not independent of each other, but connected by relations usually written down as *equations* or *inequalities*. *E.g.*, the amount of a product manufactured has influence on the number of trucks needed to transport it, and the size of the trucks has an influence on the maximal dimensions of individual pieces of the product.

data: All numbers needed for specifying instances of the model. E.g. the maximal forces on a building, the prices of the products, and the costs of the resources.

2.4 The Modeling Process

As every textbook on modeling describes, the process of building a model from a real-world problem is a tedious task. It often involves many iterations in a cycle like in Fig. 2.4.1.

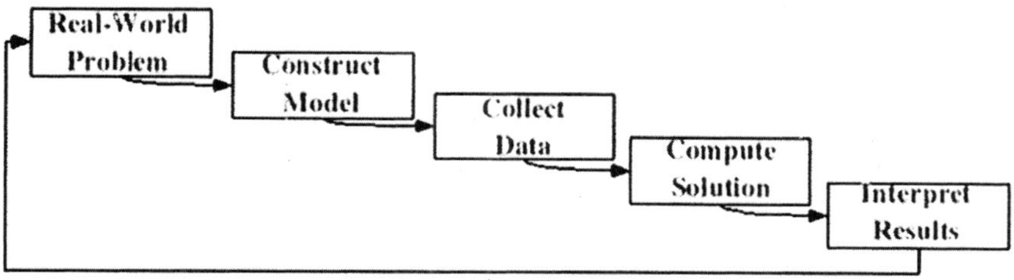

Figure 2.4.1. Modeling cycle

This is the traditional description of the modeling process, however as we will see in Chapter 3, the various stages of the *modeling cycle* appear interconnected, demanding even more interaction between the subtasks.

Several of these modeling steps require the help of the end user. For complex models, it is widely accepted that computers are needed in the *compute solution* process. For huge data sets or data which has to be retrieved from different sources, computer support for the *collect data* step is accepted, as well.

In this section we won't go into the details of the modeling process itself. This is done in the next chapter. We will rather focus on two aspects which have a strong influence on the structure of the model itself.

Until the first third of the twentieth century most of the mathematical models were used to describe phenomena and to make qualitative statements about the real-world problems described. Since then the situation has changed dramatically. The tremendous increase in computing power has shifted the interest in mathematical models more and more from problem *description* towards problem *solving*.

This has an important consequence for mathematical modeling itself and for the structure of the models: If the most important conclusions which can be drawn from a model are qualitative statements, which are derived by analytic means using a lot of mathematical theory, it is important to formulate the model in a very concise way, especially tailored to the analytical methods available. In contrast to that, the urge for numerical solutions in the last decades, made it necessary to change the model structure, to adapt it to the solution algorithms available. Section 2.4.1 will focus on the impact solution methods have on the structure of the model. The adaption of the model and the connection with a computer based solution method made it further necessary to make the model machine accessible. This will be discussed in Section 2.4.2.

2.4.1 The Importance of Good Modeling Practice

Most important for the applicability of a model in real-life situations is, whether it can be used to solve problems of industry-relevant sizes. Whether this is possible greatly depends on the solution time of the algorithm. Although modern algorithms have made a bigger number of model classes solvable, it still makes a difference which model formulation is chosen, especially if solution time is considered, even if several mathematically equivalent formulations are compared. Choosing the "right" *structure* is a matter of experience and distinguishes good from bad modellers.

When we consider MIP or nonlinear problems, the solution time can often be reduced significantly by appropriate modeling. This is important since in contrast to ordinary LPs, effective solution of MIP or nonlinear problems depends critically upon good model formulation, the use of high level branching constructs, control of the Branch-and-Bound strategy, scaling and the availability of good initial values. The solution times can be greatly influenced by observing a few rules which distinguish "bad" from "good" modeling.

Good formulations in MIP models are those whose LP relaxation is as close as possible to the MILP relaxation, or, to be precise, those whose LP relaxation has a feasible region which is close to the *convex hull*, that is the smallest polyhedron including all feasible MILP points. In practice, this means, for example, that upper bounds should be as small as possible. If α_1 and α_2 denote integer variables, the inequality $\alpha_1+\alpha_2 \leq 3.7$ can be bound-tightened to $\alpha_1+\alpha_2 \leq 3$. Another example: Expressions containing products of k binary variables can be transformed to MILP models according to

$$\left\{\delta_p = \prod_{i=1}^{k} \delta_i\right\} \Longleftrightarrow \left\{\begin{array}{cc} \delta_p \leq \delta_i,\ i=1,...,k \quad ; & \\ \sum_{i=1}^{k} \delta_i - \delta_p \leq k-1 \quad ; & \delta_i \in \{0,1\} \end{array}\right\} . \tag{2.4.1}$$

However, the formulation (2.4.1) is superior to the alternative, and algebraically equivalent set of only two inequalities

$$\left\{\delta_p = \prod_{i=1}^{k} \delta_i\right\} \Longleftrightarrow \left\{\begin{array}{c} k\delta_p \leq \sum_{i=1}^{k} \delta_i \quad ; \\ \frac{1}{k}\sum_{i=1}^{k} \delta_i - \left(k - \sum_{i=1}^{k} \delta_i\right) \leq \delta_p \end{array}\right\} \tag{2.4.2}$$

because the $k+2$ inequalities in (2.4.1) couple δ_p and the individual variables δ_i directly.

In many models, in addition, it might be possible, to derive *special cuts*, *i.e.*, *valid inequalities* cutting off parts of the LP relaxation's feasible set but leaving the convex hull unchanged. If we detect inequalities of the form $x + A\alpha \geq B$

with constants A and B, and variables $x \in \mathbb{R}$ and $\alpha \in \mathbb{N}$ in our model, we can enhance and tighten our model by the cuts

$$x \geq [B - (C - 1)A](C - \alpha) \quad , \quad C := \left\lceil \frac{B}{A} \right\rceil = ceil\left(\frac{B}{A}\right) \quad , \tag{2.4.3}$$

where C denotes the rounded-up value of the ratio B/A. If a semi-continuous variable σ, *e.g.*, $\sigma = 0$ or $\sigma \geq 1$, occurs in the inequality $x + A\sigma \geq B$ it can be shown that $\frac{x}{B} + \sigma \geq 1$ is a valid inequality tightening the formulation.

All these examples show that the quality of the model greatly depends on the care the modeler takes when designing its structure.

Preprocessing can also improve the model formulation. Preprocessing methods introduce model changes to speed up the algorithm. They apply to both continuous and MIP problems but they are much more important for MIP problems. Some common preprocessing methods are: *presolve* (logical tests on constraints and variables, bound tightening), *disaggregation of constraints*, *coefficient reduction*, and *clique and cover detection* (see [140] and references therein). However, it is important that preprocessing is used carefully, because for some classes of problems, especially if rigorous computing is needed (*e.g.*, global optimization, see Section 4.2), the roundoff errors introduced by certain preprocessing techniques effectively change the model.

In models of industrial relevance, the number of variables and constraints which can be solved effectively usually makes the difference between success and failure of a model.

For very large problems, a second difficulty "lurks behind the corner". Somebody has to put together all relevant variables, constraints, and the data, without making errors. If the model is designed once, then fixed, and only the data changes, this can be done by normal database tools. However, if the model itself changes from problem instance to problem instance the overall structure of the model has to be clearly arranged.

For example, optimization problems coming from resource optimization [229] tend to be very large and highly structured. However, from application to application neither the variables nor the constraints are fixed, not even parametrizable. In spite of that, all constraints can be generated from building blocks, and the data can be retrieved from databases.

There is a hierarchical system of models of increasing complexity and increasing non-linearity. Constraints for different resources are similar in structure but not identical, so they cannot be modeled by using loops or simple indexing.

Complicated resource optimization problems can have several 100,000 variables. A typical problem looks like in Fig. 2.4.2. There the s_i describe the resource streams needed for the amount p of end-product. Depending on the complexity of the materials these are linear or non-linear mixed integer problems. Most constraints can be constructed by general principles like resource

$$
\begin{aligned}
\min\ & B(z_1,\ldots,z_q) = \sum_{i\in I} b_i(s_i, z_1,\ldots,z_q) \\
\text{s.t.}\ & s_i = f_i(p) \quad \forall i \in I_0 \subseteq I \\
& s_i = S_i(p, z_1,\ldots,z_q) \quad \forall i \in I_p \\
& s_i \le 0 \quad \forall i \in I_{\text{in}} \subseteq I \setminus (I_0 \cup I_p) \\
& s_i \ge -M_i y_{n_i} \quad \forall i \in I_{\text{in}} \cap I_d \\
& s_i \ge 0 \quad \forall i \in I_{\text{out}} \subseteq I \setminus (I_0 \cup I_p) \\
& s_i \le M_i y_{n_i} \quad \forall i \in I_{\text{out}} \cap I_d \\
& g_j(p, s_{j_1},\ldots,s_{j_n}) \le 0 \quad \forall j \in J_u,\ j_1,\ldots,j_n \in I \\
& h_k(p, s_{k_1},\ldots,s_{k_m}) = 0 \quad \forall k \in J_g,\ k_1,\ldots,k_n \in I \\
& G_j(z_1,\ldots,z_q) \le 0 \quad \forall j \in J_u^z \\
& H_k(z_1,\ldots,z_q) = 0 \quad \forall k \in J_g^z \\
& \sum_{i=1}^{m_r} y_{n_i} \le e_r \quad \forall r \in R \\
& p, s_i, z_j \in \mathbb{R}\ (\text{or } \mathbb{Z}) \quad \forall i \in I,\ j = 1,\ldots,q \\
& y_n \in \{0,1\} \quad \forall n \in E.
\end{aligned}
$$

Figure 2.4.2. A typical model for resource optimization

conservation. The other constraints depend on the resources needed, their impact on the environment, and all this information could be retrieved from a database.

For this type of problem, a modeling system would be most convenient, in which large models can be constructed from smaller building blocks, and it is important that the modeler keeps the structure in the model description. Otherwise, the model will not be applicable for more than a few simple (and in practice irrelevant) examples.

Good modeling practice ([249], [140]) takes advantages from good use of structuring, presolve, scaling, and branching choice. Modeling appears more as an art rather than a science. Experience clearly dominates. Remodeling and reformulations of problems (see, for instance, Section 10.4 in [140]) can significantly reduce the running time and factors like the *integrality gap*, *i.e.*, in a maximization problem the difference between the LP-relaxation and the best integer feasible point found. This task is still largely the responsibility of the modeler, although work has been done on automatically reformulating mixed zero-one problems [242]. For nonlinear problems this is much more difficult, though.

Especially, in nonlinear problems it is essential that the mathematical structure of the problem can be exploited fully. This is even more true if rigorous computing is involved. It is important that good initial values are made available by, for instance, exploiting homotopy techniques, or that good estimates for constraint propagation are provided by proper use of interval analytical techniques (*cf.* [142]).

2.4.2 Making Mathematical Models Accessible for Computers

As stated before, there is a second important consequence from computer assisted model solving. Somebody has to translate the model into a form accessible by the computer.

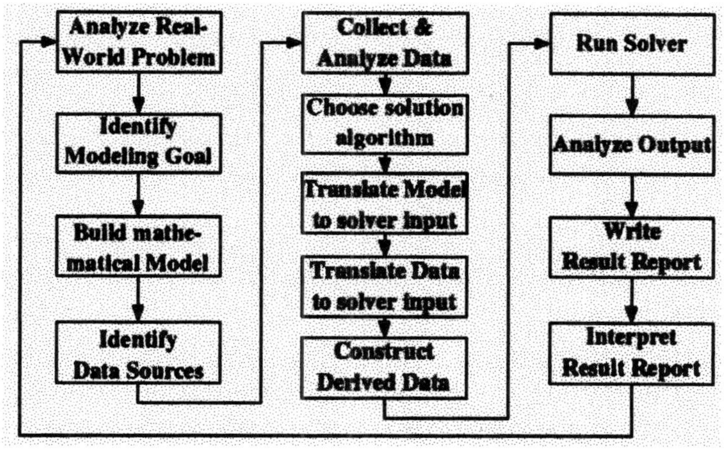

Figure 2.4.3. Detailed modeling cycle

If we consider again the simple modeling cycle in Fig. 2.4.1, we note that no step involving a computer is contained in this coarse grained view. In reality, the modeling process contains many more steps than described there. So let us consider the revised and more detailed model of the modeling process depicted in Fig. 2.4.3.

We observe that some of the additional steps in model building and solving involve translations from one format to another (*translate model to solver input format*, *translate data to solver input format*, *write status report*). These tasks are full of routine work and error prone. Furthermore, since during the various cycles sometimes the solution algorithm and the solver are changed, these "rewriting" steps have to be performed again and again.

A special case is the task *construct derived data*. Many solution algorithms need data which is a priori not part of the model, but which is also not provided by the solver. E.g. *gradients* of all functions involved are usually needed for an efficient local optimization algorithm. This not only involves routine but also mathematical work and additional translation steps.

As we have seen in Section 1.3 about the history of modeling languages, in the beginning all these translation steps were performed by hand and ended in the writing of *programs*, which would represent the mathematical model in an algorithmic way, almost like in Babylonian mathematics everything was described by means of algorithms. Data was stored in hand-written files. This method was very inflexible, error prone, and not re-usable. Maintenance of the models was close to impossible, and the models were not very scalable. This made the modeling process very expensive. In addition, a lot of expert knowledge in mathematics, modeling, and software engineering was necessary for building applicable models.

Many of the important models for military applications and for the industry were *optimization problems*, so methods were designed to reduce the modeling costs and to reduce the error rate in the translation steps, so many self-written modeling support systems were designed (and still are). However, using them still needed a lot of expert knowledge, and although scalability improved usually, flexibility did not.

Most of these problems now can be resolved by using a modeling language or modeling system, and the most important features of these will be analyzed in Chapter 4.

Acknowledgments

I want to thank Arnold Neumaier for his help and his important advice for preparing this section, and Josef Kallrath for his support and his contribution in Section 2.4.1.

Chapter 3

MATHEMATICAL MODEL BUILDING

Arnold Neumaier*

Institut für Mathematik der Universität Wien

Strudlhofgasse 4, A-1090 Wien, Austria

Arnold.Neumaier@univie.ac.at

http://www.mat.univie.ac.at/~neum

Abstract Some notes on mathematical modeling, listing motivations, applications, a numerical toolkit, general modeling rules, modeling conflicts, useful attitudes, and structuring the modeling work into 16 related activities by means of a novel modeling diagram.

Keywords: Modeling, history of modeling, model, mathematical model

3.1 Why Mathematical Modeling?

Mathematical modeling is the art of translating problems from an application area into tractable mathematical formulations whose theoretical and numerical analysis provides insight, answers, and guidance useful for the originating application.

Mathematical modeling is indispensable in many applications, is successful in many further applications, gives precision and direction for problem solution, enables a thorough understanding of the system modeled, prepares the way for better design or control of a system, and allows the efficient use of modern computing capabilities.

Learning about mathematical modeling is an important step from a theoretical mathematical training to an application-oriented mathematical expertise, and makes the student fit for mastering the challenges of our modern technological culture.

*supported by EU project COCONUT IST-2000-26063

This overview covers a vast field, and I cannot give references. For more information about mathematics, software, and applications, see, e.g., my web site at `http://www.mat.univie.ac.at/~neum/`.

3.2 A List of Applications

In the following, I give a list of *applications* whose modeling I understand, at least in some detail. All areas mentioned have numerous mathematical challenges. This list is based on my own experience; therefore it is very incomplete as a list of applications of mathematics in general. There are an almost endless number of other areas with interesting mathematical problems. Indeed, mathematics is simply the language for posing problems precisely and unambiguously (which even a stupid, pedantic computer understands).

Anthropology. Modeling, classifying and reconstructing skulls
Archeology. Reconstruction of objects from preserved fragments, Classifying ancient artifices
Architecture. Virtual reality
Artificial intelligence. Computer vision, Image interpretation, Robotics, Speech recognition, Optical character recognition, Reasoning under uncertainty
Arts. Computer animation (Jurassic Park, The Matrix)
Astronomy. Detection of planetary systems, Correcting the Hubble telescope, Origin of the universe, Evolution of stars
Biology. Protein folding, Humane genome project, Population dynamics, Morphogenesis, Evolutionary pedigrees, Spreading of infectious diseases (AIDS), Animal and plant breeding (genetic variability)
Chemical engineering. Chemical equilibrium, Planning of production units
Chemistry. Chemical reaction dynamics, Molecular modeling, Electronic structure calculations
Computer science. Image processing, Realistic computer graphics
Criminalistic science. Finger print recognition, Face recognition
Economics. Labor data analysis
Electrical engineering. Stability of electric circuits, Microchip analysis, Power supply network optimization
Finance. Risk analysis, Value estimation of options
Fluid mechanics. Wind channel, Turbulence
Geosciences. Prediction of oil or ore deposits, Map production, Earth quake prediction
Internet. Web search, Optimal routing
Linguistics. Automatic translation
Material Sciences. Microchip production, Microstructures, Semiconductor modeling
Mechanical engineering. Stability of high rise buildings, bridges, air planes; Structural optimization, Crash simulation
Medicine. Radiation therapy planning, Computer-aided tomography, Blood circulation models
Meteorology. Weather prediction, Global warming, What caused the ozone hole?
Music. Analysis and synthesis of sounds

Neuroscience. Neural networks, Signal transmission in nerves	
Pharmacology. Docking of molecules to proteins, Screening of new compounds	
Physics. Elementary particle tracking, Baryon spectrum, Laser dynamics	
Political Sciences. Analysis of elections	
Psychology. Formalizing diaries of therapy sessions	
Space Sciences. Trajectory planning, Flight simulation, Shuttle reentry	
Transport Science. Air traffic scheduling, Taxi for handicapped people, Automatic pilot for cars and airplanes	

3.3 Basic numerical tasks

The following is a list of categories containing the basic algorithmic toolkit needed for extracting numerical information from mathematical models.

Numerical linear algebra
Linear systems of equations, Eigenvalue problems, Linear programming (linear optimization), Techniques for large, sparse problems
Numerical analysis
Function evaluation, Automatic and numerical differentiation, Interpolation, Approximation (Padé, least squares, radial basis functions), Integration (univariate, multivariate, Fourier transform), Special functions, Nonlinear systems of equations, Optimization = nonlinear programming, Techniques for large, sparse problems
Numerical data analysis (= numerical statistics)
Visualization (2D and 3D computational geometry), Parameter estimation (least squares, maximum likelihood), Prediction, Classification (pattern recognition), Time series analysis (signal processing, filtering, time correlations, spectral analysis), Categorical time series (hidden Markov models), Random numbers and Monte Carlo methods, Techniques for large, sparse problems
Numerical functional analysis
Ordinary differential equations (initial value problems, boundary value problems, eigenvalue problems, stability), Techniques for large problems, Partial differential equations (finite differences, finite elements, boundary elements, mesh generation, adaptive meshes), Stochastic differential equations, Integral equations (and regularization)
Non-numerical algorithms
Symbolic methods (computer algebra), Sorting, Compression, Cryptography, Error correcting codes

Due to the breadth of the subject, this cannot be covered in a course on mathematical modeling. For a thorough education one needs to attend courses (or read books) at least on numerical analysis (which usually covers some numerical linear algebra, too), optimization, numerical methods for partial differential equations, and numerical data analysis (in particular pattern recognition).

3.4 The Modeling Diagram

The nodes of the diagram in Fig. 3.4.1 represent information to be collected, sorted, evaluated, and organized.

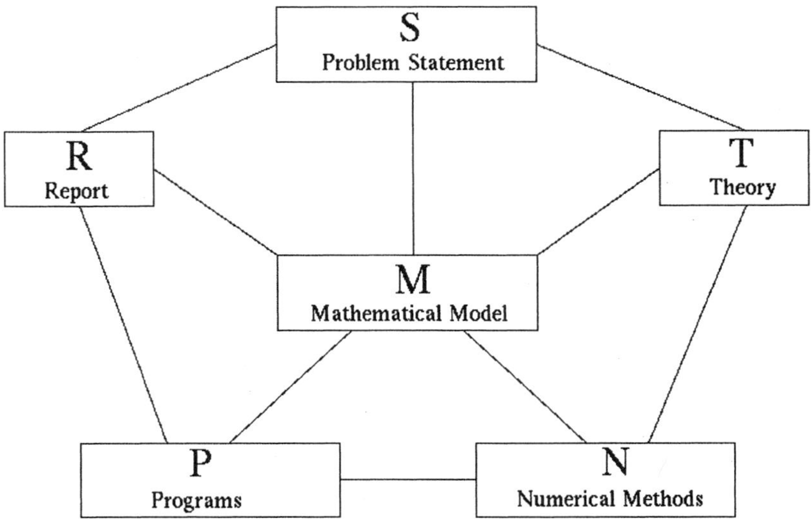

Figure 3.4.1. Modeling Diagram

The edges of the diagram represent activities of two-way communication (flow of relevant information) between the nodes and the corresponding sources of information. The following table characterizes the most important issues within each node.

S. Problem Statement
Interests of customer/boss
Often ambiguous/incomplete
Wishes are sometimes incompatible
M. Mathematical Model
Concepts/Variables
Relations
Restrictions
Goals
Priorities/Quality assignments
T. Theory
of Application
of Mathematics
Literature search

N. Numerical Methods
Software libraries
Free software from WWW
Background information
P. Programs
Flow diagrams
Implementation
User interface
Documentation
R. Report
Description
Analysis
Results
Model validation
Visualization
Limitations
Recommendations

Using the modeling diagram. The modeling diagram breaks the modeling task into $16 = 6 + 10$ different processes. Each of the 6 nodes and each of the 10 edges deserve repeated attention, usually at every stage of the modeling process. The modeling is complete only if the 'traffic' along all edges becomes insignificant. Generally, working on an edge enriches both participating nodes. If one is stuck along one edge, it is best to move to another one! Use the general rules below as a check list!

Frequently, the problem changes during modeling, in the light of the understanding gained by the modeling process. At the end, even a vague or contradictory initial problem description should have mutated into a reasonably well-defined description, with an associated precisely defined (though perhaps inaccurate) mathematical model.

3.5 General Rules

- Look at how others model similar situations; adapt their models to the present situation.
- Collect/ask for background information needed to understand the problem
- Start with simple models; add details as they become known and useful or necessary.
- Find all relevant quantities and make them precise.
- Find all relevant relationships between quantities ([differential] equations, inequalities, case distinctions).
- Locate/collect/select the data needed to specify these relationships.
- Find all restrictions that the quantities must obey (sign, limits, forbidden overlaps, etc.). Which restrictions are hard, which soft? How soft?
- Try to incorporate qualitative constraints that rule out otherwise feasible results (usually from inadequate previous versions).
- Find all goals (including conflicting ones)
- Play the devil's advocate to find out and formulate the weak spots of your model.
- Sort available information by the degree of impact expected/hoped for.
- Create a hierarchy of models: from coarse, highly simplifying models to models with all known details. Are there useful toy models with simpler data? Are there limiting cases where the model simplifies? Are there interesting extreme cases that help discover difficulties?

- First solve the coarser models (cheap but inaccurate) to get good starting points for the finer models (expensive to solve but realistic)

- Try to have a simple working model (with report) after $1/3$ of the total time planned for the task. Use the remaining time for improving or expanding the model based on your experience, for making the programs more versatile and speeding them up, for polishing documentation, etc.

- Good communication is essential for good applied work.

- The responsibility for understanding, for asking the questions that lead to it, for recognizing misunderstanding (mismatch between answers expected and answers received), and for overcoming them lies with the mathematician. You cannot usually assume your customer to understand your scientific jargon.

- Be not discouraged. Failures inform you about important missing details in your understanding of the problem (or of the customer/boss) – utilize this information!

- There are rarely perfect solutions. Modeling is the art of finding a satisfying compromise. Start with the highest standards, and lower them as the deadline approaches. If you have results early, raise your standards again.

- Finish your work in time.

Lao Tse: *"People often fail on the verge of success; take care at the end as at the beginning, so that you may avoid failure."*

3.6 Conflicts

Most modeling situations involve a number of tensions between conflicting requirements that cannot be reconciled easily.

The conflicts described below are creative and constructive, if one does not give in too easily. As a good material can handle more physical stress, so a good scientist can handle more stress created by conflict.

"We shall overcome" – a successful motto of the black liberation movement, created by a strong trust in God – generalizes to other situations where one has to face difficulties, too.

Among other qualities it has, university education is not least a long term stress test – if you got your degree, this is a proof that you could overcome significant barriers. The job market pays for the ability to persist.

fast	–	slow
cheap	–	expensive
short term	–	long term
perfecting a program	–	need for quick results
collecting the theory	–	producing a solution
doing research	–	writing up
quality standards	–	deadlines
dreams	–	actual results
simplicity	–	complexity
low quality	–	high quality
approximate	–	accurate
superficial	–	in depth
sketchy	–	comprehensive
concise	–	detailed
short description	–	long description

Einstein: *"A good theory"* (or model) *"should be as simple as possible, but not simpler."*

3.7 Attitudes

Do whatever you do with love. Love (even in difficult circumstances) can be learnt; it noticeably improves the quality of your work and the satisfaction you and others derive from it.

Do whatever you do as a service to others. This will improve your attention, the feedback you'll get, and the impact you'll have.

Take responsibility; ask if in doubt; read to confirm your understanding. This will remove many impasses that otherwise would delay your work.

Jesus: *"Ask, and you will receive. Search, and you will find. Knock, and the door will be opened for you. For whoever asks will receive, who seeks will find, and the door will be opened to him who knocks."*

Chapter 4

THEORETICAL CONCEPTS AND DESIGN OF MODELING LANGUAGES FOR MATHEMATICAL OPTIMIZATION

Hermann Schichl*
Institut für Mathematik der Universität Wien
Strudlhofgasse 4, A-1090 Wien
Hermann.Schichl@esi.ac.at

Abstract In this chapter we will present the basic design features of modeling languages, turning our attention to algebraic modeling languages. Later we will introduce an important class of optimization problems — global optimization, and illustrate the difficulties in constructing models for such problems.

Keywords: Modeling, modeling language, modeling systems, modeling software, Algebraic Modeling Language, declarative language, global optimization

4.1 Modeling Languages

The development of modeling languages started in the late 1970's when `GAMS` was designed, although, as described in Section 1.3 there had been some precursors before.

Since that time all the error prone routine tasks in Fig. 2.4.3 could be performed by the computer. The process of modeling became much more convenient, and the flexibility was increased a lot.

*funded by EU project COCONUT IST-2000-26063

In a modeling language, the model can be written in a form which is close to the mathematical notation, actually a typical feature of an *algebraic modeling language* (see Section 4.1.1).

The formulation of the model is *independent of solver formats*. Different *solvers* can be connected to the modeling language, and the translation of models and data to the solver format is done automatically. This has several advantages. The formerly tedious and error prone translation steps are done by the computer, and after thorough testing of the interface errors are very unlikely. There is a clean cut between the modeling and the numerical, algorithmic part. In addition, for hard problems *different* solvers can be tried, making it more likely that a solution algorithm is found which produces a useful result.

In a modeling language, model and model data are kept separately. There is a clear cut between the model structure and the data. Thus, many different instances of the same model class with varying data can be solved. Many systems provide an ODBC (open database connectivity) interface for automatic database access and an interface to the most widely used spreadsheet systems. This relieves the user from the laborious duty of searching for the relevant data every time the model is used. A second advantage of this concept is that during the development phase of the model (in the cycle) the approach can be tested on *toy problems* with small artificial data sets, and later the model can be applied without change for large scale industry-relevant instances with real data.

Even the problem of *derived data* is solved in most modeling systems. Due to the development of *automatic differentiation* (see *e.g.*, [91]) the modeling languages can generate derived information (gradients, sparse Hessians,...) from the model description without assistance of the user.

Today, there are several modeling systems in use, as described in Section 1.3. At first we will analyze the biggest class, the algebraic modeling languages, in Section 4.1.1. In the later sections we will shed light on the other language classes, ending with languages.

4.1.1 Algebraic Modeling Languages

This is the biggest class of modeling languages. Typical representatives are `AIMMS` [24] (see Chapter 6), `GAMS` [32] (see Chapter 8), `AMPL` [76] (Chapter 7), `mp-model` [6], `LINGO` [213] (Chapter 9), `MPL` (Chapter 13), `NOP` [182], `NOP-2` [207] (Chapter 15), `Numerica` [238], and `MINOPT` [218] (Chapter 11).

In [120] and [122] Tony Hürlimann describes why algebraic modeling languages can be viewed as a new paradigm of programming. Usually programming languages are divided into three different classes:

Imperative Languages These languages are also called *procedural languages*, typical representatives are `C`, `C++`, `Pascal`, `FORTRAN`, `Java`. With these languages a computer is programmed in a way which has a strong connection to the von Neumann concept of a computer. The stage of the computation can always be described by the state of the computer's memory locations, its instruction pointer, and some status registers. At each step this state changes until the computation reaches a terminal state, usually when the desired result is computed. The ability to assign values to variables and change the corresponding memory location in that way, and the explicit sequential execution are characteristics of an imperative language. An extension to that concept, as HÜRLIMANN expresses it: the ultimate consequence, is *object oriented programming* where computation proceeds by changing the local state of objects.

Functional Languages This paradigm is based on the concept that every computation can be viewed as a function $f : X \longrightarrow Y$ translating an input from X to a unique output in Y. Since every value can be represented as a function, no variables are necessary, and there is no need for an assignment operation. There is a distinction between function definition and application of a function, and most important, functions are viewed as values themselves. A typical representative of functional programming languages is `LISP`.

Logic Programming Languages This paradigm was developed in the 1960s during the construction of theorem provers, programs which were designed to prove mathematical theorems. In this development process it was detected that every mathematical proof can be regarded as a computation following specific rules. In the reverse, every computation can as well be regarded as a proof, and so the most important representative of this language category, the programming language `Prolog` was designed.

Every program in `Prolog` consists of a non-empty set of goals and a set of *(Horn)-rules*[1]. Using the backtracking algorithm, or more modern resolution methods (*Constraint Logic Programming*, see [129]) the program tries to reach a goal obeying the rules. As a short summary, one could say that a Logic Programming language can be seen as a "notational system for writing logical statements together with specified algorithms for implementing inference rules" [159, p. 426]. Typical Logic Programming languages are `Prolog V` and `ECLiPSe`.

All languages from these classes specify a problem in an algorithmic way. Like in Babylonian mathematics, it is not specified *what the problem is* but rather

A *Horn-rule* or *Horn-clause*is a disjunction of literals (i.e. atomic statement) with at most one positive literal. If the clause contains only the positive literal, it is called a *fact*.

how to solve the problem. For this reason we call these languages *algorithmic languages*.

In contrast to that, modeling languages store the *knowledge* about a model, they *define the problem* and usually do not specify how to solve it. They are *declarative languages*.

The important question here is: How do we describe what the problem is? The answer is: By specifying the problem's properties.

We start with a state space X (usually $\mathbb{R}^n \times \mathbb{Z}^m$). On this state space we define a set of *constraints* by mathematical formulas (usually *equations*, *inequalities*, and *optimality requirements*), which together define a relation $R : X \to \{\text{true}, \text{false}\}$ deciding about the *admissibility* of a member $x \in X$. We say x is *admissible*[2] if no constraints are violated, *i.e.*, if $R(x) = \text{true}$, and we define the *mathematical model* $M := \{x \in X \mid R(x)\}$ for the problem we are considering. This formulation is very close to the modern mathematical approach, and it has an important consequence:

Since no solution algorithm is specified, there is a priori no reason, why a problem defined in the declarative way should be solvable at all, and there is definitely no reason why there should be an algorithm for finding a solution.

Fortunately, however, there are large classes of problems (like *linear programming*) for which algorithms to explicitly find the set M exist, and the number of solution algorithms for various problem categories grows steadily.

If we direct our attention back to declarative languages, we see their three most important aspects:

Problems are represented in a declarative way.
There is a clear separation between problem definition and the solution process.
There is a clear separation between the problem structure and its data.

Algebraic modeling languages are a special class of declarative languages, and most of them are designed for specifying *optimization problems*. Usually they are capable of describing problems of the form

$$\begin{aligned} \min\ & f(x) \\ \text{s.t.}\ & F(x) = 0 \\ & G(x) \leq 0 \\ & x \in \mathbf{x}, \end{aligned} \tag{4.1.1}$$

or something slightly more general. Here $\mathbf{x}$ denotes a subbox of $X = \mathbb{R}^m \times \mathbb{Z}^n$.

[2] I have chosen the word *admissible*, since in optimization the word *feasible* is used to describe all points which obey the equality- and inequality constraints but not necessarily the optimality requirement.

The problem is flattened, *i.e.*, all variables and constraints become essentially one-dimensional, and the model is written in an *index-based* formulation, using *algebraic expressions* in a way which is close to the mathematical notation. Typically, the problem is declared using *sets*, *indices*, *parameters*, and *variables*.

Conceptually similar entities are grouped in sets. The entities in the sets are later referenced by indices to the elements of those sets. Groups of entities (variables, constraints) can then be compactly represented and used in algebraic expressions. Here it must be noted that one of the most important characteristics of modern algebraic modeling languages is their ability to describe *non-linear models*.

In AMPL, for instance, the expression

$$\sum_{i \in S} x_i y_i$$

would be written as

```
sum { i in S } x[i]*y[i];
```

This leads to a problem formulation which is very close to a mathematical formulation, and translation of mathematical models to declarations in a modeling language usually only involves syntactic issues.

The algebraic modeling language is then responsible for creating a *problem instance* that a solution algorithm can work on. This is done by expanding the compact notation by indexing all the sets and adding the model data; this is often called the *set indexing ability* of algebraic modeling languages. Very often the modeling language provides an additional *presolve* phase before transferring the problem to the solution algorithm. In this phase the model is analyzed and certain simplification techniques are applied. This flattened and simplified model is then passed on to the solver.

As a small example we consider the problem

$$\begin{aligned} \min\ & x^T Q x + c^T x \\ \text{s.t.}\ & Ax \leq b \\ & \|x\| \geq 1 \\ & x \in [l, u], \end{aligned} \tag{4.1.2}$$

with an $N \times N$-matrix Q and an $M \times N$-matrix A.

Writing down all matrix products in index form we transform problem (4.1.2) into the "flat" model

$$\begin{aligned}
\min \quad & \sum_{i=1}^{N}\Big(\sum_{j=1}^{N} Q_{ij}x_j + c_i\Big)x_i \\
\text{s.t.} \quad & \sum_{j=1}^{N} A_{ij}x_j \le b_i \quad \forall i = 1,\dots,M \\
& \sqrt{\sum_{i=1}^{N} x_i^2} \ge 1 \\
& x_j \in [l_j, u_j] \quad \forall j = 1,\dots,N.
\end{aligned} \tag{4.1.3}$$

This model can then easily be "programmed" in *e.g.*, AMPL:

```
### PARAMETERS ###
param N>0 integer;
param M>0 integer;
param c {1..N};
param b {1..M};
param Q {1..N,1..N};
param A {1..M,1..N};
param l {1..
N};
param u {1..N};
### VARIABLES ###
var x {1..N};

### OBJECTIVE ###
minimize goal_function:
  sum {i in 1..N} (sum {j in 1..N} Q[i,j]*x[j] + c[i]) * x[i];

### CONSTRAINTS ###
subject to linear_constraints {j in 1..M}:
   sum {i in 1..N} A[j,i]*x[i] <= b[j];
norm_constraint:  sqrt(sum {j in 1..N} x[j]^2) >= 1;
box_constraints {j in 1..N}: l[j]<=x[j]<=u[j];

################ DATA #####################
data sample.dat;

###########################################
solve; display x;
```

By reading through the definition, we can easily identify the various parts of the flat model, only complemented by declarations of parameters and variables. In the data section we define that the model data should be read from the file *sample.dat*, and the last line contains the only *procedural statements*, `solve` which calls the solver, and `display x` which prints the solution.

Before we consider different approaches to modeling languages, here is a short summary on the most important design features of algebraic modeling languages:

Variables, constraints with arbitrary names,
Sets, indices, algebraic expressions (possibly non-linear),
Notation which is close to the mathematical formulation,
Problem structure is data independent,
Models can be scalable, by a change of the parameters a switch can be made from *toy-problems* to *real-life problems*.
Most statements are declarative, except for *conditionals*, *loops*, and very few procedural statements,
Flexibility in the types of models that can be specified,
Convenient for the modeler (little overhead in model formulation...),
Simple interface between modeling language and solver

- It must be easy to connect a solver to the modeling language and to have easy access to the model, the data, and the derived information like derivatives.
- Sending data back from the solver to the modeling language like progress reports, solutions, or error messages should be straightforward.

Simple and powerful data handling (ODBC, spreadsheet interfaces, data files,...),
Automatic differentiation.

For a further analysis see [78].

4.1.2 Non-algebraic Modeling Languages

In certain areas, algebraic modeling languages have not been very successful.

Sometimes, writing a model with variables and constraints is just too tedious, especially if the model is highly structured.

For chemical engineering there are two modeling systems `gPROMS` [7] and `ASCEND` [187] which build models in an object oriented way. For manufacturing processes a system called `EXTEND` [197] is available. All models are built from primitive entities (*e.g.*, reactors or tanks), and these can be assembled in compound entities with certain attributes. The modeler only needs to specify the parts and their connections, and the modeling system builds the mathematical model automatically. It introduces variables and derives constraints from physical and chemical principles like the law of mass conservation.

A very recent development combining these ideas came from the ESPRIT project "Simulation in Europe Basic Research Working Group (SiE-WG)" led by a group of simulation experts. The combined effort of the members of this project led to the development of the new *object-oriented modeling language* `Modelica` [63].

In *Constraint Logic Programming* there is a different reason for the lack of interest in algebraic modeling languages. For most of the difficult discrete problems there is no generic algorithm available which can solve the problems without hints from the modeler.

Hence, for solving these hard problems, it is necessary to store at least partial algorithmic knowledge together with the problem definition. Most algebraic modeling languages do not provide this possibility. Even more, discrete problems usually need model descriptions with very specialized constraints (`all_diff` — in a set of variables all take different values, `cardinality` — exactly N of a given set of variables have the value *true*). These discrete constraints are usually not provided by algebraic modeling languages. There they have to be reformulated using special mixed-integer linear constraints (see [138, Chap. 5]), which sometimes is inefficient and hides the model structure.

In these areas the modern Constraint Logic Programming languages like `ECLiPSe` are usually used. A sample file is given below. It can be used to solve the famous n-queens problem using various search and backtracking strategies.

```
% ECLiPSe SAMPLE CODE
%
% AUTHOR:       Joachim Schimpf, IC-Parc
%
% The famous N-queens problem, using finite domains
% and a selection fo different search strategies
%
:- set_flag(gc_interval, 100000000).
:- lib(lists).
:- lib(fd).
:- lib(fd_search).
%--------------------
% The model
%--------------------
queens(N, Board) :-
    length(Board, N),
    Board :: 1..N,
    ( fromto(Board, [Q1|Cols], Cols, []) do
        ( foreach(Q2, Cols), param(Q1), count(Dist,1,_) do
            noattack(Q1, Q2, Dist)
        )
    ).

noattack(Q1,Q2,Dist) :-
    Q2 #\= Q1,
    Q2 - Q1 #\= Dist,
    Q1 - Q2 #\= Dist.
%-----------------------
% The search strategies
%-----------------------
labeling(a, AllVars) :-
    ( foreach(Var, AllVars) do
        count_backtracks,
        indomain(Var)              % select value
    ).

% functions labeling(b,.. to labeling(d,.. deleted here !

labeling(e, AllVars) :-

\   middle_first(AllVars, AllVarsPreOrdered), % static var-select
    ( fromto(AllVarsPreOrdered, Vars, VarsRem, []) do
%      search_space(Vars, Size), writeln(Size),

\       delete(Var, Vars, VarsRem, 0, first_fail),   % dynamic var-select
        count_backtracks,
```

```
        indomain(Var, middle)          % select value
    ).

% reorder a list so that the middle elements are first
middle_first(List, Ordered) :-
        halve(List, Front, Back),
        reverse(Front, RevFront),
        splice(Back, RevFront, Ordered).
%-----------------------------------
% Toplevel code
%
% all_queens/2 finds all solutions
% first_queens/2 finds one solution
% Strategy is a,b,c,d or e
% N is the size of the board
%-----------------------------------
all_queens(Strategy, N) :-            % Find all solutions
    setval(solutions, 0),
    statistics(times, [T0|_]),
    (
       queens(N, Board),
       init_backtracks,
       labeling(Strategy, Board),
%     writeln(Board),
%     put(0'.),
       incval(solutions),
       fail
    ;
       true
    ),
    get_backtracks(B),
    statistics(times, [T1|_]),
    T is T1-T0,
    getval(solutions, S),

\   printf("\nFound %d solutions for %d queens in %w s with %d backtracks%n",
           [S,N,T,B]).

first_queens(Strategy, N) :-          % Find one solution
    statistics(times, [T0|_]),
    queens(N, Board),
    statistics(times, [T1|_]),
    D1 is T1-T0,

\   printf("Setup for %d queens done in %w seconds", [N,D1]), nl,
    init_backtracks,
    labeling(Strategy, Board),
    get_backtracks(B),
    statistics(times, [T2|_]),
    D2 is T2-T1,

\   printf("Found first solution for %d queens in %w s with %d backtracks%n",
           [N,D2,B]).

%-------------------
% Utilities
%-------------------
search_space(Vars, Size) :-
    ( foreach(V,Vars), fromto(1,S0,S1,Size) do
         dvar_domain(V,D),
         S1 is S0*dom_size(D)
    ).
:- local variable(backtracks), variable(deep_fail).
init_backtracks :-
         setval(backtracks,0).
get_backtracks(B) :-
         getval(backtracks,B).
count_backtracks :-
```

```
        setval(deep_fail,false).
count_backtracks :-
        getval(deep_fail,false),        % may fail
        setval(deep_fail,true),
        incval(backtracks),
        fail.
```

Looking at this file two facts become immediately obvious. First, it is big although the model description needs only about 15 lines near the beginning. Second, the declarative information is only a very small part of the model. The remaining file contains a lot of algorithmic knowledge: two solution strategies, one for searching all solutions and one for finding only one.

Finally, there are modeling languages which directly interface to spreadsheet systems. This is especially convenient for users who do not want to bother about learning new systems. They can enter their data into the spreadsheets as usual, and afterwards they add a model by specifying spreadsheet formulas and bounds for certain cells. Then an add-on program for the spreadsheet system solves the so-entered model and stores the result back into the "variable cells". For MS Excel there is a *spreadsheet modeling system* by FRONTLINE Systems.

4.1.3 Integrated Modeling Environments

A more graphical approach to model building is taken by the so called integrated modeling environments. exploiting *graphical user interfaces* (GUI) for various tasks. There are GUI's

A) to support the model construction phase, *i.e.*, model building is done in a menu-driven way,

B) to visualize the solution process or to illustrate the solution,

C) to help build complete end-user interfaces (to allow for both model output and end-user input through multiple linked pages).

`AIMMS`, `MPL` and `OPL Studio` offer A), `MPL` by Maximal Software being the first modeling system with a graphical user interface, and `AIMMS` (see Chapter 6) by Paragon Decision Technology B.V. being the most advanced by allowing multiple views of the same model.

`MPL`, `OPL Studio` and `Mosel` offer B), `MPL` by showing a graph of the matrix and the objective function, and `OPL Studio` (see Chapter 17) by ILOG and `Mosel` by Dash Optimization being the most advanced by allowing the visualization of search trees.

`AIMMS` offers C), and this puts `AIMMS` apart from the other modeling systems, such as `Mosel`, `MPL` and `OPL Studio` which offer instead APIs and component libraries that allow linking to other programming language.

4.1.4 Model-Programming Languages

Recently, several language providers try to bridge the gap between the algorithmic languages used in constraint programming and the declarative languages mainly used in mathematical optimization.

Many modeling languages nowadays have built-in scripting capabilities (see Chapters 7 and 8), but these new languages provide elements for declaring models and for describing solution algorithms in a much more integrated way.

The `Mosel` [101] (Chapter 12) approach overall looks much like a programming language, even the model declaration part has to be programmed.

The language `LPL` [121] (Chapter 10), on the other hand, provides a declarative part which still looks similar to mathematical notation, and the algorithmic part has the look and feel of an imperative programming language.

Also `AIMMS` (Chapter 6) can be seen as a modeling system which combines declarative and algorithmic elements. There, however, the GUI is the main focus, and declarations and programming are based on it.

Basically, a trend can be seen that modeling languages provide the possibility to record algorithmic knowledge in addition to purely declarative modeling. This can be an advantage for the modeler because of increased flexibility, but the future will show whether this approach makes models more or less maintainable, and whether it increases or decreases the shelf-life time of the models.

The approach that the modeling system provides just the models and that the solver interface contains the algorithms is still widely used. This has the advantage that different solution algorithms can get different algorithmic "hints". However, it is more difficult to document the whole model solution process with this approach, especially if we consider that, *e.g.*, `LPL` provides automatic document generation capabilities.

4.1.5 Other Modeling Tools

Apart from modeling systems and modeling languages there are tools for analyzing already existing models.

The `ANALYZE` system [88] provides various tools for linear programming problems. An LP can examined w.r.t. *constraint effectiveness*, *objective effect*, and *grouping of constraints* (the system can also make suggestions on better constraint sorting).

For non-linear models there is a successor product. `MPROBE` [40] provides in addition to the aforementioned methods also a shape analysis tool for non-linear model parts.

4.2 Global Optimization

It was already outlined in Section 1.3 that the development of modeling languages started with the invention of matrix generators. At that time, linear programs were the only real-life models which could be effectively solved.

As the computing power increased and the solution algorithms became more effective, higher and higher dimensional nonlinear programs were made accessible. The modeling systems had to keep up with the development, and ultimately this was one reason for the emergence of algebraic modeling languages. The algorithms available solved *local optimization* problems in an approximate way. So most of the now available modeling systems provide very good connectivity for linear solvers and local optimization algorithms.

Modeling and computer aided model analysis can serve many different needs. As was pointed out in [138] various levels of rigor can be distinguished for models, depending on the modeling goal.

$$\text{Validation} \leq \text{Verification/Falsification} \leq \text{Mathematical Proof}$$

The highest level of rigor is necessary for *mathematical proofs*. There are real-world applications which need a similar level of rigor, *e.g.*, in chemical engineering and robotics. Most of these problems are global optimization problems. They are of the form

$$\begin{aligned} \min\ & f(x) \\ \text{s.t.}\ & F(x) \in \mathbf{F} \\ & x \in \mathbf{x}, \end{aligned} \tag{4.2.4}$$

where $\mathbf{F}$ and $\mathbf{x} \subseteq X$ are both boxes (multidimensional intervals), and $X = \mathbb{R}^n \times \mathbb{Z}^m$. In contrast to local optimization, the task is to find *all globally best* solutions to this problems and to do it in a rigorous way, *i.e.*, to compute the solution set S of (4.2.4) and a mathematical proof that S is the solution set.

When solving such models harmful approximations must not be performed, and a lot of techniques have to be applied which are not usually necessary for local optimization. The most important tool is *interval arithmetic* with *directed rounding*.

During the last few years an increasing number of algorithms were developed for global optimization (*e.g.*, BARON [203], αBB [4], GLOBSOL [142], and the COCONUT project [46]). Some of the solvers are now reaching a strength which enables them to solve some industry-relevant problems.

In this section we will glance at global optimization and pinpoint the difficulties which arise when interfacing algebraic modeling languages with global optimization software.

4.2.1 Problem Description

Let us consider again problem (4.2.4). Here we describe the most general problem class GLOBAL (see Section 1.2.1). The task is to find *all* $x \in \mathbf{x}$ with $F(x) \in \mathbf{F}$ and the property that

$$\forall y \in \mathbf{x} \text{ with } F(y) \in \mathbf{F} : \; f(y) \geq f(x).$$

The variables x can come in different flavours

continuous: The variable is real or complex valued, and usually bounds are provided for its domain (*e.g.*, the amount of fluid flowing through a pipe in a chemical plant).

integer: These are discrete variables which can take all integers (or complex integers) as values. Sometimes they are in addition bound constrained (*e.g.*, the number of television sets produced by some production site).

binary: The value is either 0 or 1. These variables are also called *decision variables* (*e.g.*, decide whether or not a certain chemical reactor is to be included when building a new plant).

discrete variables: These can take a finite number of values which do not need to be numbers.

semi-continuous: They are either zero or bigger as some threshold (*e.g.*, we either do not start production in facility A at all or we have to manufacture a minimum amount of goods).

partial integer: These variables take either integer values in $0, \ldots, N$ or continuous values in $[N, M]$.

There are several types of constraints F

continuous: These constraints mainly involve continuous variables. Usually they are piecewise differentiable and composed from elementary functions. They can be linear, quadratic, convex, or nonconvex.

discrete: They only take discrete values. Classification constraints are of this type.

logical: Constraints of this type involve logical operators and binary variables.

global: The number of variables involved in a single constraint is small for most real-life models. There are, however, some constraints which involve all or just a lot of variables. These are called global constraints. Typical representatives are *all-different*, *cardinality*[3], and *histogram* constraints.

special ordered sets: there are two categories:

- **SOS-1** At most one from a set of variables is non-zero, all the others vanish.
- **SOS-2** At most two variables from a given ordered list of variables are non-zero, and if two are non-zero, they have to be neighbors in the list.

exclusion regions: these are constraints of the form $x \notin \mathbf{x}^e$.

Compared to models with low requirements in rigor, for global optimization much more care has to be taken in data handling. For mathematical proofs all *data* items must be *known exactly*. *Round-off* in the translation of the input data from modeling system to the solver can destroy important model properties. E.g. the very simple system

$$\begin{aligned} 0.4x + 0.6y &\geq 1 \\ x, y &\in [0, 1] \end{aligned} \tag{4.2.5}$$

has exactly one solution $x = y = 1$. The system becomes infeasible when rounded to IEEE floating point representation.

For some problems the *data* might be *uncertain*, see Section 1.2.8. For example the elasticity module for steel bars can vary by more than 15%. For rigorous problem solving it is important that all parameters can be entered together with their amount of uncertainty.

Sometimes constraints are less relevant than others or they may be slightly violated in some circumstances. E.g. the radiation patterns of radiotherapy for cancerous tumors can be optimized. There are constraints on the minimal dose needed for the tumor cells, and maximum doses for the remaining tissue types. Sometimes there is no solution, depending on the size of the tumor and the geometric shapes. Then the physician might say: "Well, I don't mind if the liver gets a higher dose than allowed, as long as this excess radiation is confined to a small part of the tissue." In these cases we call the constraint a *soft constraint*.

4.2.2 Algebraic Modeling Languages and Global Optimization

Most modeling systems around have originally been developed for nonlinear local optimization and linear programming. When trying to connect a solver for global optimization problems to such a modeling language, the programmer usually runs into one or all of the following problems. Therefore, the global optimization community has developed its own input languages or modeling languages, like `Numerica` [238], `NOP` [182], or `NOP-2` [207] (see Chapter 15).

The difficulties start with the fact that most modeling systems pass a presolved problem, and all input data is rounded to floating point representation. Sometimes it is not even possible to decide whether a datum the solver sees is integer or only some real number very close to an integer. This involves round-off and a loss of important knowledge and makes mathematical rigor impossible.

The second point is that the performance of global optimization algorithms usually depends on whether specific structural knowledge about the problem can be exploited. Unfortunately, the problem which is passed from the modeling system to the solver is "flat" and has lost most of its mathematical structure, except for the sparsity pattern.

There is almost no modeling language around in which uncertain data can be specified. This makes it, *e.g.*, almost impossible to model safety programs (see Section 1.2.7).

There is only restricted support for soft constraints.

Many modeling languages only support the first three variable categories (continuous, integer, binary), and there is almost no support for global constraints (all different, cardinality, histogram,...) and exclusion regions.

Automatic differentiation is not sufficient for constructing all derived information needed for a global optimization algorithm. There is no inherent support for *interval analysis*, and hence essential data such as implied bounds cannot be generated automatically (except in `Numerica`).

There is no support for matrix operations, so all matrix operations have to be flattened, hence the structure is lost (consider, *e.g.*, the constraint $A^T DAx = b$). Some matrix operations cannot be used inside a model at all, like $\det(A) \geq c$, or A is *positive semidefinite*, specifying these constraints is impossible in all modeling languages I know.

Finally, many global optimization problems can only be solved by using several *mathematically equivalent formulations* of the same model at the same time. A very nice example for that principle from an application in robotics can be found in [153]. There is no modeling language which supports equivalent formulations.

4.3 A Vision — What the Future Needs to Bring

In this section we will make a few technical suggestions about extensions to the existing modeling languages which are necessary in order to be able to write global and stochastic optimization problems and multi-stage problems in a convenient way.

4.3.1 Data Handling

From the declarative part, which specifies the model structure, the modeling system generates the *problem instance* by adding the model data. This principle was already described in Section 4.1.1.

Since data is a very important part in the modeling process and later during the use of the model, it is important that knowledge about the data is recordable as well.

- The *source of the data* is important, and how much it suits the model.
- *Accuracy* or estimates on data accuracy should be recordable in the modeling system.

- Directed rounding, interval enclosures, and exact values should be supported for rigorous computing (see Chapter 15).
- The date of data construction is probably interesting, as well, and how long it can be considered *up-to-date*.
- Data *validity* and *completeness* is also relevant.

Data handling should be *convenient*. An appropriate interface to automatic data retrieval and data management tools should be included in a modeling system. There should be convenient ways to access *databases* (ODBC) and *spreadsheet* data.

4.3.2 Solver Views

Constraint Logic Programming (CPL) and their languages teach that for the processing of very complex models the declarative knowledge usually is not sufficient for solving the problem, as explained in Section 4.1.4.

Depending on the solver used, additional hints must be specified to guide the algorithm. These hints are neither part of the model structure nor part of the model's data. Different solvers will need different hints and guides for handling the same model instance. Useful additional information includes *search and selection strategies*, *scenarios for stochastic programs*, *starting points for local optimization*, *convex underestimating functions*, *valid cuts*, *bisection and backtracking hints*, *switching between mathematically equivalent model formulations*.
These hints can be kept separate for every solver used, and the modeling system would be responsible to construct a *solver's view of the model instance*, which is a mixture of declarative and algorithmic knowledge.

4.3.3 GUI

The introduction of GUI front-ends, which is, *e.g.*, done in `AIMMS` (see Chapter 6), `MPL` (see Chapter 13), and {OPL (see Chapter 17) provides the modeler with additional possibilities.

- With graphical capabilities and additional fonts, there could be an even *more mathematical notation*. It looks similar to the front-end of Mathematica 4, like in Fig. 4.3.1,
- Provide *model structuring* with coarse and fine grained views.
- Offer *syntax based editing* capabilities with *online help*.
- Visualize the connection between the model parts, cross referencing between variables and constraints.
- *Visualize Progress* during the solution process.
- *Visualize Result*.

$$\min \sum_{i \in X} (x_i - x_{0\,i})^2$$

$$\text{s.t. } \bigwedge j \in J : \sum_{i \in X} A_{ij}\, x_i \le b_j$$

$$\text{s.t. } \bigwedge i \in I : x_i \in [l_i, u_i]$$

Figure 4.3.1. Optimization model displayed in the Mathematica GUI front-end

4.3.4 Object Oriented Modeling — Derived Models

In the same way that object oriented programming is the ultimate consequence of imperative programming (*cf.* Section 4.1.1), the declarative languages need the same kind of extension: *derived models*.

This mimics the way mathematicians define new objects. The algebraic structure of a group is defined to be a monoid in which every element has an inverse. The remaining properties like the law of associativity and the existence of a unit element are "inherited" from the monoid structure. As a further example, we take the definition of an algebra as a vector space, which is also a ring with the same additive group, and additional compatibility laws are valid.

The same could be done in modeling. A derived model would be a model which has all properties of all the models it is an epigone of plus additional variables and constraints.

For example, a *screw* could be represented by variables and constraints and a few parameters like its size. *Steel* is a material, and all its properties are recorded with variables and constraints, and in the same way a model for *wood* is produced.

The derived modeling method now would make it straightforward to construct models for steel screws and for wooden screws.

4.3.5 Hierarchical Modeling

As described in Section 1.2.4, multi-stage models often arise naturally. For instance, in the first step we just want a method of proving that a given robot design is free of catastrophic singularities. We build a model A for that and solve it. The strict separation of model structure and data enables us to prove design correctness for many robots, and at some stage we decide that we want to compute an optimal design which is cheap and free of structural deficiencies.

The resulting problem is a two-stage design problem, where the first stage consists of model A, the only difference is that the design parameters become free variables instead of being fixed by the model data.

For making such a hierarchical approach to modeling possible, all *models* in a modeling language *should be reusable as "functions"* in bigger models.

4.3.6 Building Blocks

Many huge optimization problems are built from a number of smaller (almost) identical building blocks, especially in *chemical engineering*, *manufacturing processes*, *resource optimization*, *optimal scheduling*,...

Adding support for building blocks (with configurable automatic constraints when connected together) and enhancing the building block management by providing GUI support will *decrease the number of modeling errors* for big problems and *reduce the modeling costs*.

A good example for such a system is, *e.g.*, `BoFit` by `ProComm` [190], a system for the daily resource planning for electrical energy providers.

4.3.7 Open Model Exchange Format

It would be great if all modeling system providers could agree upon a common exchange interface for mathematical models. Until now it is a tedious task to translate a model from one modeling system to another.

The Open Model Exchange Format could be XML- and MathML-based. This would have the advantage that it would provide a convenient interface for model presentation on the Worldwide Web, it would need a consistent interface for floating point numbers which does not produce read/write mismatches due to rounding errors, and it should be easily extendable.

It would be useful to have ASCII and binary interfaces and encryption to protect intellectual properties. The most important feature would be that it is *open*, *public*, and *widely supported*.

Acknowledgments

I want to thank Arnold Neumaier for his help and his important advice for preparing this section, and Oleg Shcherbina for his support and the many references he provided.

Chapter 5

THE IMPORTANCE OF MODELING LANGUAGES FOR SOLVING REAL-WORLD PROBLEMS

Josef Kallrath
BASF-AG, GVC/S-B009 (Scientific Computing) & Dept. of Astronomy, University of Florida
D-67056 Ludwigshafen, Germany *Gainesville, FL 32611*
josef.kallrath@t-online.de kallrath@astro.ufl.edu

Josef Liesenfeld
MaBOS GmbH
D-68723 Schwetzingen, Germany
josef.liesenfeld@mabos.com

Abstract Algebraic modeling languages play an important role in solving real-world problems. They speed up the modeling process itself and the process of coding the model, they increase the efficiency of the solution process, and they may extend the shelf-life of a model. In short: modeling languages increase the productivity and maintainability of optimization projects.

The model coded in a modeling language reflects the structure of the real-world problem. The efficiency of the solution process benefits at least in two different aspects from a modeling system: first, by supporting good modeling practice and techniques and passing appropriate parameters to the solvers and, second, the time needed to realize and finish an optimization project depends on the capabilities of the modeling software. The shelf-life of a model is extended, as the model coded in a modeling language provides a valuable easy-to-read documentation.

From a practitioner's point of view it would be very useful if the modeling languages and modeling systems designers develop their software packages in close synchronization with the evolution of the solvers. It would be desirable to see features such as optimization under uncertain data, stochastic optimization, or robust optimization, as well as multi-criteria optimization being supported by providing appropriate language elements and syntax structures.

Keywords: Real-world problems, modeling software, optimization under uncertainty, practitioners

5.1 Modeling Languages and Real World Problems

Users of mathematical optimization benefit in three major ways when models are developed and problems are solved. First, there is the gain through the greater understanding of the problem. The very act of working through a model formulation with its builder can be of considerable benefit to a client. This knowledge is manifested in some mathematical documentation but also in the coded model itself. Second, there is the decision support system to be developed using the model and its solution capability. Using such a system can lead to substantial savings. Third, there is the availability of a model for future experimental purposes. It is possible to test ideas for future planning on the model that could not be conducted on actual processes themselves.

Declarative algebraic modeling languages support all three aspects. It is usually easy to learn it within a short time, and thus even for newcomers in industrial OR groups it can be used for quick prototyping and to get a model up and running as quickly as possible without having to focus on programming. Then come the phases of reformulation. Matrix generators written in FORTRAN, C, or C++ do not easily support this task. In a modeling language it takes only minutes to add or to reformulate a constraint. Thus, we clearly see an increase in productivity. After a while the model might have been validated and there is enough confidence that it produces reliable results. Then it already provides a first level of documentation and great flexibility in testing new scenarios.

In Section 2.4.1 is was pointed out that good modeling practice can be vital to the success of an optimization project. An algebraic modeling language can help enormously in testing different model formulations. A modeling language remains flexible when the client asks for new model features or when reformulations become necessary. It provides stability when it comes to re-engineering of software many years after its first installation. At MaBOS we have seen customers who had run, for instance, `Xpress-MP` in a mainframe environment for many years without problem. Then, the client wanted to shift the software from the mainframe platform to PCs. The only piece of software which did not cause any problem was `Xpress-MP` with its model and solver environment, while the data interfacing from the mainframe to other sources of data were subject to all sorts of disaster. If such a model were coded in COBOL or another ancient language, the disaster would have been complete.

Last but not least: If a real-world problem is mapped to a mathematical model coded in a modeling language, the chances are much higher that the model can still be understood in five or ten years. If there is a project documentation which holds all mathematical relations and the basic assumptions of the model, and if the model has been coded according to this documentation in a modeling language, there is not too much effort to maintain the software, or to understand even after many years what the original developers had in mind when the model was built. Some packages, such as LPL even produce the documentation automatically once the code is entered in the modeling language environment. In short, using a modeling language in an optimization project increases maintainability.

5.2 Requirements from Practitioners towards Modeling Languages and Modeling Systems

Long years experience from many projects tells us that a modeling language or modeling system can be checked against or should fulfill the following criteria:

Syntax (complete, easy-to-learn, easy-to-read). Complete means that it covers all the language features and structures supported by the state-of-the art solvers, *i.e.*, all nonlinear functions, all sorts of variables such as binary, integer, semi-continuous or partial-integer variables, but also special ordered sets. It would be highly desirable for modeling languages to have better support for optimization under uncertainty. Some modeling language providers support stochastic optimization for LP problems. Although, there are promising approaches [a detailed discussion on various algorithms for stochastic integer optimization can be, for instance, found in the survey of Klein-Haneveld & van der Vlerk (1999,[145]), and, more recently, in Schultz (2003,[216])] and successfully solved real world problems for stochastic MILP problems [*cf.*, Schultz (1995, [215]) or Carøe & Schultz (1999, [37])], this technology is not yet supported by current modeling language systems. Although one might argue that it is not strictly part of the syntax, we want to stress at this place that it helps a lot and can increase the readability of the code, if the modeling language allows one to define aliases or macros.

Solver Suite (LP, MILP, NLP, MINLP, CP, metaheuristics, global optimization techniques). For difficult combinatorial problems or NLP problems different solvers can demonstrate very different behaviours. Thus, it is very helpful if a modeling system offers a great variety of solvers, or allows users to connect their own solvers to the modeling system. Technically advanced users often develop their solvers themselves. It would be useful if the modeling systems provide an easy way to call solvers users wrote themselves in FORTRAN, C, C++, or Dynamic Link Libraries (DLLs).

Infeasibility Tracing. During the early stage of an optimization project one might see the solver returning the statement *problem is infeasible*. It is a very difficult task to identify potential reasons for these infeasibilities. It can be a great help if the modeling system helps to support the analyst in this phase, for instance, by the method of independent infeasible sets (IIS).

Multiple platforms/platform independence (Windows, UNIX, LINUX, Macintosh). A modeling system should not only support multiple platforms but should also be completely independent of the platforms and the operating systems. It is also very desirable to keep command line orientation as part of modeling languages. There are users, especially in academia, who prefer this, as very often this approach is faster and also somewhat more resistant to changes of the operating system.

Open Design (multiple solvers/databases). There are a great variety of public domain, free and commercial solvers available to solve LP, MILP, NLP or MINLP problems. The modeling system should provide links to such solvers as mentioned above (solver suite). The other part of openness refers to access the input data. The list of ways of storing data ranges from plain text files, over spreadsheets and databases to huge supply chain management systems such as SAP.

Indexing (powerful sparse index and data handling). Real world problems, especially supply chain optimization models, can easily require ten or more indices. The software needs to allow that many indices, but it also needs to handle the resulting data structures efficiently. Part of the indexing is how index sets are supported. Here it is very useful that the concept of subsets of index sets is efficiently implemented, or that operations such as intersection or union of index sets are supported. Efficient implementations of subsets of sets allow to construct efficiently to define a subset of all feasible combinations of site, plants and products.

Scalability (millions of constraints/variables). It is one thing to generate an LP matrix involving a few thousands of variables and constraints. It is a completely different task to do this with a few millions of objects. It is very important to be sure that the modeling languages has been designed in such a way to guarantee linear scale-up.

Memory Management. The model generation phase can be very demanding in terms of memory management when treating problems with several millions of variables and constraints. The relevant point here is how many number or non-zero coefficients can be stored per Megabyte, and also how efficiently this memory is accessed.

Performance (10 million nonzeros in less than a minute). There have been cases, in which the solver produced useful MILP-solutions within a two or three minutes, but it took 20 minutes to generate the matrix. This is certainly not ideal. The model generation phase should be very, very fast.

Robustness (very stable code, years of testing). Most modeling languages are around for a long while. It can be said that they fulfill this criterion. If we take a decade as the typical time scale to develop a modeling language and turn it into a robust system tested by many customers, it is foreseeable when the newer developed languages will fulfill the robustness criterion. Of course, the robustness is connected to the level of continuous development, *i.e.*, integrating the feedback into new releases and thus eliminating existing problems. Thus, the decision to obtain a new modeling language might be supported by the vendor's information on how much effort they put into development.

Deployment (embed into applications). This can be a major task. It would be very helpful, if the development of a GUI or a user-friendly application could be done in a short time.

Synchronism (keeping up with the state-of-the-art solvers). One often observes that the solvers develop faster than the modeling systems. It would be very positive if both areas could be kept in close synchronization. For instance, specialized MILP solvers can support column generation, but modeling languages usually do not support this feature. From a practitioner's point of view it would be very useful if the modeling languages and modeling systems develop their software packages in close synchronization with the evolution of the solvers. Otherwise, the full capabilities of the solvers cannot be used.

In addition to these points modeling systems could be checked against the following features are often required in solving real world problems but are not yet fully supported:

1 Optimization under uncertain data or stochastic optimization, robust optimization as well as multi-criteria optimization being supported by providing appropriate language elements and syntax structures.

2 In the area of global optimization it would be a great advantage if for certain nonlinear terms, associated convex underestimators or other structural information would be provided in addition.

3 Metaheuristics (Simulated Annealing, Tabu Search, Genetic Algorithms) adjusted to solve the standard optimization problem (1.2.1). This requires one to automatically generate appropriate variable neighborhood concepts. Of course, it is not expected that the metaheuristics compute the global optima, but at least that they should, especially for hard combinatorial problems, generate some useful feasible points. It is very positive to see that systems such as `GAMS` and `Mosel` now allow to call solvers based on metaheuristics such as Tabu Search.

4 A Graphical User Interface (GUI) might be used to enter and code a model. Some software packages in the chemical process industry and in the energy service industry allow the modelers to develop not in an algebraic but rather a graphical way. A graphical representation of optimization models could be used to explain or capture models from non-experts, similar to a *data flux diagram* (DFD), used in systems analysis, or an *entity-relationship diagram* (ERD) used in database modeling.

Two aspects, whose resolution is a nontrivial matter, may hopefully will find their way in to modeling systems:

1 *Version management*: Modeling systems should offer facilities to manage version of models so that it is easier to track the evolution of a model over time. This does not only address differences, but also a way to document the evolution for use by others.

2 *Distributed model development*: Modeling systems should offer facilities to support the development of a large model by a group of developers that want to work *simultaneously*. Typical problems have to do with maintaining consistency and avoiding name clashes.

Last but not least we would mention a point less related to the modeling language but rather to the philosophy of modeling language providers. Whatever progress is made in a modeling language, the providers should strictly keep *backward compatibility*, *i.e.*, previous model files should fully function under the new system. This puts a long binding constraint on a language provider company. Any feature supported by the language can never be taken out. It is very frustrating for a user if a model developed suddenly does not run under a new release of the modeling system. It is especially problematic if a company offers both a modeling system and a solver, and if these parts depend on the operating system. The costs to adjust are enormous if properly counted over the whole set of customers. Often projects have been finished years ago and the people who developed the model are not part of the company anymore or have been external consultants anyway. Thus, a new budget has to be applied for, a new project has to be set up whose task is to convert from the old modeling system to the new one and the consultants have to be hired again. Modeling system providers may argue that this is too costly for them to follow this strict rule. They might even offer a converter from the old to the new system, but still there are remaining costs externalized to their customer base because such converters usually do not function perfectly and even then a lot of testing is still required. If a modeling system provider has sold several thousand licences, we strongly believe that the total costs (costs inside the company plus all externalized costs) are smaller when all maintenance is properly done and centralized in the modeling system provider's place. And there is one great advantage for modeling providers following the strict principle of backward compatibility: they retain the loyalty of their customers.

Acknowledgments

I want to thank Bjarni Kristjansson (Maximal Software Inc.) for taking the time to discuss some of these requirements in great detail with me. Some of the requirements listed in this chapter stem from his Bad Honnef presentation. The comments provided by Johannes Bisschop (Paragon Decision Technologies B.V.) are also thankfully appreciated.

II

THE MODELING LANGUAGES IN DETAIL

Chapter 6

THE MODELING LANGUAGE AIMMS

Johannes Bisschop
Paragon Decision Technology B.V.
P.O. Box 3277
2001 DG Haarlem, The Netherlands
Jan.Bisschop@paragon.nl

Marcel Roelofs
Paragon Decision Technology B.V.
P.O. Box 3277
2001 DG Haarlem, The Netherlands
Marcel.Roelofs@paragon.nl

Abstract AIMMS[1] is an allround development environment for modeling experts to create fully functional, often optimization-based, decision support applications ready for use by end-users. In addition to a basic overview of the philosophy behind AIMMS, the paper will focus on advanced facilities for units of measurement, the modeling of time, and an open approach to solve Mixed-Integer NonLinear Programs (MINLP).

Keywords: AIMMS, Mixed Integer Nonlinear Programming, Outer Approximation, multi-agent system, deployment, unit-based scaling

In this paper, we will give an overview of the functionality provided by the AIMMS modeling system. First, in Section 6.1 we will provide a brief overview of the design goals behind AIMMS, as well as a brief description of the available modeling tools to accomplish these goals. A full description, also covering a more detailed account of the basics, can be found in the AIMMS user manuals,

[1]General information about AIMMS is available on the internet:
E-mail: info@paragon.nl, Web: www.aimms.com

which can be obtained from our web site. The remainder of the paper will focus on a number of more in-depth descriptions of selected advanced topics.

- Section 6.2 describes an open approach to solve MINLP models using the well-known outer approximation algorithm, and discusses possible further uses of such an open approach.
- Section 6.3 gives a detailed account of the units of measurement facilities in AIMMS.
- Time-based modeling is discussed in Section 6.4 relating planning horizons and calendars.
- AIMMS offers a powerful link to Excel which is described in Section 6.5.
- Section 6.6 provides an account of the multi-agent technology in AIMMS which allows a modeler to build distributed applications across a local area network.

Finally, a short indication of future developments in AIMMS is stated in the last section, followed by an appendix containing a complete feature summary of the present AIMMS release.

6.1 AIMMS Design Philosophy, Features and Benefits

The acronym AIMMS stands for Advanced Integrated Multi-dimensional Modeling Software. At its heart, AIMMS offers an algebraic modeling language, which, in its basic functionality, is comparable to other modeling languages such as GAMS or AMPL, but which also provides advanced modeling concepts not seen in these languages. In addition, AIMMS offers a graphical environment around its language, constructed to support two primary user groups: *modelers* (application developers) and *end-users* (decision makers). By allowing modelers to create a graphical end-user interface around their model, AIMMS provides the ability to place all of the power of the most advanced mathematical modeling techniques directly into the hands of the people who need these to make decisions.

AIMMS overview. As a development environment, AIMMS possesses a unique combination of advanced features and design tools which allow a modeler to build complex decision support applications which are easily maintainable—in a fraction of the time required with conventional programming languages. Fig. 6.1.1 provides a top-level overview of the components available in AIMMS. The modeling tools referred to in the figure will be described briefly in relation to the functional aspects outlined in this section.

AIMMS design philosophy. In practice, most operational real-life models are complex, detailed and large- to very large-scale. AIMMS has been designed

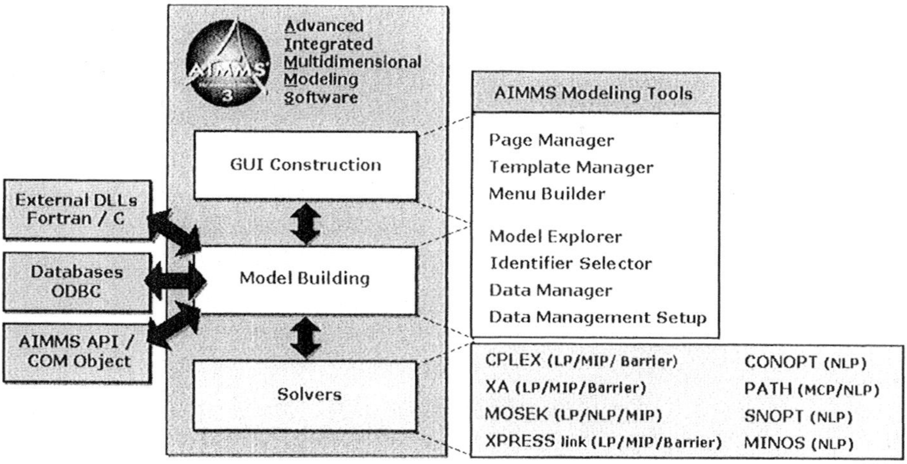

Figure 6.1.1. Graphical overview of AIMMS components

to optimally support modelers in building and maintaining such models, by offering excellent tools and advanced features for

- design, implementation and maintenance of a model,
- visualization of results, and
- deployment of a model.

In this section, we will discuss the features available in AIMMS for each of these areas in more detail.

Multi-dimensional modeling language. The multi-dimensional modeling language in AIMMS offers a powerful index notation which enables a modeler to capture the complexity of real-world problems in an intuitive manner. In addition, the language allows one to express very complex relationships in a compact manner without the need to worry about memory management or sparse data storage considerations. The combined declarations and procedures using these multi-dimensional structures can be organized, edited and displayed using an advanced interactive model editor.

Model explorer. Through our experience with the large-scale models of our customers over the past decade, we've come to realize that the traditional model building facilities—basically one or more ASCII files and an editor to operate on them—do not optimally support the modeler of an extensive, large-scale model. Not only is it often hard to remember where you placed a particular identifier declaration, but even worse, that location may have been dictated by the first use of the identifier in the remainder of your model. Therefore, AIMMS uses a *model database* to store all declarations and procedures within a model in a location independent manner, and provides access to these declarations through a (graphical) *model tree*. As the tree may also be extended with additional (named) organizational nodes, a modeler can impose any logical (directory-like) structure on his model through the *model explorer* tool. An example of a model tree is illustrated in Fig. 6.1.2. Note that each identifier type is displayed using a type-specific icon in the tree structure, making it easy to recognize identifiers either as a set, parameter, variable or constraint, etc. In the end, this greatly helps not only the modeler but also others reviewing the model in finding one's way around in large and complex models.

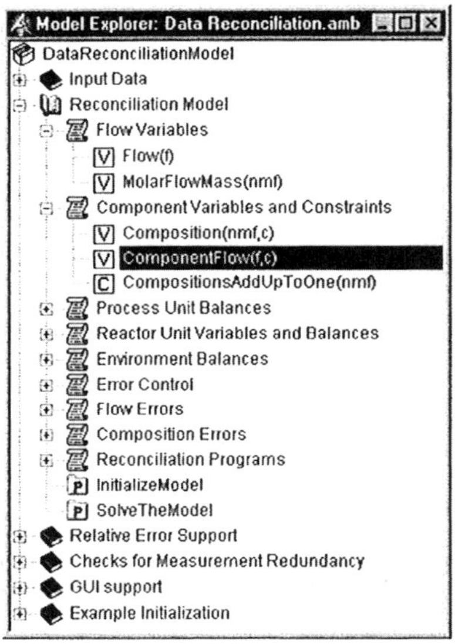

Figure 6.1.2. Example of a model tree

Declaration attributes. All attributes of an identifier declaration or procedure in AIMMS are presented in a graphical form as illustrated in the figure 6.1.3. For each identifier type, AIMMS only displays the relevant attributes, and provides attribute wizards wherever possible. In addition, within these forms, AIMMS supports convenient features such as automatic name completion, or can directly pop up the declaration form of an identifier occurring in an attribute through the context menu, allowing a modeler to make changes to it or look up its exact declaration.

Simultaneous overview. While an attribute form may be great for editing a single declaration, it effectively destroys the possibility to look at several declarations simultaneously, which was possible in the good old ASCII world.

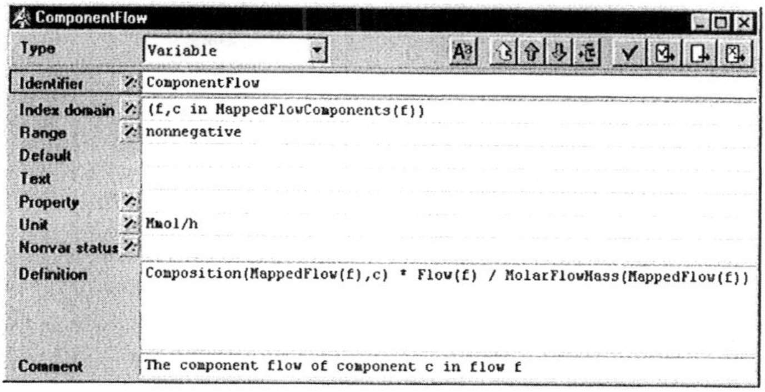

Figure 6.1.3. Declaration attribute window

For this reason, AIMMS also offers a view window in which the attributes of multiple identifiers can be viewed simultaneously, together with an *identifier selector* tool to flexibly construct an identifier selection based on one or more criteria (*e.g.*, all constraints referring to a particular variable). Through this tool, it is again possible to look at, and even change, multiple declarations of interest simultaneously, even if they are declared in completely different parts of the model.

Ready for use by domain experts. Often, in practice, the translation of real-life problems as understood by industry or domain experts in terms of working optimization models represents a challenge, due to incommonalities in the formulations used by domain and optimization experts. AIMMS helps to solve this problem by providing a common platform that can be understood by both domain and optimization experts, and can be used to interact and agree on the specification of a problem to be modeled. In many cases, the AIMMS model can serve as a complete replacement of a more formal problem specification on paper, cutting down the time and effort it takes to formulate and tackle a problem. Moreover, because AIMMS automatically translates optimization problems into a form that can be communicated with a solver and also supports a high-level multi-dimensional data manipulation language, AIMMS is an ideal tool to try different problem formulations and solution approaches, allowing you to quickly and easily find a solution that works for the problem at hand. Diagnostic tools such as a debugger and a profiler are available to give further insight into problems that may occur during model development or execution.

Declarative versus procedural language. Besides a declarative component for specifying multi-dimensional parameters, variables and constraints of a mathematical program, AIMMS also supports a powerful functional and procedural language. Parameters can have a definition, which is automatically updated whenever the data of identifiers referred to in the defining expression are changed. Procedures allow the implementation of complex data manipulations and data checks, or the implementation of algorithms around optimization models in very little statements. Together, they form an essential addition to almost any real-life modeling application.

Advanced language features. In addition, AIMMS also offers advanced language features for several commonly used modeling concepts such as *time-based modeling*, with calendars, (rolling) horizons and (semi-)automatic aggregation and disaggregation, or *units of measurement*, which almost always occur in real-life applications but are hardly addressed by modeling languages. Both topics will be discussed in more detail in Sections 6.4 and 6.3.

Visualization of results. Another important topic that has always been an integral part of the design of AIMMS, is the ability to visualize input data and results. Visualization is not only a essential feature for end-users of optimization models, but can also be of great help to modelers to interpret a solution, or to find a modeling mistake by visually interpreting the (faulty) result. Because of the importance for both types of users AIMMS has tightly integrated data visualization with its modeling language. AIMMS supports several types of graphical controls that can be placed on (end-user) pages, ranging from common controls such as tables, grids and 2D/3D charts, to more advanced controls such as a Gantt chart control and a network flow control in support of time- and network-based models.

Populating controls with data. As the focus within a modeling system like AIMMS should be on modeling a problem rather than programming the behavior of a GUI, populating a control with data almost never requires programming. It suffices to simply place the control in an area on an end-user page and select the model identifier(s) that hold the relevant data to be displayed through an identifier selection dialog box.

Advanced GUI tools. Usually, real-life models involve a lot of input and output data. As result, a graphical end-user application around such a model, will typically have a very large number of screens for both entering the input data and visualizing the results. To support building such applications in a quick and maintainable manner, AIMMS offers features such as page templates, a page tree in which pages can be organized in the *page manager* and *template manager* tools in a similar fashion as model declarations can be organized in the model tree, as well as special navigational controls, which can automatically navigate through the end-user following the organizational structure imposed by the page tree.

Case management. AIMMS offers an advanced scheme for storing model results and dealing with multiple scenarios through the use of cases and datasets. With the special *data manager* tool, a modeler can manage the entire collection of cases and datasets constructed for a particular AIMMS application. In addition, a set of cases can be selected to initiate a batch run, or graphical controls on an end-user page can be configured to to view model data for multiple cases simultaneously.

Deployment features. When the development a modeling application has been completed, the next phase is the successful deployment of the model in real life. One can distinguish basically two different deployment scenarios of an AIMMS model, namely

- as a stand-alone end-user application, or
- as an (optimization-based) component with another application.

Each of these scenarios poses its own requirements. AIMMS possesses a collection of advanced connectivity features, which allows an AIMMS model to be deployed in each of them.

Data connectivity. Whether deployed as an end-user application or as an optimization component, any model run on a daily basis must be fed with up-to-date data. AIMMS has features for reading from or writing to various data sources, such as

- ASCII data files,
- ODBC- and OLE DB-compliant databases,
- Excel spreadsheets, and
- XML documents.

For each of these data sources, it is possible to flexibly map the data present in a data source onto the corresponding identifiers in a model.

Calling AIMMS externally. AIMMS also provides C++ API as well as a COM object for external access to a model. This allows, for instance, the use of an AIMMS model from within another application, or as an optimization component underneath a web service. Based on the COM object, AIMMS comes with a ready-to-use Excel add-in in which all data transfer between the spreadsheet and AIMMS can be configured through a convenient configuration dialog box. This add-in allows the use of Excel as a the user interface to an underlying AIMMS model. The Excel add-in is discussed in more detail in Section 6.5.

Summary of benefits. The table in Fig. 6.1.4 summarizes the key benefits obtained from using AIMMS from three different perspectives, namely from a developer, an end-user, and a management perspective.

For Developers	For End-Users	For Management
• **Quick time-to-market:** –Modeling rather than programming –Integrated modeling environment –Easy to use: Point & click GUI builder –No rework to build 'end-user' production version from modeling prototype –High-quality productivity tools –Easy upgrade from GAMS • **Robust:** –Enforces structure and generic building –Best of breed execution engine –Linked to best solvers available (Cplex, Conopt, Xpress, ...) • **Flexible:** –Incremental application development –One implementation for modeling and production • **Transferable:** –Facilitates code sharing & transfer between programmers	• **Industry proof application:** –Robust –Stable • **Easy to use application:** –GUI Based –Fully functional Windows style • **Application tailored to real-life business requirements:** –Fast but flexible modeling approach assures that application can be build to cover actual business requirements (often unlike packaged solutions) –Flexible on changing topologies –Connectible to existing systems and/or data sources	• **Creates competitive edge:** –Fast time-to-market –Easily to up-and-downscale –Easily adaptable/extendable on changing environment • **Low Cost:** - One package deal (Solver, Modeler and GUI) - High maintainability reduces running costs - Fast Prototyping (POC) - Code sharing / transfer between programmers • **Fast and better insight in complex problems:** –Facilitates a mutual understanding between business side and OR professionals • **Brings structure to modeling and application building:** –Giving fast and improved insight in applications during (re)developing and auditing

Figure 6.1.4. Key benefits of AIMMS from different perspectives

6.2 AIMMS Outer Approximation (AOA) Algorithm

Outer approximation is a basic and well-known approach to solving Mixed-Integer Nonlinear Programming (MINLP) models. The underlying algorithm is an interplay between two solvers, namely one for solving mixed-integer models and one for solving nonlinear models. Even though the overall basic algorithm is remains unchanged, a user of the algorithm may want to specify the individual algorithmic steps in a customized fashion to obtain better performance and/or a better solution for a particular problem. To support such problem-specific optimizations, the outer approximation algorithm in AIMMS has been implemented as an open approach (a white box) instead of the usual closed (black box) implementation.

This section. This section describes and motivates an open approach to solving MINLP models based on outer approximation. Following a brief introduction to the problem statement and the basic algorithm, an open approach implemented using a single control procedure written in the AIMMS language is described. The control procedure uses external functions that are designed to execute some of the fixed algorithmic steps of the outer approximation algorithm, as well as some more general bookkeeping functions. The section concludes with a discussion how the open approach can be applied to modify the basic outer approximation algorithm, as well as how the open approach outlined here, could be applied to solve other complex problem types.

6.2.1 Problem Statement

MINLP. The mixed-integer nonlinear programming models to be solved can be stated as follows.

Minimize:

$$f(x, y)$$

Subject to:

$$\begin{aligned} h(x,y) &= 0 \\ g(x,y) &\leq 0 \\ Cy + Dx &\leq d \\ x \in \mathcal{X} &= \{x \in \mathbb{R}^n | x^L \leq x \leq x^U\} \\ y \in \mathcal{Y} &= \mathbb{Z}^m \end{aligned}$$

Usual Assumption. The usual assumption is that the nonlinear submodel (i.e. the model in which all integer variables are fixed) is convex. This assumption is to guarantee that each locally optimal solution of the nonlinear submodel is also a globally optimal solution. In practice, however, this assumption does not always hold, but the algorithm can still be applied. Convergence to a global optimum of the MINLP using the outer approximation algorithm is then no longer guaranteed.

6.2.2 Basic Algorithm

The algorithm consists of solving an alternating sequence of mixed-integer linear models and nonlinear models. First, the entire model is solved as a nonlinear program with all integer variables relaxed as continuous variables between their bounds. Then a linearization is made around the optimal solution, and the resulting constraints are added to the linear constraints that are already present. This new linear model is referred to as the master model, and is solved as an integer program. The integer part of the resulting optimal solution is then temporary fixed, and the original MINLP model with fixed integer variables is solved as a nonlinear submodel. Again, a linearization around the optimal solution is made and the new linear constraints are added to the master model. In addition, one or more constraints are added to cut off the previously-found integer solution of the master model. This iterative process is repeated until one of the termination criteria is satisfied.

Termination criteria. Several termination criteria are used in practice. The most frequently-used criterion is the iteration limit, because the size of the underlying integer program tends to grow significantly each time linearization constraints are added. In addition, practical experience has shown that a good solution is found during the first few iterations. A second criterion is the worsening of the objective value of two successive nonlinear submodels. Such worsening is not expected, but can be observed when an optimal solution of the nonlinear program is not a global optimum. A third termination criterion is insufficient improvement in the objective value of the mixed integer program in relation to the objective value of the previously-solved nonlinear submodel. Such insufficient improvement indeed happens when the optimal solution of the nonlinear submodel is also optimal or near-optimal for the overall MINLP model. However, insufficient improvement can also happen when the optimal solution of the nonlinear program is not a global optimum.

Final solution. Upon termination of the algorithm the best solution found at that point is declared as the final solution. This solution need not be optimal. The final solution is optimal whenever the nonlinear subproblems are always solved to global optimality and the algorithm is not stopped as a result of any of the above three termination criteria. In most practical applications, however, the search is terminated without finding a proven optimal solution.

Linearizations. The term 'outer approximation' refers to the linear approximation of the convex nonlinear constraints at selected points along the boundary of the convex solution region. The accumulation of such (inequality) constraints forms an outer approximation of the solution region, and this approximation can be used in the optimization instead of the nonlinear constraints from which it is derived. The formula for the linearization of a scalar nonlinear inequality $g(x, y) \leq 0$ around the point $(x, y) = (x^0, y^0)$ is as follows.

$$g(x^0, y^0) + \nabla g(x^0, y^0)^T \begin{bmatrix} x - x^0 \\ y - y^0 \end{bmatrix} \leq 0$$

The non-convex case. The linear approximation is no longer an outer approximation in case the solution region is not convex. In that case there is the possibility that portions of the solution region are cut off as illustrated in Fig. 6.2.5, where the linearization cuts off a portion of the nonlinear feasible region.

Loosening inequalities. This cutting off of the feasible region of the master model can be avoided by allowing the solution of the master model to slightly move away from the linear feasible region. The implementation is to allow deviations through artificial nonnegative variables and penalizing them while

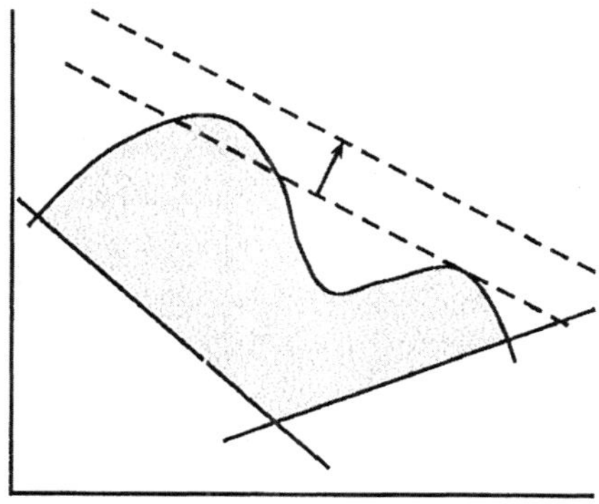

Figure 6.2.5. A non-convex linearization cutting off part of the feasible region

solving the master model. Fig. 6.2.5 illustrates how such an approach effectively enlarges the linear feasible region. Even though such an approach is not valid on theoretical grounds, in practice it turns out to be a workable heuristic to obtain better final solutions. Such loosening of linear approximations becomes especially of interest after several iterations of the outer approximation algorithm. At that time, there are linearizations at different points of the solution space, and these linearizations are all considered simultaneously in the master model. Loosening the corresponding inequalities has the desirable effect of preventing unwanted infeasibilities.

6.2.3 Open Solver Approach

The general communication pattern between `AIMMS` and any solver linked to AIMMS follows a predefined layout, defined by the AIMMS Open Solver Interface. The three main solver components that should be present in all solvers are as follows.

- A solver interface that should at least implement a `solve` function and some functionality to return the solution of the model.
- A solver data block that contains the problem instance as it is generated by `AIMMS`. By using generic callback functions, the solver will request `AIMMS` for the problem data once `AIMMS` has issued a `solve` statement.
- The solver then uses one or more algorithms (*e.g.*, the simplex algorithm, a branch-and-bound method) that are needed to solve the model. During the solution algorithm it is possible to execute some callback functions when installed from with the `AIMMS` model (*e.g.*, an `AIMMS` procedure that is run after every iteration of the solution algorithm).

AOA specific solver communication. In the specific case of the AOA solver, the solver returns control back to AIMMS as soon as the model has issued a `solve` statement, and the solver data block has been read by the solver. AIMMS will then run the AOA control procedure that has been assigned from within the model text. Fig. 6.2.6 illustrates the process. In this control procedure all functions that are exported by the AOA model are available to build a customized AOA solution algorithm. The basic outer approximation algorithm is just a predefined instance of such a customized solution algorithm.

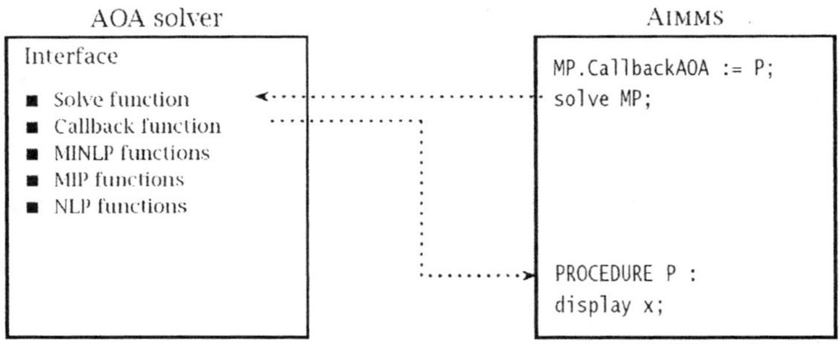

Figure 6.2.6. Solver communication illustrated

AOA module. All functionality concerning the AIMMS Outer Approximation (AOA) algorithm can be found in a single AIMMS module, called `OuterApproximation`, that is provided as part of the AIMMS system. This module contains not only the basic algorithm written in AIMMS, but also a description of all interface functions that are exported by the AOA Solver. In general terms, there are functions related to the manipulation of solutions and there are functions related to the generation and solution of the master MIP models and the NLP submodels. Selected examples of specific functions are

- functions to save, delete, retrieve and replace intermediate solutions,
- functions to add and delete linearizations of nonlinear constraints,
- functions to add and delete a cut to eliminate a particular integer solution,
- functions to solve the master MIP model and the NLP submodel,
- functions to indicate whether a particular type of model has been solved to optimality,
- functions for getting the value of the objective function, etc.

With these functions it is straightforward to write procedures in the AIMMS language to capture the logic of each step of the overall outer approximation solution algorithm.

Basic control procedure. The control procedure outlined below implements the basic outer approximation algorithm, and interacts with the Aimms Outer Approximation solver interface to solve and modify the NLP and master MIP submodels. The implementation of the control procedure can be modified as the user sees fit, or the user may create a completely different solution procedure. The control procedure refers to five procedures that execute the various steps. These steps are not further explained, but their reference is self-explanatory.

```
Initialize_All_Algorithmic_Identifiers ;
Solve_The_Relaxed_MINLP_Model_As_An_NLP ;

Repeat
    Add_Linear_Approximations_And_Solve_The_Master_MIP_Model ;
    Solve_The_NLP_Submodel_With_Fixed_Integer_Variables ;
    Terminate_Or_Prepare_For_Next_Iteration ;
Endrepeat ;
```

6.2.4 Alternative Uses of the Open Approach

With the open approach to implementing the outer approximation algorithm for MINLP models, it is possible to add to the existing control procedures or write alternative control procedures to optimize the solution process for particular classes MINLP problems. This section describes some less trivial examples of modifications, and also describes, in general, how the open approach may be useful for other problem types.

Solve more NLP's. Practical experience has shown that it is sometimes difficult to get a feasible solution to the initial relaxed NLP model. Based on the particular application at hand, the user may specify how multiple starting values can be found, and then modify the algorithm to solve multiple NLP's to get a feasible and/or a better solution. While doing so, it is also possible to specify how the algorithm should switch between different solvers. Such extensions could then also be applied to the NLP submodel inside the `Repeat` statement of the basic control procedure outlined above.

Retain integer solutions. It is possible to activate a MIP callback procedure whenever the MIP solver has found an integer solution. Even though these intermediate solutions are not optimal, the user may want to save the integer portion of these solutions for later evaluation. Once the main algorithm has terminated, all these integer solutions can be retrieved and evaluated by solving the corresponding nonlinear submodel. In some instances, one of these extra solutions may be a better solution of the original MINLP model than the one produced by the main algorithm.

Adjust penalties. Setting the penalties on the deviations of the linear approximation constraints in the master MIP submodel is a delicate manner, and has an effect on the solution quality in case the nonlinear submodels are nonconvex. The user may think of several problem-dependent strategies to adjust the penalty values, and implement them inside the basic AOA algorithm.

Application to other problem types. The outer approximation is an example of a more general class of decomposing solution algorithms, i.e. algorithms that decompose an entire optimization problem into two or more smaller optimization problems and use the interaction between these submodels to find a solution to the overall problem. Currently, several research groups investigate the use of various decomposition algorithms to tackle difficult combinatorial optimization problems. Possibly, an open approach as outlined in this section for the outer approximation algorithm may serve as a natural extension to modeling languages in support of such decomposition algorithms.

Conclusion. An open approach to solving MINLP models with an outer approximation algorithm allows for more variety in solution approaches, and thus offers the possibility to obtain better solutions for particular applications through customization of the underlying solution algorithm. `AIMMS` is the first modeling system to offer such an open interface.

6.3 Units of Measurement

Measurement plays a central role in observations of the real-world. Most observed quantities are measured in some unit (*e.g.*, dollar, hour, meter, etc.), and the magnitude of the unit influences the mental picture that a modeler may have of an object (*e.g.*, ounce, kilogram, ton, etc.). When such objects are combined in a numerical relationship, the corresponding units must be *commensurable*. Without such consistency, the mathematical relationships become meaningless. In spite of the central role of units in real-world problem descriptions, modeling languages do usually not offer support for units of measurement other than the possibility to mention these as a comment in the model text.

Why units in models. There are, however, several good reasons to offer first class support for units of measurement throughout a model formulation as provided by AIMMS. The explicit mentioning of units can enhance the readability of a model, which is especially helpful when others read and/or maintain a model. Units also provide the `AIMMS` compiler with additional checking power to find errors in model formulations. Finally, through the use of units a modeler can let `AIMMS` perform the job of unit conversion and scaling.

Standard units. The model editor in `AIMMS` will give a modeler access to a large number of predefined quantities and units, and in particular to those of the International System of Units (referred to as SI from the French "Systeme

Internationale"). The SI system is an improved metric system adopted by the Eleventh General Conference of Weights and Measures in 1960. The entire SI system of measurement is constructed from the atomic base units associated with the following nine basic quantities.

Quantity	Atomic Base Unit	Text
length	m	meter
mass	kg	kilogram
time	s	second
temperature	K	kelvin
amount of mass	mol	mole
electric current	A	ampere
luminous intensity	cd	candela
angle	rad	radian
solid angle	sr	steradian

Table 6.1. Basic SI quantities and their base units.

Derived quantities and units. All quantities which are not one of the nine basic SI quantities are called *derived* quantities. Each such quantity has a derived base unit which can be expressed in terms of the atomic base units of the basic SI quantities. Optionally, a compound unit symbol can be associated with such a derived base unit, like the symbol `N` for the unit `kg*m/s^2`. The following table illustrates some of the more well-known derived quantities and their corresponding derived base units. Note that five of them have an associated compound unit symbol. Many other derived quantities are available in `AIMMS` as predefined unit symbols.

Quantity	Derived Base Unit	Text
area	`m^2`	square meter
volume	`m^3`	cubic meter
force	`N = kg*m/s^2`	newton
pressure	`Pa = kg/m*s^2`	pascal
energy	`J = kg*m^2/s^2`	joule
power	`W = kg*m^2/s^3`	watt
charge	`C = A*s`	coulomb
density	`kg/m^3`	kilogram per cubic meter
velocity	`m/s`	meter per second
angular velocity	`rad/s`	radian per second

Table 6.2. Selected derived SI quantities and their base units.

Related units. Aside from the base unit that must be associated with every quantity, it is also possible to specify a number of *related* units. Related units are those units that can be expressed in terms of their base unit by means of a linear relationship. A typical example is the unit `km` which is related to the base unit `m` by means of the linear relationship x `km = 1000*`x `m`. Similarly, the unit `degC` (degree Celsius) is related to the base unit `K` through the formula x `degC = (`x `+ 273.15) K`.

Standard unit prefix notation. Frequently, related units are a multiple of their own base unit, which is reflected through a prefix notation that indicates the level of scaling. AIMMS supports the the standard SI prefix symbols and their corresponding scaling factor. Familiar examples are `kton`, `MHz`, `kJ`, etc.

Flexible specification. To give a modeler maximum freedom to choose his own quantities, units and naming conventions, `AIMMS` is not exclusively committed to any particular standard. However, modelers are encouraged to use the standard SI units and prefix symbols to make the model as readable and maintainable as possible.

6.3.1 Unit Analysis

Unit consistency. By associating a unit with every relevant identifier in the model, `AIMMS` will automatically verify whether all terms in the assignments and constraints of the model are unit consistent. When `AIMMS` detects unit inconsistencies, this may help a modeler to solve conceptual problems in the model, which could otherwise have remained undetected for a long time.

Atomic units. With every derived unit or compound unit symbol, it is possible to associate a unique unit expression consisting of a constant scale factor and atomic units only. All assignments and definitions in `AIMMS` are interpreted as formulas expressed in terms of these atomic unit expressions, and unit consistency checking is based on this interpretation. While ignoring the constant scale factors, `AIMMS` will verify that the atomic unit expression for every term in either an assignment statement or a constraint is identical. If the resulting unit check identifies an inconsistency, an error or warning will be generated.

An example. Consider the identifiers `a`, `b`, and `c` having units `[m]`, `[km]`, and `[10*m]` respectively, all with `[m]` as their corresponding associated atomic unit expression, and scale factors 1, 1000 and 10, respectively. Then the assignment `c := a + b` is *unit consistent*, because all terms share the same atomic unit expression `[m]`.

6.3.2 Unit-Based Scaling

With each identifier for which a UNIT attribute is specified, AIMMS associates two values:

- the *scaled* value (i.e. expressed in terms of the unit specified), and
- the *unscaled* value (i.e. expressed in terms of the associated atomic unit expression).

The transformation between scaled and nonscaled values is completely determined by the product of explicit and implicit scale factors associated with the various quantity and unit definitions.

When used. AIMMS uses internally unscaled values for all data storage and arithmetic computations. This guarantees automatic scale consistency. However, for external use, scaled values are more natural when exchanging data with components outside the AIMMS execution system. Specifically, AIMMS uses scaled values when

- displaying the data of an identifier in the (end-)user interface,
- exchanging data for a particular identifier with files and databases using the READ and WRITE statements,
- passing arguments to external procedures and functions,
- storing the result value(s) of an external function, and
- communicating the variables and constraints of a mathematical program to a solver.

Automatic scaling for solvers. During communications with a solver, AIMMS will scale all variables and constraints (including variable definitions) in accordance with the scale factor associated with the UNIT attribute in their declaration. This choice is based on the assumption that the specified units reflect the expected order of magnitude of the numbers associated with the variables, parameters and constraints, and that the resulting numerical values will neither be very large nor very small. As a result, the values of all rows and columns in the generated mathematical program are expected to be of the same, reasonable, order of magnitude. Especially nonlinear solvers may greatly benefit from this choice.

6.3.3 Unit Conventions

AIMMS allows a modeler to *globally* override the default format for data exchange using READ and WRITE, DISPLAY and SOLVE statements by selecting an appropriate *unit convention*. A convention offers a global medium to specify alternative (scaled) units for multiple quantities, units, and identifiers.

An example. The following declaration illustrates the use of a CONVENTION to define the more common units in the Anglo-American unit system at the quantity level, the unit level and the identifier level.

```
CONVENTION:
    identifier     : AngloAmericanUnits
    per identifier : GasolinePurchase : gallon,
                     PersonalHeight   : feet
    per quantity   : Velocity    : mph,
                     Temperature : degF,
                     Length      : mile
    per unit       : cm  : inch,
                     m   : yard,
                     km  : mile
```

Application order. For a particular identifier, AIMMS will select a unit from a convention in the following order.

- If a unit has been specified for the identifier, AIMMS will use it.
- If the identifier can be associated with a specific quantity in the convention, AIMMS will use the unit specified for that quantity.
- In all other cases AIMMS will apply the convention to an atomic unit directly, or to all composing atomic units used in a compound unit.

6.4 Time-Based Modeling

Time plays an important role in various real-life modeling applications. Typical examples are found in the areas of planning, scheduling, and control. The time scale in control models is typically seconds and minutes. Scheduling models typically refer to hours and days, while the associated time unit in planning models is usually expressed in terms of weeks, months, or even years. In spite of the importance of time in many application areas, most modeling languages do not provide special concepts to deal with time conveniently. To facilitate time-based modeling, AIMMS provides a number of unique language features to relate abstract time periods in a model with real calendar time.

Use of time periods. Time-dependent data in a model is usually associated with time periods. Some data items associated with a period index can be interpreted as taking place *during* the period, while others take place *at a particular moment.* For instance, the stock in a tank is usually measured at, and associated with, a specific moment in a period, while the flow of material into the tank is usually associated with the entire period.

Use of time as a continuous quantity. Time-dependent data in a model can also represent continuous time values. For instance, consider a parameter containing the starting times of a number of processes. Even though this representation is not ideal for constructing most time-based optimization models, it allows time to be expressed to any desired accuracy.

Calendar periods versus model periods. A large portion of the data in time-dependent models originates from the real world where quantities are specified relative to some calendar. Optimization models usually refer to abstract model periods such as p_1, p_2, p_3, etc., allowing the optimization model to be formulated independent of real time. This common distinction makes it essential that quantities associated with real calendar time can be converted to quantities associated with model periods and vice versa.

Rolling horizon. In many planning and scheduling applications, time-dependent models are solved repeatedly as time passes. Future data becomes present data and eventually becomes past data. Such a moving time span is usually referred to as a "*rolling horizon*". By using the various features that AIMMS offers, it is fairly straightforward to implement models with a rolling horizon.

Calendars and horizons. AIMMS offers two special data types for time-based modeling applications, namely CALENDAR and HORIZON. Both are index sets with special features for dealing with time. CALENDARS allow a modeler to create a set of time slots of fixed length in real time, while HORIZONS enable him to distinguish past, planning and beyond periods in the model.

Timetables. In addition, AIMMS offers support for automatically creating *timetables* (represented through indexed sets) which link model periods in a HORIZON to time slots in a CALENDAR in a flexible manner. Based on a timetable, AIMMS provides functions to *aggregate* data defined over a CALENDAR to data defined over the corresponding HORIZON and vice versa. Fig. 6.4.7 illustrates an example of a timetable relating a horizon and a calendar.

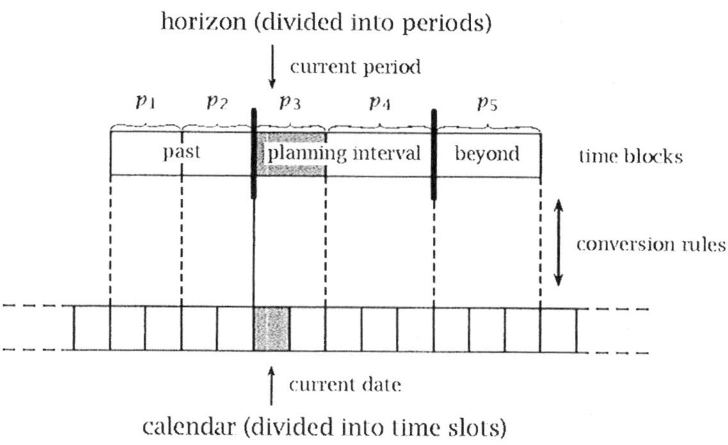

Figure 6.4.7. Timetable relating calendar and horizon

Explanation. The horizon consists of periods divided into three time blocks, namely a past, the planning interval, and beyond. There is a current period in the horizon which can be linked to a current date in the calendar. The calendar consists of time slots and its range is defined by a begin date and an end date. When a modeler constructs a mathematical program, it will typically be in terms of periods in the planning interval of the horizon. However, the input data of the model will typically be in terms of calendar periods. The conversion of calendar data into horizon data and vice versa is done on request by AIMMS in accordance with pre-specified conversion rules.

6.4.1 Calendars

A *calendar* is defined as a set of consecutive *time slots* of unit length covering the complete time frame from the calendar's begin date to its end date. A calendar can be used to index data defined in terms of calendar time.

An example. The following example is a declaration of a daily calendar and a monthly calendar

```
CALENDAR:
   identifier       : DailyCalendar
   index            : d
   element          : CurrentDay
   text             : "A work-week calendar for production planning"
   begin date       : "1996-01-01"
   end date         : "1997-06-30"
   unit             : day
   timeslot format  : "%d/%m/%y" ;

CALENDAR:
   identifier       : MonthlyCalendar
   index            : m
   begin date       : CalendarBeginMonth
   end date         : CalendarEndMonth
   unit             : month
   timeslot format  : "%m/%y" ;
```

Varying number of time slots. The calendar `DailyCalendar` thus declared will be a set containing the elements `'01/01/96'`,...,`'06/30/97'` for every day in the period from January 1, 1996 through June 30, 1997. When the `BEGIN` and `END DATE` attributes are specified as string parameters containing the respective begin and end dates (as in `MonthlyCalendar`), the number of generated time slots can be changed dynamically.

6.4.2 Horizons

A horizon in AIMMS is basically a set of planning periods. The elements in a horizon are divided into three groups, also referred to as time blocks. The main group of elements comprise the *planning interval*. Periods prior to the planning interval form the *past*, while periods following the planning interval form the *beyond*. When variables and constraints are indexed over a horizon, AIMMS automatically restricts the generation of these constraints and variables to periods within the planning interval.

Effect on constraints and assignments. Whenever a horizon is used to construct a time-dependent model, AIMMS has the following features:

- constraints are excluded from the past and beyond periods,
- variables are assumed to be fixed for these periods, and
- assignments and definitions to variables and parameters are, by default, only executed for the periods in the planning interval.

An example. Given a scalar parameter MaxPeriods, the following example illustrates the declaration of a horizon.

```
HORIZON:
   identifier     : ModelPeriods
   index          : h
   parameter      : IntervalStart
   current period : IntervalStart
   definition     : ElementRange( 1, MaxPeriods, prefix: "p-" ) ;
```

If, for instance, the scalar MaxPeriods equals 10, this will result in the creation of a period set containing the elements 'p-01',...,'p-10'. The start of the planning interval is fully determined by the value of the element parameter IntervalStart, which for instance could be equal to 'p-02'. This will result in a planning interval consisting of the periods 'p-02',...,'p-10'.

An example. Consider the parameter Demand(h) together with the variables Production(h) and Stock(h). Then the definition of the variable Stock can be declared as follows.

```
VARIABLE:
   identifier     : Stock
   index domain   : h
   range          : NonNegative
   definition     : Stock(h-1) + Production(h) - Demand(h) ;
```

When the variable Stock is included in a mathematical program, AIMMS will only generate individual variables with their associated definition for those values of h that correspond to the current period and onwards. The reference Stock(h-1) refers to a fixed value of Stock from the past whenever the index h points to the current period. The values associated with periods from the past (and from the beyond if they were there) are assumed to be fixed.

6.4.3 Data Conversion of Time-Dependent Identifiers

When a modeler is working with time-dependent data, it is usually not sufficient to provide and work with a single fixed-time scale. The following examples serve as an illustration.

- Demand data is available in a database on a day-by-day basis, but is needed in a mathematical program for each horizon period.
- Production quantities are computed per horizon period, but are needed on a day-by-day basis.
- For all of the above data weekly or monthly overviews are also required.

The procedures 'Aggregate' and 'Disaggregate'. With the `Aggregate` and `Disaggregate` procedures `AIMMS` can perform an aggregation or disaggregation step from one time scale to another. Both procedures perform the aggregation or disaggregation of a single identifier in one time scale to another identifier in a second time scale, given a timetable linking both time scales and a predefined aggregation type. The syntax is as follows.

- `Aggregate`(*timeslot-data*, *period-data*, *timetable*, *type*[, *locus*])
- `Disaggregate`(*period-data*, *timeslot-data*, *timetable*, *type*[, *locus*])

Aggregation types. Possible values for the *type* argument of the `Aggregate` and `Disaggregate` procedures are given by the elements of the predefined set `AggregationTypes`:

- `summation`,
- `average`,
- `maximum`,
- `minimum`, and
- `interpolation`.

Illustration of aggregation. Table 6.3 demonstrates the aggregation and disaggregation taking place for each conversion rule. The conversion operates on a single period consisting of 3 time slots in the calendar.

Conversion rule	Calendar to horizon			Horizon to calendar		
	3	1	2	3		
`summation`	6			1	1	1
`average`	2			3	3	3
`maximum`	3			3	3	3
`minimum`	1			3	3	3

Table 6.3. Conversion rules for "during" quantities.

Illustration of interpolation. Table 6.4 demonstrates aggregation and disaggregation of a horizon of 3 periods, each consisting of 3 time slots, for loci of 0, 1, and 0.5 (i.e. at the beginning, at the end and in the middle of a period). The underlined values are the values determined by the reverse conversion.

Locus	Horizon data								
	0			3			9		
0	0	1	2	3	5	7	9	9	9
1	0	0	0	1	2	3	5	7	9
0.5	0	0	1	2	3	5	7	9	9

Table 6.4. Conversion rules for interpolated data.

6.5 The AIMMS Excel Interface

There are situations where it would be useful to be combine the optimization power of AIMMS with other applications. For example, end-users might insist on using the common spreadsheet interface of Microsoft Excel for data entry, while an the underlying optimization model could be more easily specified in AIMMS. In this section we will discuss the communication features between AIMMS and Excel.

Why combining AIMMS and Excel. Although Microsoft Excel does offer some optimization functionality itself through its Solver add-in, there are several reasons why using the AIMMS Excel interface would be a much better choice for integrating optimization in a spreadsheet:

- Optimization models can be specified much easier and more intuitively in AIMMS.
- Optimization models (and especially large-scale or complex models) in Excel cannot be maintained very easily because the formulation is hidden in the cell formulas.
- Excel optimization models suffer from "dimensional arthritis", i.e. the dimension of a multi-dimensional collection of variables can hardly be modified once the model is formulated.
- Optimization models in AIMMS solve much faster in general.

Separation of data and algorithms. When an AIMMS model is combined with a Microsoft Excel spreadsheet, only the data needs to be represented in Excel, together with a specification of how Excel must interact with AIMMS. This means that the data and the algorithms for optimization can be separated, which allows the Excel user only to focus on the part that is important to him, namely the data. The algorithms for optimization should be contained in the AIMMS model.

Improving existing applications. There are Excel applications that make use of the relatively weak solving capabilities of Excel. These applications can now be improved by using the optimization power of AIMMS.

Two major modes. When specifying the interaction between AIMMS and Excel two major modes of communication can be distinguished.

- Excel is the main application.
- AIMMS is the main application.

6.5.1 Excel as the Main Application

In this first mode a modeler will typically use the Excel program as a front end to access an optimization application. To accommodate this mode the AIMMS system provides a special *Excel add-in* that extends Excel with some new commands that allow a modeler to specify the communication between AIMMS and Excel in a point-and-click manner. The Excel add-in offers the possibility to exchange data between the two applications and to execute AIMMS-procedures from within Excel. These processes can be controlled through the provided interface of the Excel add-in by defining so-called *execution sequences*.

Additional advantages. The fact that AIMMS and Excel are being combined has the added advantage that the developer of the AIMMS model and the Excel spreadsheet can decide which data is accessible to the end user. Such data can be put in the spreadsheet, while data that is not considered to be of use to the end user can remain hidden inside the AIMMS model. This offers the advantage that end users are only confronted with data that's important and meaningful to them.

An Example. Fig. 6.5.8 shows the AIMMS Interface Setup dialog box. All actions that are part of an execution sequence can be specified in a point- and-click manner.

6.5.2 AIMMS as the Main Application

In this mode a modeler will typically use AIMMS to build an end-user interface for his application, and use Excel as a data source providing some or all of the model data. To access Excel data from within AIMMS it is possible to transfer data using the common ODBC or OLE DB interface of Excel, or to use the special *AIMMS-supplied Excel functions* that allow for the direct communication of data in various tabular formats within Excel.

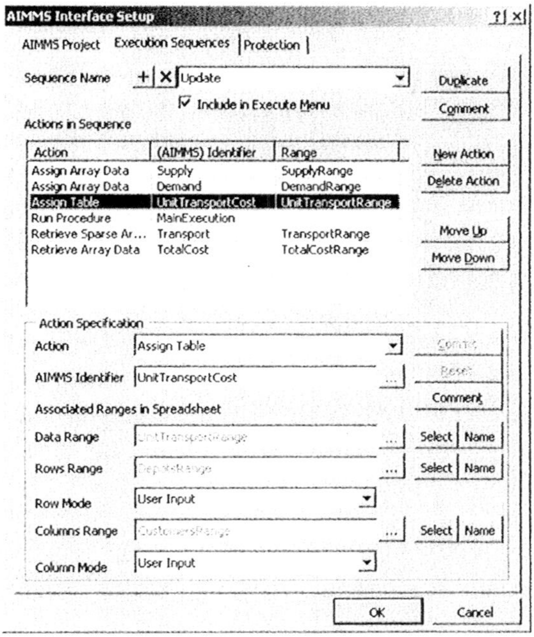

Figure 6.5.8. The AIMMS Interface Setup dialog box

Supported Functions. AIMMS supports functions for reading data from and writing data to Excel workbooks in various tabular formats. It is even possible to run Excel macros from within AIMMS and to performs several sheet and workbook manipulation functions (*e.g.*, creating, deleting, printing)

An Example. The following advanced example uses the ExcelAssignTable function to copy the contents of a two-dimensional AIMMS identifier Transport to a sheet named 'Total' in the 'Result.xls' Excel workbook. The destination range as well as the ranges that will contain the row and column labels are dynamically constructed using AIMMS string manipulation functions and the function 'ExcelColumnName' which maps column numbers to the characters Excel uses to identify the columns in a sheet.

```
ExcelAssignTable( "Results.xls",
    Parameter     : Transport(c,d),
    DataRange     : "B2:" + ExcelColumnName(1+Card(Depots)) +
                    FormatString("%i", 1+Card(Customers))
    RowsRange     : "A2:A" + FormatString("%i", 1+Card(Customers))
    ColumnsRange : "B1:" + ExcelColumnName(1+Card(Depots)) + "1"
    Sheet         : "Total"
);
```

6.6 Multi-Agent Support

AIMMS offers extensive facilities to build multi-agent applications across a local area network, and is the first modeling system to do so. This section covers some basic concepts to give the reader an idea about agent applications in general, their application in optimization modeling, and the support provided by AIMMS to construct such applications.

6.6.1 Basic Agent Concepts

Agents. An *agent* is a software program that independently performs its task on behalf of its user. As software programs all perform tasks on behalf of users, the distinguishing feature is to be found in the word "independently". Even though contact between an agent and its user is possible, the typical agent operates autonomously.

A multi-agent system and application. A *multi-agent system* is a community of agents that communicate together to accomplish individual and/or common goals. These agents are sometimes referred to as intelligent agents or decision agents to reflect their decision-making role within the community. A *multi-agent application* is a specific application of a multi-agent system to accomplish a specified set of individual and/or common goals.

Agent characteristics. In addition to being *autonomous* and *intelligent*, as indicated above, agents are also *reactive* and *proactive*. They are reactive in that they respond to other agents. They are proactive in that they work to accomplish a goal.

Agent communication. Agents communicate by sending messages. Even though messages can be very complex structures in certain multi-agent languages, in this context it suffices to think of messages as a medium to transfer data in the form of numbers and strings from one agent to another. Since agents can be proactive and reactive, messages can either be initiated or be responded to. Messages play a central role in the overall design of a multi-agent system.

Communication network. Agents in a multi-agent system do not necessarily reside on the same host computer. As shown in Fig. 6.6.9, agents can be distributed over several computers, and each agent has its own associated message queue. Two-way communication is indicated through the words "request" and "response", and both requests and responses are implemented through messages. Whenever an agent is idle, it will read the next message on its message queue, and react accordingly. Such communication between agents takes place asynchronously, because it is not known when an agent will be available to react to a particular message.

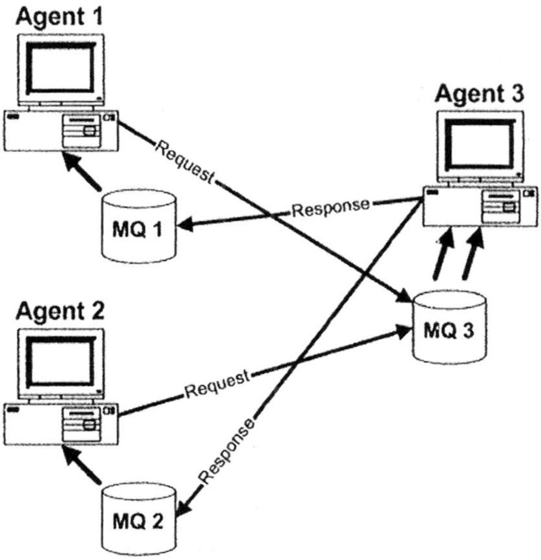

Figure 6.6.9. Communication between agents

Synchronization. There are applications in which the communication between agents has to be synchronized in order to be meaningful, such as a market clearing agent that wants to receive all bids before clearing the market. Typically, synchronization facilities are introduced through special message types or wait mechanisms.

Computing paradigm. The computing paradigm underlying multi-agent systems is that of *parallel computing* and *non-deterministic computing*: agents can reside on different processors, and the order of execution is not predetermined due to the asynchronous communication between agents.

6.6.2 Examples of Motivation

For the purpose of brevity, the examples int this subsection are presented in a rather abstract and introductory format.

A purchasing agent. A purchasing agent is a software program that independently checks inventories at regular moments, determines order quantities, requests price information from suppliers, selects a supplier on the basis of several criteria, places orders, traces deliveries, and updates the inventory data when goods have arrived. This is a typical example of an individual and proactive decision agent.

Auctions. In an auction run by autonomous agents there may be bidding agents and a market clearing agent. The bidding agents supply their bid each round in accordance with well-specified market rules. The market clearing agent informs the bidding agents each round of how the market is cleared. The market rules also determine when the auction is over. Such an auction is a typical example of a multi-agent system containing all agent characteristics.

Control systems. In an advanced plant control system there will be a hierarchy of agents. At the lowest level there are multiple agents observing, registering, and analyzing plant measurements. At a higher level there are coordinating agents that gather status information from agents below them in the hierarchy. Such a control system is a typical example of distributed computing with asynchronous decision making.

Algorithms. The computational time required in several solution approaches, in the areas of simulation and optimization, can be improved through course-grain parallelization. Worker agents can be introduced to execute work that can be done in parallel. Typical examples are Monte Carlo simulations, second stage computations in two-stage stochastic programming, partitioned branch-and-bound computations, and submodels in multi-commodity networks following the allocation of arc capacity to commodities. In such multi-agent applications it is common to introduce a dispatcher agent to keep track of idle worker agents. The dispatcher will then ensure that all computational work is divided effectively among all the workers.

Distributed problem solving. Assume that a large decision problem, such as an extensive supply-chain planning model, can be divided up into distinct decision phases. In addition, assume that each phase has its own model, algorithm and parameter settings, and that there are separate constraints across the various phases. Then a distributed solution approach may be appropriate for obtaining a good-quality solution to the overall problem, which could be tackled using agent technology. The basic concept is to have multiple agents solving phases using the different models, algorithms and parameter settings, and letting the output of one phase be the input to the next. The evolving tree of solutions can be observed by special agents to verify global constraints, and to give feedback to try alternative parameter settings based on their observations. These special agents can also eliminate partial solutions based on their poor overall quality. Humans can also watch the functioning of such a multi-agent application, and can also provide input to steer the overall approach.

Massive parallelization. By using many agents at once, and letting each agent run on its own processor, it may be possible to find one or more good alternative solutions to a very large problem in a matter of hours. The same approach, but without the use of agents, could take weeks or even months before the same overall solution quality is obtained.

6.6.3 Agent-Related Concepts in AIMMS

AIMMS agents. It may ease understanding to view an AIMMS agent as an AIMMS session running an instance of a specific AIMMS project. When there are multiple such agents, their corresponding AIMMS sessions will be running on one or more computers within a local area network, and their corresponding project files will be stored at a specific network location.

Agent roles. In a multi-agent system it is quite natural to distinguish between several types of agents. The types are indicated through the *role* that each agent plays in the overall underlying application. Some examples of roles such as purchasing agent, bidding agent, market clearing agent, worker agent, and dispatcher agent were encountered in the previous subsection. Agent roles play a crucial part in constructing and understanding multi-agent systems as implemented in AIMMS.

Messages between agents. In AIMMS, messages make up the vocabulary to be used between agents. An AIMMS agent can send a message to another agent by calling a send procedure with arguments as illustrated in Fig. 6.6.10. The first two arguments are always the names of the sending and the receiving agents. The remaining arguments are the data items to be transferred. In this example, the two data items are input arguments.

Messages en route. Once an AIMMS agent makes a call to a send message, the underlying AIMMS technology automatically packs the input data to be sent to the addressee. Once sent, the data package will arrive in the message queue of the addressee. As soon as the receiving agent is idle, the underlying AIMMS technology takes the first message from the queue, unpacks the data and calls the handling procedure that corresponds to the particular message.

Message handlers. In AIMMS, a message handler is a procedure with the same data arguments as the corresponding send procedure. The underlying AIMMS technology ensures that the incoming data are passed to the arguments of the message handler. Inside the body of the message handler the modeler can specify all actions necessary to process the incoming message.

Synchronous versus asynchronous. AIMMS supports two types of message transfer, namely asynchronous and synchronous

Asynchronous transfer. When a message only involves input arguments, AIMMS will send the message in a asynchronous manner, i.e. the sending

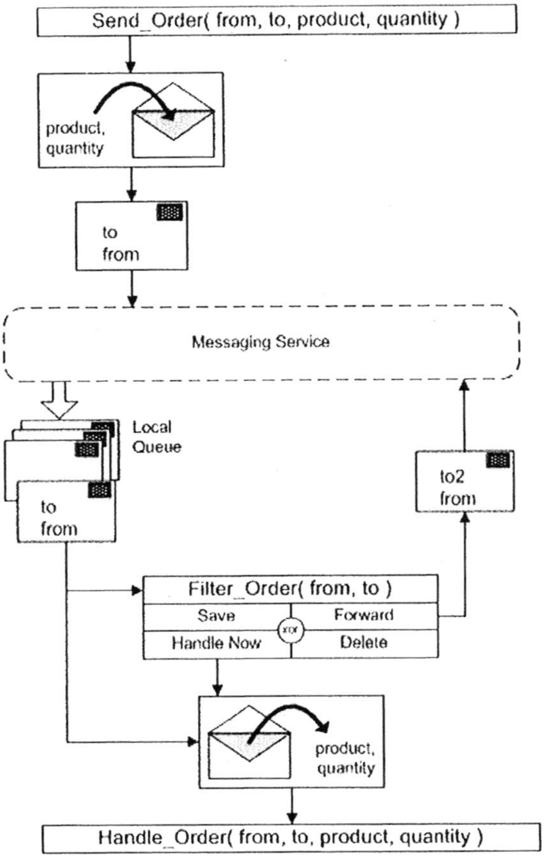

Figure 6.6.10. Messages with only input arguments

agent does not wait for the receiving agent to complete handling the message. This makes it possible for the sending agent to send messages to several agents without delay. Any response from a receiving agent to an incoming message, should be send back via another (asynchronous) message to the original agent.

Synchronous transfer. Whenever messages involve output arguments, AIMMS will transfer the message in a synchronous manner, i.e. the sending agent will wait until the receiving agent has completely handled the message, and will assign the returned output data to the corresponding output arguments. Synchronous message transfer is subject to a time-out, to avoid sending agents to block indefinitely when the message does not arrive with, or cannot be handled by the receiving agent for whatever reason.

Synchronization mechanisms. If the underlying application (for example, an auction) requires the sending agent to collect one or more responses before continuing execution, this can be implemented by using the special waiting functions offered by AIMMS. With these functions one can block the current execution of the sending agent until one or more specified messages are received and handled. Naturally, these synchronization functions require a time-out value that determines the maximum period that such a function will wait for the responding message(s).

6.6.4 Agent Construction Support

AIMMS agent design components. AIMMS provides several facilities for designing and running a multi-agent system using the concepts described above. The design facilities consist of two extensive dialog boxes: one for the overall designer of a multi-agent application, and one for the designer of a specific agent with its own particular agent role.

Community setup. The overall designer of a multi-agent application in AIMMS must specify all the necessary ingredients such as agent roles and agent messages, which, together, are also referred to as the *community setup*. For each role he must indicate which messages can be sent or received, and for each message he must indicate its from-agents and its to-agents. AIMMS offers a special dialog box to support a modeler in this design task.

Agent setup. The designer of an individual agent project in AIMMS must select its specific role from the roles in the community setup. In accordance with this role, he then needs to specify the send procedures for all messages to be sent, and the message handlers for messages to be received. Again, AIMMS offers a dialog box to support a modeler in this design task.

Other facilities. Other facilities to assist in running a multi-agent system in AIMMS consist of being able to log on to an existing network of agents, to start a remote agent, and to log off. In addition, there is a whole layer inside AIMMS, not visible from within the model, that ensures that the AIMMS agent functionality works properly across the local area network.

6.7 Future Developments

Development input. Modeling systems such as AIMMS are continuously developed as a result of customer requests, evolving external technologies, new solvers, and last but not least, the many ideas created by the developers of modeling systems. Several of the topics mentioned below will be implemented in

AIMMS during the years to come. Some new features will become available in the AIMMS 3.x series within the next few years. Others may become operational only in the AIMMS 4.x series to be first released in years thereafter.

Customer requests. Requests from customers vary from small extensions to the GUI builder in AIMMS (such as multi-language support and new graphical objects) to major extensions in the modeling language (such as object-oriented and component-based modeling). Another example of a major extension is a new facility to construct modeling applications for use via the internet. Such applications should be able to serve several users at once with facilities to share common data. Think of an application such as portfolio optimization with access to client-specific data as well as an extensive common financial database to be run as a web service. These and other requests will directly contribute to the future releases of AIMMS.

External technologies. External developments such as new operating systems, new windowing systems, new hardware architectures, and new software communication standards have a direct impact on the development of modeling systems. AIMMS will grow along with these emerging technologies, and several new software standards such as XML and COM have already been incorporated into AIMMS.

New solvers. During the last decennium there has been an explosion of new solvers for all kinds of mathematical programs. Examples are solvers for quadratic programming, mixed-integer quadratic programming, cone programming, mixed complementarity programming, mixed-integer nonlinear programming, and nonconvex nonlinear programming. Some of these new solvers are already available in AIMMS, and several other ones will be added over time. The open approach to solve mixed-integer nonlinear programming models is a good example of how new hybrid algorithms could be made available in the future.

Internal ideas. The developers of AIMMS have an extensive wish list with new features to be added to AIMMS. Some of these newly planned extensions provide extensive support for logic modeling, scheduling, and stochastic modeling. Other new features to be added are novel extensions to the model construction tools, the graphical interface builder, and the multi-agent technology. In addition, there will be several additions to the internal implementation of AIMMS to allow for further improvements in sparse data storage and overall execution efficiency.

Appendix

A AIMMS Features Overview

In this appendix several features of `AIMMS` are briefly mentioned to give the reader an overview of the complete functionality of the `AIMMS` system.

A.1 Language Features

The following summary of language features provides a good overview of the current modeling language in `AIMMS`.

- Powerful multi-dimensional modeling language, including a rich set of numeric, statistical and financial functions, as well as advanced set-, element- and string-valued manipulations.
- Integrated multi-dimensional programming language for specifying both data manipulations and algorithms around optimization models.
- Automatically updated multi-dimensional definitions for specifying global relationships between entities in a model.
- Time-based modeling through calendars and horizons, including support for rolling horizons and automatic aggregation and disaggregation.
- Support for units of measurement, including unit consistency checking, unit-based scaling and unit conventions.
- Support for calling external C/FORTRAN procedures within the `AIMMS` language.
- Integrated case management facilities to support extensive what-if analysis.
- Advanced graphical tools to support and ease model building and maintenance.

A.2 Mathematical Programming Features

The following summary provides a good overview of mathematical programming related features in `AIMMS`.

- Mathematical program types: LP, MIP, QP, MIQP, NLP, MINLP, MCP.
- Supported solvers: CPLEX, Xpress-Optimizer, XA, Mosek, CONOPT, MINOS, SNOPT, PATH, AOA (highly customizable, `AIMMS`-based, outer approximation algorithm for MINLP).
- NLP allows the use of external C/FORTRAN functions in constraints, including support for estimated or user-supplied derivative evaluation.
- Support for callbacks to the modeling environment during a solve.
- Complete set of matrix manipulation functions to modify and re-solve a once generated mathematical program.
- Open solver interface to allow easy solver integration with `AIMMS`.

A.3 End-User Interface Features

Several of the end-user interface features in `AIMMS` can be summarized as follows.

- Integrated GUI builder for constructing end-user interfaces around a model.
- Data-driven graphical controls such as tables, charts, network flow object, Gantt chart, tightly linked to multi-dimensional model identifiers.
- Support for adding custom controls, which will behave as the built-in controls to the interface builder and end-user, through a fixed set of interfaces.
- Page templates to enable a uniform look and feel.

- Menu builder for customized end-user menus and toolbars.
- Page manager for specifying (tree-based) page order, as well as end-user navigation through special navigation controls and menus.

A.4 Connectivity and Deployment Features

AIMMS has several facilities to interact with other software components, and they are summarized below.

- Database connectivity through ODBC/OLE DB, with built-in mapping capabilities between database columns and model identifiers.
- Support for reading/writing XML files, with extensive mapping capabilities between the XML schema and model identifiers.
- Excel add-in, allows data transfer with and execution of AIMMS model from within Excel.
- AIMMS API and COM object, allows AIMMS applications to be used as an optimization component within other applications.

B Application Examples

AIMMS applications are applied in a wide range of industries. Some examples are:

- downstream oil & gas, chemical, and process industry, *e.g.*, crude oil scheduling, catalytic cracking optimization, refinery scheduling and blending, production simulation, brewery optimization
- upstream oil & gas industry, *e.g.*, long-term oil/gas exploration decision support
- electricity, *e.g.*, energy market clearing, unit commitment,
- airline industry, *e.g.*, revenue management
- forestry, *e.g.*, strategic forestry management
- financial industry, *e.g.*, option evaluation and portfolio risk management tools.

Chapter 7

DESIGN PRINCIPLES AND NEW DEVELOPMENTS IN THE AMPL MODELING LANGUAGE

Robert Fourer
Department of Industrial Engineering and Management Sciences
Northwestern University
Evanston, IL, USA
4er@iems.northwestern.edu

David M. Gay
AMPL Optimization LLC
New Providence, NJ, USA
dmg@ampl.com

Brian W. Kernighan
Department of Computer Science
Princeton University
Princeton, NJ, USA
bwk@cs.princeton.edu

Abstract The design of the AMPL modeling language stresses naturalness of expressions, generality of iterating over sets, separation of model and data, ease of data manipulation, and automatic updating of derived values when fundamental values change. We show how such principles have guided the addition of database access, complementarity modeling, and other language features.

Keywords: AMPL, modeling language, optimization, Mathematical Programming, model-data separation, sets, iteration, set manipulation, automatic updates, database access, complementarity, sparsity, nonlinearity

Introduction

AMPL is a little language that grew. Originally designed to express linear programming problems, it gradually expanded to encompass nonlinear programming problems — possibly with complementarity constraints and some integer variables — and it acquired a command environment for manipulating such problems. Although AMPL's presolve phase occasionally determines the solution to a problem, AMPL was never meant to solve problems itself; rather, it works with separate solvers that may even run on different machines.

The rest of this chapter provides more detail on AMPL's design and capabilities. AMPL's historical background has strongly affected its design, so section 7.1 gives more detail about AMPL's history. Section 7.2 presents a simple example, a variant of the venerable diet problem, to illustrate some aspects of AMPL's design. One of AMPL's strengths lies in the generality of its indexing and set expressions; section 7.3 demonstrates some of this by discussing an example of airline fleet assignment that uses sets of quadruples and slices.

In the decade between the appearances of the first and second editions of the AMPL book [76], we extended AMPL in various ways. Three further sections summarize these extensions, including some that are still underway. Section 7.4 deals with iterative schemes and other flow-of-control issues; section 7.5 considers new kinds of models — complementarity, combinatorial, and stochastic; and section 7.6 discusses communications with other systems — databases, Internet services, and solvers. Section 7.7 summarizes differences between the first and second AMPL book editions, and a brief conclusion appears in section 7.8.

7.1 Background and Early History

Several threads converged in the initial design of AMPL. In the early 1980s little languages were a topic of interest in the Computing Science Research Center of Bell Laboratories. These are special-purpose languages designed to simplify and facilitate computing in well focused application areas. Two nice examples are `grap` [16], a `troff` preprocessor that makes it easy to typeset various kinds of graphs, and `chem` [15], a `troff` preprocessor for typesetting chemical formulae.

Another aspect of AMPL's background was the hoopla surrounding Narendra Karmarkar's announcement of an efficient polynomial-time linear programming algorithm [141]. It was clear to us in the Computing Science Research Center that one needs more than just a powerful algorithm; one also needs a problem-formulation language that is natural for use by people yet suitable for processing by a computer, and one needs a convenient way to provide the relevant data and data structures to solvers. We believed that a mathematically natural language would help to make the audience for optimization technology much broader than it would otherwise be.

Bob Fourer and David Gay had first met when both worked at the National Bureau of Economic Research's Computer Research Center in Cambridge, Massachusetts, where Fourer worked on documentation for `sesame`, an LP solver that William Orchard-Hayes was developing for NBER. Fourer subsequently attended graduate school at Stanford, then joined the faculty at Northwestern University. Among other things, he wrote about the advantages of modeling languages over matrix generators [71]. Meanwhile Gay had moved to Bell Labs. When Gay and Fourer happened to meet at a conference, Fourer mentioned he would be coming up for sabbatical. In that way it developed that he was invited to visit Bell Labs for the 1985-86 academic year, with the expectation that the three of us (Fourer, Gay, Kernighan) would work on a modeling language for linear programming.

Our initial work led to a report [73] about the first version of AMPL. Models acceptable to that version would still work today, but commands would not: the first implementation did not recognize any commands, not even "`solve`". It simply read a model and subsequent data and wrote out a translated problem in a simple text format (that we no longer use). Much of the current data-specification syntax was provided by a separate preprocessor. The initial AMPL report eventually became our 1990 AMPL paper [74] in *Management Science*.

Many improvements ensued, as the following sections will show.

7.2 The McDonald's Diet Problem

As an introduction, imagine a student who must decide what foods to buy at a certain popular fast-food establishment so as to minimize cost while meeting some nutritional requirements. For concreteness, suppose the 9 foods and 7 nutrients shown in Table 7.1 are relevant. Suppose further that the food costs,

Table 7.1. McDonald's Diet Problem foods and nutrients.

Foods:		*Nutrients:*	
QP	Quarter Pounder	Prot	Protein
FR	Fries, small	Iron	Iron
MD	McLean Deluxe	VitA	Vitamin A
SM	Sausage McMuffin	Cals	Calories
BM	Big Mac	VitC	Vitamin C
1M	1% Lowfat Milk	Carb	Carbohydrates
FF	Filet-O-Fish	Calc	Calcium
OJ	Orange Juice		
MC	McGrilled Chicken		

Table 7.2. McDonald's Diet Problem data.

	QP	MD	BM	FF	MC	FR	SM	1M	OJ	
Cost	1.84	2.19	1.84	1.44	2.29	0.77	1.29	0.60	0.72	Need:
Protein	28	24	25	14	31	3	15	9	1	55
Vitamin A	15	15	6	2	8	0	4	10	2	100
Vitamin C	6	10	2	0	15	15	0	4	120	100
Calcium	30	20	25	15	15	0	20	30	2	100
Iron	20	20	20	10	8	2	15	0	2	100
Calories	510	370	500	370	400	220	345	110	80	2000
Carbo	34	35	42	28	42	26	27	12	20	350

Table 7.3. Concrete McDonald's Diet Problem.

Minimize

$$1.84x_{QP} + 2.19x_{MD} + 1.84x_{BM} + 1.44x_{FF} + 2.29x_{MC} + 0.77x_{FR} + 1.29x_{SM} + 0.60x_{1M} + 0.72x_{OJ}$$

Subject to

$$28x_{QP} + 24x_{MD} + 25x_{BM} + 14x_{FF} + 31x_{MC} + 3x_{FR} + 15x_{SM} + 9x_{1M} + x_{OJ} \geq 55$$

$$15x_{QP} + 15x_{MD} + 6x_{BM} + 2x_{FF} + 8x_{MC} + 4x_{SM} + 10x_{1M} + 2x_{OJ} \geq 100$$

$$6x_{QP} + 10x_{MD} + 2x_{BM} + 15x_{MC} + 15x_{FR} + 4x_{1M} + 120x_{OJ} \geq 100$$

$$30x_{QP} + 20x_{MD} + 25x_{BM} + 15x_{FF} + 15x_{MC} + 20x_{SM} + 30x_{1M} + 2x_{OJ} \geq 100$$

$$20x_{QP} + 20x_{MD} + 20x_{BM} + 10x_{FF} + 8x_{MC} + 2x_{FR} + 15x_{SM} + 2x_{OJ} \geq 100$$

$$510x_{QP} + 370x_{MD} + 500x_{BM} + 370x_{FF} + 400x_{MC} + 220x_{FR} + 345x_{SM} + 110x_{1M} + 80x_{OJ} \geq 2000$$

$$34x_{QP} + 35x_{MD} + 42x_{BM} + 38x_{FF} + 42x_{MC} + 26x_{FR} + 27x_{SM} + 12x_{1M} + 20x_{OJ} \geq 350$$

nutrients, and nutrient requirements are as given in Table 7.2 (derived from the establishment's literature).

Since the expressions for total food cost and resulting nutrients are linear, this problem has the form of a general linear programming problem,

$$\begin{array}{ll} \text{Minimize} & c^T x \\ \text{Subject to} & Ax = b \\ & x \geq 0 \end{array}$$

but such a formulation is too general for convenient manipulation. The problem also has the concrete form shown in Table 7.3, but this form is too specific —

Table 7.4. Abstract model for the McDonald's Diet Problem.

Given
$\mathcal{F}$, a set of foods,
$\mathcal{N}$, a set of nutrients,

$a_{ij} \geq 0$, the units of nutrient i in one serving of food j,
for each $i \in \mathcal{N}$ and $j \in \mathcal{F}$,

$b_i \geq 0$, the units of nutrient i required in the diet, for each $i \in \mathcal{N}$,

$c_j \geq 0$, the cost per serving of food j, for each $j \in \mathcal{F}$,

Define $x_j \geq 0$ to be the number of servings of food j to be purchased,
for each $j \in \mathcal{F}$.

Minimize $\sum_{j \in \mathcal{F}} c_j x_j$,

Subject to $\sum_{j \in \mathcal{F}} a_{ij} x_j \geq b_i$ for each $i \in \mathcal{N}$.

Table 7.5. Diet Model in AMPL (`mcdiet.mod`).

```
set NUTR;      # nutrients
set FOOD;      # foods

param amt {NUTR,FOOD} >= 0;     # amount of nutrient in each food
param nutrLow {NUTR} >= 0;      # lower bound on nutrients in diet
param cost {FOOD} >= 0;         # cost of foods

var Buy {FOOD} >= 0 integer;    # amounts of foods to be purchased

minimize TotalCost: sum {j in FOOD} cost[j] * Buy[j];

subject to Need {i in NUTR}:
      sum {j in FOOD} amt[i,j] * Buy[j] >= nutrLow[i];
```

too hard to set up and maintain.

Between these extremes lies the general mathematical model for the diet problem shown in Table 7.4. AMPL was designed to permit easy transcription of mathematical models in such a form. An AMPL model for the diet problem is shown in Table 7.5. This model represents a whole class of diet problems; to obtain the particular instance corresponding to Tables 7.2 and 7.3, we must supply the relevant data. There are various ways to do this, but the simplest for small examples is to provide a file of AMPL data statements, such as the one in Table 7.6. With files `mcdiet.mod` and `mcdiet1.dat` containing the text shown in Tables 7.5 and 7.6, we obtain a continuous-variable solution to the problem if we use AMPL's default solver, MINOS, which ignores integrality restrictions on variables. This is shown in Table 7.7.

Table 7.6. AMPL data statements for McDonald's Diet Problem (`mcdiet1.dat`).

```
data;
param: FOOD:             cost :=
  "Quarter Pounder"      1.84        "Fries, small"       .77
  "McLean Deluxe"        2.19        "Sausage McMuffin"   1.29
  "Big Mac"              1.84        "1% Lowfat Milk"      .60
  "Filet-o-Fish"         1.44        "Orange Juice"        .72
  "McGrilled Chicken"    2.29 ;

param: NUTR: nutrLow :=
   Prot  55   VitA 100   VitC  100
   Calc 100   Iron 100   Cals 2000   Carb 350 ;

param amt (tr):          Cals  Carb  Prot  VitA  VitC  Calc  Iron :=
   "Quarter Pounder"      510    34    28    15     6    30    20
   "McLean Deluxe"        370    35    24    15    10    20    20
   "Big Mac"              500    42    25     6     2    25    20
   "Filet-o-Fish"         370    38    14     2     0    15    10
   "McGrilled Chicken"    400    42    31     8    15    15     8
   "Fries, small"         220    26     3     0    15     0     2
   "Sausage McMuffin"     345    27    15     4     0    20    15
   "1% Lowfat Milk"       110    12     9    10     4    30     0
   "Orange Juice"          80    20     1     2   120     2     2;
```

Table 7.7. Continuous-variable solution of McDonald's Diet Problem.

```
ampl: model mcdiet.mod;
ampl: data mcdiet1.dat;
ampl: solve;
MINOS 5.5: ignoring integrality of 9 variables
MINOS 5.5: optimal solution found.
7 iterations, objective 14.8557377
ampl: option omit_zero_rows 1;
ampl: display Buy;
Buy [*] :=
   '1% Lowfat Milk'  3.42213
     'Fries, small'  6.14754
  'Quarter Pounder'  4.38525
;
```

Table 7.8. Integer solution of McDonald's Diet Problem.

```
ampl: option solver cplex; solve;
CPLEX 8.0.0: optimal integer solution; objective 15.05
27 MIP simplex iterations
15 branch-and-bound nodes
ampl: display Buy;
Buy [*] :=
  '1% Lowfat Milk'  4
       Filet-o-Fish  1
    'Fries, small'  5
 'Quarter Pounder'  4
;
```

Table 7.9. McDonald's Diet for 63 foods, 12 nutrients.

```
ampl: reset data;
ampl: data mcdiet2.dat;
ampl: solve;
CPLEX 8.0.0: optimal integer solution; objective 0
0 MIP simplex iterations
0 branch-and-bound nodes
ampl: display Buy;
Buy [*] :=
              'Barbeque Sauce'  50
                      Croutons  55
           'Hot Mustard Sauce'  50
;
```

In reality, one cannot order fractional servings, so it is better to use a solver that respects integrality. This is shown in Table 7.8.

Solving with a larger data set is easy. Table 7.9 illustrates this and reveals a weakness in the model: one can satisfy all the nutritional requirements in the bigger data set by using free condiments. A more palatable solution requires additional constraints linking condiment amounts to purchases of related foods.

7.3 The Airline Fleet Assignment Problem

The McDonald's Diet Problem provides an introduction to many fundamental aspects of AMPL, but it illustrates only a few of the set indexing features that are essential in a good modeling language. AMPL's facilities for constructing and manipulating sets and for iterating over sets make it a powerful language that can readily express complex models. Whereas the diet example involves only simple one-dimensional sets (FOOD and NUTR), our example in this section — the Airline Fleet Assignment Problem — also relies on higher-dimensional sets and on *slices* through these sets, as well as indexed collection of sets. Multi-dimensional sets can be viewed as having members that are pairs, triples, or higher-order "tuples" of elements, in which case slices are subsets in which certain components of the tuples have prescribed values. A model for the Fleet Assignment Problem appears in Tables 7.10 and 7.11.

The model begins with declarations of three simple sets: FLEETS of airplanes, CITIES served by the airplanes, and discrete TIMES at which the airplanes take off or land. The TIMES set is circular, meaning that its members are ordered, with the first member considered to follow the last when computing the "next" member. Set FLEGS describes the schedule to be flown; it consists of 5-tuples (f,c1,t1,c2,t2) with f in FLEETS, c1, c2 in CITIES, and t1, t2 in TIMES, such that fleet f can provide an airplane that departs city c1 at time t1 and arrives in city c2 at time t2. Parameters leg_cost and fleet_size are indexed over these sets.

The model's first four sets are fundamental: their values must be provided before AMPL can generate a problem instance. The next three sets to be declared — LEGS, SERV_CITIES, and OP_TIMES — are all derived from the fundamental sets. LEGS is the set of all quadruples (c1,t1,c2,t2) such that (f,c1,t1,c2,t2) appears in set FLEGS for at least one f. SERV_CITIES is an indexed collection of sets, that is, a collection of similar sets that are distinguished from one another by subscripts (in square brackets): for each f in FLEETS, SERV_CITIES[f] is the set of cities served by fleet f. OP_TIMES is another indexed collection of sets: for each f in FLEETS and c in SERV_CITIES[f], OP_TIMES[f,c] is the ordered set of times at which an airplane from fleet f may take off or land at city c. The phrase circular by TIMES defines each set OP_TIMES[f,c] to be ordered in the same way as the fundamental set TIMES. Both setof expressions in the declaration for OP_TIMES iterate over slices of set FLEGS, the first slice being the set of all triples (t1,c2,t2) such that for a given pair (f,c), (f,c,t1,c2,t2) is in FLEGS, and the second similarly being the set of all triples (c1,t1,t2) such that for a given (f,c), (f,c1,t1,c,t2) is in FLEGS.

AMPL allows one to specify a model in either of two ways:

- "row-wise" — declaring each objective or constraint all at once, through an algebraic expression; or
- "column-wise" — using each variable's declaration to indicate its contributions to various constraints and objectives.

AMPL's node and arc declarations are the most often used column-wise notation: a node declaration sets up a network balance-of-flow constraint to which variables declared with arc can contribute. Since airline fleet assignment has the

Table 7.10. Airline Fleet Assignment Model, part 1 — column-wise specification of network flow costs and balance constraints.

```
set FLEETS;
set CITIES;
set TIMES circular;

set FLEGS within
   {f in FLEETS, c1 in CITIES, t1 in TIMES,
                 c2 in CITIES, t2 in TIMES: c1 <> c2 and t1 <> t2};

          # (f,c1,t1,c2,t2) represents the availability of fleet f
          # to cover the leg that leaves c1 at t1 and
          # whose arrival time plus turnaround time at c2 is t2

param leg_cost {FLEGS} >= 0;
param fleet_size {FLEETS} >= 0;
          # leg costs and sizes for each fleet

set LEGS = setof {(f,c1,t1,c2,t2) in FLEGS} (c1,t1,c2,t2);
          # the set of all legs that can be covered by some flight

set SERV_CITIES {f in FLEETS} =
   union {(f,c1,t1,c2,t2) in FLEGS} {c1,c2};

set OP_TIMES {f in FLEETS, c in SERV_CITIES[f]} circular by TIMES =
       setof {(f,c,t1,c2,t2) in FLEGS} t1  union
       setof {(f,c1,t1,c,t2) in FLEGS} t2;

          # for each fleet and city served by that fleet,
          # the set of active arrival & departure times at that city

minimize Total_Cost;

node Balance {f in FLEETS, c in SERV_CITIES[f], OP_TIMES[f,c]};

          # for each fleet and city served by that fleet,
          # a node for each possible time

arc Fly {(f,c1,t1,c2,t2) in FLEGS} >= 0  <= 1
   from Balance[f,c1,t1]  to Balance[f,c2,t2]
   obj Total_Cost leg_cost[f,c1,t1,c2,t2];
          # arcs for fleet/flight assignments

arc Sit {f in FLEETS, c in SERV_CITIES[f], t in OP_TIMES[f,c]} >= 0
   from Balance[f,c,t]  to Balance[f,c,next(t)];
          # arcs for planes on the ground
```

Table 7.11. Airline Fleet Assignment Model, part 2 — row-wise specification of flight coverage and fleet size limitations.

```
subject to Service {(c1,t1,c2,t2) in LEGS}:
   sum {(f,c1,t1,c2,t2) in FLEGS} Fly[f,c1,t1,c2,t2] = 1;
          # each leg must be served by some fleet

subject to Capacity {f in FLEETS}:
       sum {(f,c1,t1,c2,t2) in FLEGS:
             ord(t2,TIMES) < ord(t1,TIMES)} Fly[f,c1,t1,c2,t2]
     + sum {c in SERV_CITIES[f]} Sit[f,c,last(OP_TIMES[f,c])]
   <= fleet_size[f];

          # number of planes used is the number in the air
          # at day's end (arriving "earlier" than they leave)
          # plus the number on the ground in each city at day's end
```

form of a network flow problem plus "side" constraints, it is convenient to use a mixture of row-wise and column-wise specifications. Table 7.10 shows the objective, `Total_Cost`, and the `Balance` constraints expressed column-wise; this part of the model sets up a separate network flow problem for each fleet. Table 7.11 shows the side constraints expressed row-wise. The `Capacity` constraints restrict the number of airplanes in each fleet network, and the `Service` constraints insure that exactly one fleet is assigned to each flight in the schedule. Slice notations again appear, this time in summations in these constraints.

7.4 Iterative Schemes

Many applications require solving sequences of problems. AMPL has various facilities that are useful in such applications. The following subsections discuss flow of control expressions and commands, named subproblems, and debugging facilities.

7.4.1 Flow of Control

AMPL offers two sorts of conditional computations: if-then-else expressions and flow-of-control commands. The former are occasionally useful in declarations, such as the following from a model in our first paper on AMPL [74]:

```
param minv 'minimum inventory' {p in prd, t in time}
   = dem[p,t+1] * (if pro[p,t+1] then pir else rir);
```

Here `pro[p,t+1]` is a logical parameter, that is, a parameter that takes only the value 0 (for false) or 1 (for true).

More generally, it is often convenient to solve sequences of problems or to compute certain sets and parameters using sequences of commands. For this purpose, AMPL provides several flow-of-control commands,

```
if lexpr then command [ else command ] ;
for  [ loopname ] { indexing } command
repeat [ loopname ] [ when lexpr ] { command } [ when lexpr ] ;
break [ loopname ] ;
continue [ loopname ] ;
```

where square brackets enclose optional parts, *lexpr* represents a logical expression, and *when* is either `while` or `until`. Each *command* is either a simple command or a sequence of commands within braces.

Table 7.12 is a simple sensitivity analysis script that uses a `for` loop to solve a sequence of diet problems, each with a larger limit on the amount of sodium allowed, until the constraint on sodium consumed is no longer binding. Table 7.13 shows output from running this script.

Table 7.12. Simple script for sensitivity analysis.

```
model diet.mod;
data diet2a.dat;

set NALOG default {};  # starts out empty
param NAobj {NALOG};
param NAdual {NALOG};

for {theta in 52000 .. 70000 by 1000} {
   let n_max["NA"] := theta;
   solve;
   let NALOG := NALOG union {theta};
   let NAobj[theta] := Total_Cost;
   let NAdual[theta] := Diet["NA"].dual;
   if Diet["NA"].dual > -1e-6 then break;
}
```

7.4.2 Named Subproblems

AMPL has commands `drop` to temporarily ignore specified constraints and objectives and `restore` to honor them again; and analogously `fix` to freeze specified variables at their current values and `unfix` to let them vary again. These commands are sometimes useful in solving sequences of related problems. But where it is desirable to switch between solving substantially different problems, usually it is clearer to give names to the problems to be solved. This

Table 7.13. Output from the sensitivity analysis script.

```
ampl: option solver_msg 0;
ampl: commands diet.run;
ampl: display NAobj, NAdual;
:        NAobj       NAdual     :=
52000    113.428    -0.0021977
53000    111.23     -0.0021977
54000    109.42     -0.00178981
55000    107.63     -0.00178981
56000    105.84     -0.00178981
57000    104.05     -0.00178981
58000    102.26     -0.00178981
59000    101.082    -0.000155229
60000    101.013    -5.27818e-19
;
```

is done via an AMPL `problem` declaration that lists the variables to be varied, the constraints to be enforced, and the objectives to be considered; these lists may involve iteration over sets.

As an example, we show how a cutting-stock problem would be solved in AMPL via Gilmore-Gomory column generation. Table 7.14 presents a model for optimizing over a given list of patterns, and Table 7.15 exhibits a model for generating a new pattern given dual prices on the desired widths. Associated `problem` declarations and commands to construct initial data appear in Table

Table 7.14. Model for cutting optimization with given patterns.

```
param roll_width > 0;           # width of raw rolls
set WIDTHS ordered;             # set of widths to be cut
param orders {WIDTHS} > 0;      # number of each width to be cut

param nPAT integer >= 0;        # number of patterns
set PATTERNS = 1..nPAT;         # set of patterns
param nbr {WIDTHS,PATTERNS} integer >= 0;

var Cut {PATTERNS} integer >= 0;    # rolls cut using each pattern

minimize Number:                    # minimize total raw rolls cut
   sum {j in PATTERNS} Cut[j];

subject to Fill {i in WIDTHS}:
   sum {j in PATTERNS} nbr[i,j] * Cut[j] >= orders[i];
```

Table 7.15. Model for new pattern generation.

```
param price {WIDTHS} default 0.0;
var Use {WIDTHS} integer >= 0;

minimize Reduced_Cost:
   1 - sum {i in WIDTHS} price[i] * Use[i];

subject to Width_Limit:
   sum {i in WIDTHS} i * Use[i] <= roll_width;
```

Table 7.16. Script for cutting-stock problem declarations and initialization.

```
option solver cplex;            # need an integer solver
option display_transpose -10;   # for nicer formatting
problem Cutting_Opt: Cut, Number, Fill;
option relax_integrality 1;

problem Pattern_Gen: Use, Reduced_Cost, Width_Limit;
option relax_integrality 0;

let nPAT := card(WIDTHS);
let {w in WIDTHS} nbr[w,ord(w)] := floor(roll_width/w);
let {w in WIDTHS, i in WIDTHS: i <> w} nbr[w,ord(i)] := 0;
```

Table 7.17. Script switching problems for cutting-stock optimization.

```
repeat {
   solve Cutting_Opt;
   let {i in WIDTHS} price[i] := Fill[i].dual;

   solve Pattern_Gen;
   if Reduced_Cost < -0.00001 then {
      let nPAT := nPAT + 1;
      let {i in WIDTHS} nbr[i,nPAT] := Use[i];
   }
   else break;
};
display nbr, Cut;
option Cutting_Opt.relax_integrality 0;
solve Cutting_Opt;
```

7.16, and Table 7.17 shows a script that uses the named problems to carry out the column-generating scheme.

7.4.3 Debugging

AMPL has a single-step mode that provides for stopping and examining data (perhaps by a display command) in the middle of a script. Relevant commands include step *n* to execute the next *n* commands (or just the next command if *n* is omitted), next *n* to execute the next *n* commands with compound commands (such as loops and if-then-else commands) counting as a single command, skip *n* to skip the next *n* commands, and cont to continue until the current compound command is finished. Table 7.18 illustrates some of these commands.

Among the output commands useful for debugging is expand, which shows portions of the problem instance that AMPL has generated: either the linear parts of specified constraints and objectives, or the linear contributions of specified variables to the constraints and objectives. Table 7.19 demonstrates this command.

Table 7.18. Using single-step mode with the cutting-stock script.

```
ampl: model cut.mod; data cut.dat;
ampl: option single_step 1;
ampl: commands cut.run;
cut.run:1(0) option ...
<2>ampl: step 6;
cut.run:12(337) repeat ...
<2>ampl: display nbr;
nbr [*,*]
:  1 2 3 4 5 :=
20 5 0 0 0 0
45 0 2 0 0 0
50 0 0 2 0 0
55 0 0 0 2 0
75 0 0 0 0 1
;
<2>ampl: step
cut.run:13(349) solve ...
<2>ampl: cont
```

7.5 Other Types of Models

AMPL has long dealt with linear, piecewise-linear, and nonlinear models. Linear models are the most frequently used, but nonlinear models are not uncommon and are easily expressed by means of nonlinear algebraic expressions.

Table 7.19. Expanding the cutting-stock constraints.

```
ampl: display nbr;
nbr [*,*]
:  1 2 3 4 5 6 7 8 :=
20 5 0 0 0 0 1 1 3
45 0 2 0 0 0 2 0 0
50 0 0 2 0 0 0 0 1
55 0 0 0 2 0 0 0 0
75 0 0 0 0 1 0 1 0
;
ampl: expand Fill;
subject to Fill[20]:
        5*Cut[1] + Cut[6] + Cut[7] + 3*Cut[8] >= 48;

subject to Fill[45]:
        2*Cut[2] + 2*Cut[6] >= 35;

subject to Fill[50]:
        2*Cut[3] + Cut[8] >= 24;

subject to Fill[55]:
        2*Cut[4] >= 10;

subject to Fill[75]:
        Cut[5] + Cut[7] >= 8;
```

Piecewise-linear terms are sometimes of interest, and AMPL offers special syntax for them. More recently, with Michael Ferris, we extended AMPL to handle complementarity constraints [64]. Lately we have been working to extend AMPL to handle some more general combinatorial [72] and stochastic programming problems. This section discusses some of these extensions.

7.5.1 Piecewise-Linear Terms

Piecewise-linear terms arise naturally in economic contexts, such as with graduated income taxes. They also may be useful as approximations for nonlinear terms; piecewise-linear approximations may permit using a linear rather than a nonlinear solver. AMPL offers the syntax

<< *bkpoint-list* ; *slope-list* >> *variable*

to indicate that a piecewise-linear term "multiplies" a variable. The term is zero when the variable is zero and has slopes given in the *slope-list* when the variable has a value between corresponding breakpoints in the *bkpoint-list*.

Piecewise-linear terms can be converted to equivalent linear expressions. In simple cases, such as a convex piecewise-linear term in an objective that is to be minimized or on the left-hand side of a $\leq$ constraint, it suffices to replace the term with suitable linear inequality constraints. Otherwise, linearization also involves the use of auxiliary constraints and binary variables, or equivalent Special Ordered Sets of Type 2 [9] that are recognized by many integer programming solvers. AMPL automatically provides the auxiliary constraints and variables if necessary, along with information that enables solvers that handle SOS2 sets to ignore these extra constraints and variables.

In general, AMPL permits entities in the lists of breakpoints and slopes to be iterated. For instance, Kallrath has posed a problem [139] with a nonlinear objective that could be expressed by the AMPL fragment shown in Table 7.20. He proposes approximating the nonlinear objective by a piecewise-linear one, to permit using an integer programming solver. Table 7.21 shows how AMPL would express the piecewise-linear objective, with a number Nbkpts[r] of breakpoints that depends on the reactor r. (Auxiliary linear constraints insure that Vol[r] is zero whenever binary variable Active[r] is zero, and that Vol[r] lies between the first and last breakpoints otherwise.)

Table 7.20. Nonlinear objective in Kallrath's problem.

```
set Reactors;
param costFix {Reactors};
param costInv {Reactors};

var Active {Reactors} binary;
var Vol {Reactors} >= 0;

minimize TotalCosts:
   sum {r in Reactors}
      (sqrt(costInv[r]*Vol[r]) + costFix[r]) * Active[r];
```

7.5.2 Complementarity Problems

A complementarity constraint specifies that a given pair of inequalities must be satisfied, at least one of them with equality. Among the applications that can be stated as collections of complementarity constraints — so-called complementarity problems — are equilibrium problems in economics and engineering, mechanical contact problems, and optimality conditions for nonlinear programs, bi-level linear programs, and bimatrix games.

As an example, consider the first-order necessary conditions for a smooth nonlinear programming problem involving inequality constraints: either an inequality is slack and its Lagrange multiplier is zero, or the inequality is tight

Table 7.21. Piecewise-linear objective in Kallrath's problem.

```
set Reactors;
param costFix {Reactors};
param costInv {Reactors};

var Active {Reactors} binary;
var Vol {Reactors} >= 0;

param Nbkpts {Reactors} integer > 0;
param Bpoint {r in Reactors, 1 .. Nbkpts[r]};

minimize TotalCosts:
  sum {r in Reactors} (
    << {q in 1 .. Nbkpts[r]-1} Bpoint[r,q] ;
       sqrt(costInv[r]*Bpoint[r,1]) / Bpoint[r,1],
       {q in 1 .. Nbkpts[r]-1}
         (sqrt(costInv[r]*Bpoint[r,q+1]) - sqrt(costInv[r]*Bpoint[r,q]))
           / (Bpoint[r,q+1] - Bpoint[r,q]) >> Vol[r]
  + costFix[r] * Active[r] );

subj to UseLo {r in Reactors}: Active[r] * Bpoint[r,1] <= Vol[r];
subj to UseHi {r in Reactors}: Vol[r] <= Active[r] * Bpoint[r,Nbkpts[r]];
```

and its multiplier must have the correct sign. All conventional (smooth) mathematical programming problems are thus complementarity problems, but the converse is not true: some complementarity problems cannot be expressed as constrained optimization problems.

Table 7.22 shows an AMPL model (from [76]) for a production cost minimization problem stated as an equilibrium model. The keyword `complements` separates each pair of inequalities that are complementary in the sense that at least one must hold with equality.

Every complementarity condition involves exactly two inequalities. In the example of Table 7.22, there is one inequality on either side of the complements operator, and at least one of these inequalities must hold with equality. In a *mixed-complementarity* condition, the `complements` keyword separates an expression from a pair of inequalities, as in

expr1 <= *expr2* <= *expr3* `complements` *expr4*.

This condition is interpreted according to the following rules:

$$\begin{array}{ccc} expr1 = expr2 & \Longleftrightarrow & expr4 \geq 0 \\ expr1 < expr2 < expr3 & \Longleftrightarrow & expr4 = 0 \\ expr2 = expr3 & \Longleftrightarrow & expr4 \leq 0 \end{array}$$

As an example where mixed complementarity conditions arise, Table 7.23 is a variant of Table 7.22 corresponding to a minimization problem with lower and upper bounds on the production-level variables.

Table 7.22. Production cost minimization as an equilibrium model.

```
set PROD;    # products
set ACT;     # activities

param cost {ACT} > 0;         # cost per unit of each activity
param demand {PROD} >= 0;   # units of demand for each product
param io {PROD,ACT} >= 0;   # units of each product per unit of activity

var Price {i in PROD};
var Level {j in ACT};

subject to Pri_Compl {i in PROD}:
   Price[i] >= 0 complements
      sum {j in ACT} io[i,j] * Level[j] >= demand[i];

subject to Lev_Compl {j in ACT}:
   Level[j] >= 0 complements
      sum {i in PROD} Price[i] * io[i,j] <= cost[j];
```

Table 7.23. Bounded-variable cost minimization as an equilibrium model.

```
set PROD;    # products
set ACT;     # activities

param cost {ACT} > 0;         # cost per unit of each activity
param demand {PROD} >= 0;   # units of demand for each product
param io {PROD,ACT} >= 0;   # units of each product per unit of activity

param level_min {ACT} > 0; # min allowed level for each activity
param level_max {ACT} > 0; # max allowed level for each activity

var Price {i in PROD};
var Level {j in ACT};

subject to Pri_Compl {i in PROD}:
   Price[i] >= 0 complements
      sum {j in ACT} io[i,j] * Level[j] >= demand[i];

subject to Lev_Compl {j in ACT}:
   level_min[j] <= Level[j] <= level_max[j] complements
      cost[j] - sum {i in PROD} Price[i] * io[i,j];
```

Complementarity conditions are mainly of interest when they are part of problems that cannot be stated easily or at all as conventional optimization problems. For example, it may make sense to view the demand for each product as a decreasing function of the product's price, as in Table 7.24.

Some complementarity solvers require a "square" system, with

$$\text{\# of variables} = \text{\# of complementarity constraints} + \text{\# of equality constraints}$$

Table 7.24. Economic equilibrium with price-dependent demands.

```
set PROD;    # products
set ACT;     # activities

param cost {ACT} > 0;       # cost per unit of each activity
param io {PROD,ACT} >= 0;   # units of each product per unit of activity

param demzero {PROD} > 0;   # intercept and slope of the demand
param demrate {PROD} >= 0;  # as a function of price

var Price {i in PROD};
var Level {j in ACT};

subject to Pri_Compl {i in PROD}:
   Price[i] >= 0 complements
      sum {j in ACT} io[i,j] * Level[j]
         >= demzero[i] - demrate[i] * Price[i];

subject to Lev_Compl {j in ACT}:
   Level[j] >= 0 complements
      sum {i in PROD} Price[i] * io[i,j] <= cost[j];
```

For the convenience of such solvers, AMPL translates complementarity conditions into a canonical mixed-complementarity form of *constraint* `complements` *variable.* To help catch mistakes, AMPL complains by default if a complementarity problem is not square.

More generally, mathematical programs with equilibrium constraints, often called MPECs, have no restriction on the numbers of variables and constraints and may involve objectives. When working with MPECs, one can specify

```
option compl_warn 0;
```

to suppress warnings about nonsquare complementarity systems.

7.5.3 Combinatorial Optimization

Combinatorial optimization problems arise in many contexts and are of considerable practical interest. Some are readily expressed as integer programs, but formulations in other terms are often more convenient. Constructs convenient to general-purpose combinatorial optimization can be added to conventional algebraic modeling languages in a way that lets one exploit conventional facilities, such as iteration over sets, while permitting links to new kinds of solvers.

Solvers incorporating the approach known as constraint programming [169], [161] are of particular interest in this context. They can operate directly on a variety of constructs that can be introduced into algebraic modeling languages, and they employ a tree-search approach (much like branch-and-bound for integer programming) that is optimal at least in principle. The tree search may fail in practice to find an optimum in a reasonable amount of time, but it can often be made practical through a judicious choice of search strategies. (Providing convenient ways to specify the search strategy is thus an important challenge.)

As an example, we contrast an AMPL integer programming formulation with a prospective constraint programming formulation for a problem of job sequencing with setups, simplified from [133]. We are given a set of jobs to be done, with a processing time and due time for each job. Jobs must be completed by their due times, and preferably as close to their due times as possible; hence there is also an early completion penalty for each job. A single machine carries out all jobs, one at a time, and readying the machine for a particular job involves a specified setup cost and time. The problem is to sequence the jobs so as to minimize the total setup costs plus earliness penalties.

In either an integer programming or a constraint programming formulation of this problem, it makes sense to define a variable `ComplTime[j]` for the completion time of job `j`. Since the constraints will prevent any job from finishing late, the total earliness penalty can be written as

```
sum {j in Jobs} duePen[j] * (dueTime[j] - ComplTime[j])
```

A typical integer programming formulation would introduce, for each pair of jobs `i` and `j`, a binary (zero-one) variable `Seq[i,j]` that is to equal one if and only if job `i` immediately precedes job `j`. The setup cost would then be

```
sum {i in Jobs, j in Jobs} setupCost[i,j] * Seq[i,j]
```

A more natural expression for the setup cost would introduce an array named `JobForSlot` of job-valued variables, with `JobForSlot[k]` to equal the `k`th job handled by the machine. This kind of formulation, with fewer variables each having a larger domain, is particularly amenable to constraint programming techniques. It permits the setup cost to be written

```
sum {k in Slots} setupCost[JobForSlot[k],JobForSlot[k+1]]
```

with a special `JobForSlot[0]` initialized to represent the initial state of the machine. This expression's use of variables as "subscripts" to the `setupCost` parameter is another natural formulation feature that would not be accepted by an integer programming solver but that is handled directly by solvers for constraint programming.

In the model's constraints, an integer programming formulation would require that completion times respect the due times, and that completion times be consistent with the setup and processing times:

For each job i,

```
CompleTime[i] <= dueTime[i]
```

For each job pair (i,j),

```
ComplTime[i] + setupTime[i,j] + procTime[j] <=
ComplTime[j] + BIG * (1 - Seq[i,j])
```

The use of the parameter BIG is a standard integer programming "trick" for insuring that only the sequencing constraints on adjacent jobs are significant. What we really want to say here is that a job's completion time is bounded both by its due time and by the completion time of the next job less the appropriate processing and setup times. Since the completion time is bounded by just these two amounts, it can be constrained to equal the lesser of them:

For each slot k,

```
ComplTime[JobForSlot[k]] = min (
   dueTime[JobForSlot[k]],
   ComplTime[JobForSlot[k+1]]
     - procTime[JobForSlot[k+1]]
     - setupTime[JobForSlot[k],JobForSlot[k+1]] )
```

A constraint programming solver does not attempt to bound the optimum by relaxing integer variables, and so can deal directly with operators such as min that are not strictly linear. It is the use of min that permits this to be an equality constraint; the equality has the advantage of fully defining the ComplTime variables, so that the solver can be instructed to search only over the JobForSlot variables.

It remains to constrain the variables to represent a valid sequencing. In the integer programming formulation, this is done through "assignment" constraints that allow only one Seq[i,j] to be 1 for each i and for each j:

For each job i,

```
sum {j in Jobs} Seq[i,j] = 1
```

For each job j,

```
sum {i in Jobs} Seq[i,j] = 1
```

The same restriction can be expressed more naturally as a single assertion that no job is repeated in the JobForSlot array. An AMPL iterated operator can serve to express such an assertion:

```
all_different {k in Slots} JobForSlot[k]
```

A constraint programming solver operates directly on this constraint, using an efficient matching procedure to reveal jobs that can be eliminated from the variables' domains at nodes of the search tree.

Conventional linear and integer programs are most easily conveyed to solvers, which need to see only a matrix and some vectors. Conventional nonlinear programming problems are more complicated, in that solvers must be able to evaluate nonlinear expressions and their derivatives. AMPL converts conventional linear and nonlinear programs to a standard format, a so-called `.nl` file, which contains linear parts in a sparse-matrix format and expression graphs for nonlinear expressions. AMPL's solver-interface library provides facilities to read the .nl files and to evaluate expressions and their derivatives on request from a solver. First and second derivatives are computed efficiently by backwards automatic differentiation. (See [90] for more on automatic differentiation in general, and [81] and [82] for details on AMPL's use of it.) Source for AMPL's interface library is freely available; see `www.ampl.com/hooking.html` for more details.

When combinatorial constraints are employed as described in this subsection, complications of an additional kind can arise. Some solvers need to be given the `.nl` file's actual expression graphs in particular formats. This can be achieved by means of recursive "tree-walks" of the expression graphs after they have been read from the .nl file; more details appear in [72], with illustrations from the C++ interface to ILOG Solver.

7.5.4 Stochastic Programming

Many practical planning problems involve uncertainty. Tomorrow's prices, demands, and resources are seldom known precisely, so it is generally necessary to make educated guesses. Sometimes historical data can guide those guesses and even suggest probabilities for them. Other situations may require a more Bayesian approach in which one simply guesses at probabilities for various scenarios.

Stochastic programming deals with uncertainties of several kinds. Assuming probability distributions on some input parameters, one might minimize expected cost or maximize expected profit, perhaps as penalized by a variance term. Sometimes it suffices — or is only feasible — to require that a constraint hold with some probability, for example, that the probability of a particular kind of failure be sufficiently low or that some measure of success be met with sufficiently high probability. There is a large and growing literature on stochastic programming; see for example [136] or [20].

Many kinds of algorithms have been proposed for stochastic programs. In simple cases involving discrete random variables, perhaps in the form of scenarios, one can simply state a "deterministic equivalent", a conventional mathematical programming problem equivalent to the stochastic program. If continuous distributions are involved, one can approximate the problem by sampling, and perhaps again deal with a deterministic equivalent. Unfortunately, it is all too easy to find examples where the deterministic equivalent is too large to solve. Decomposition algorithms, perhaps with dynamic sampling, such as importance sampling, may then appear to be the only reasonable computational alternatives.

In some such situations, it is desirable for the solver to be able to sample the relevant distributions as necessary.

We are working on extensions to AMPL that would facilitate dealing with at least some stochastic programming problems. One of our goals is to give suitable solvers the ability to do their own sampling. To this end, we will permit declaration of random parameters. Because of how they would be conveyed to solvers and to accord with the conventional notion of "random variables", these quantities may in fact be declared as variables with the `random` attribute. Just as conventional parameters may either be fundamental or derived, with fundamental ones being supplied values after the model has been stated and derived ones computed from expressions in the model, so random variables may be either fundamental or derived. For example, in the declarations

```
param avail_mean >= 0;
param avail_var >= 0;
var avail {1..T} random = Normal (avail_mean, avail_var);
```

`avail` is a derived random variable whose distribution is determined by the fundamental parameters `avail_mean` and `avail_var`. Fundamental random variables could be assigned distributions by the `let` command, as illustrated in Table 7.25. Both fundamental and derived random variables would be sent to solvers as expression graphs, in much the same way as AMPL's "defined variables".

Table 7.25. Assigning Distributions to Random Variables.

```
set PROD; param T integer > 0;
param mktbas {PROD} >= 0;
param grow_min {PROD} >= 0;
param grow_max {PROD} >= 0;
var Market {PROD,1..T} random;
.......
let {p in PROD} Market[p,1] := mktbas[p];
let {p in PROD, t in 2..T} Market[p,t] :=
     Market[p,t] + Uniform (grow_min[p], grow_max[p]);
```

New kinds of expressions would help in stating some problems. For example, a discrete distribution would be expressed by a sequence of (*probability*, *value*) arguments to a new `Discrete` built-in function, as in

```
Discrete (1/3, 20,  1/3, 50,  1/3, 175)
Discrete ( {s in SCEN} (prob[s],demand[s]) )
```

(In general, function arguments in AMPL can be iterated, as in the latter `Discrete` invocation.) New built-in functions `Expected_value` and `Variance` would permit use of expected values and variances in objectives and constraints, and the new built-in function `Probability` (*logical-expression*) would permit stating reliability constraints.

In recourse problems, one makes a decision now — in stage 1 — and makes corrections, i.e., takes recourse, later — in subsequent stages, after new information becomes available. For such problems, a declared suffix `.stage` could be employed to distinguish the different stages of recourse variables; see the discussion of suffixes in section 7.6.3. Solvers could learn which constraints involve random entities from the sparsity pattern of the Jacobian matrix, which is available in the `.nl` file.

`Discrete`, `Uniform`, and other (half-)bounded distributions will also offer new opportunities for AMPL's presolve phase to simplify the problem after deducing bounds on certain expressions.

7.6 Communicating with Other Systems

In the following subsections, we consider a variety of ways that AMPL can communicate with other entities, such as databases, compute servers, and solvers.

7.6.1 Relational Database Access

In many practical applications, it is convenient to maintain data — both input and certain forms of output — in the form of relational tables in database or spreadsheet files. A relatively recent addition to AMPL is its general mechanism for communicating with these external data repositories. A `table` declaration establishes connections between AMPL entities, such as sets, parameters and variables, and their external representations. Subsequent `read table` commands copy data from external repositories to the AMPL session, while `write table` commands copy values back from the AMPL session to the external repositories. The AMPL model itself remains strictly independent of the data. Special *table handlers* communicate with the external representations; an open interface makes it possible for anyone to provide additional table handlers that deal with new kinds of external data representations.

For example, the AMPL book [76] defines a variant of Table 7.5 above in which there are also lower and upper bounds `f_min[j]` and `f_max[j]` on the amount `Buy[j]` of food `j` that is bought, and similarly lower and upper bounds `n_min[i]` and `n_max[i]` on the amounts of nutrient `i` consumed. Table 7.26 is a script that obtains data for `diet.mod` from a Microsoft Access database, solves the problem, and writes some results to another Access table.

As seen in this script, the first part of a table declaration generally provides some quoted strings; these identify the table handler and provide handler-dependent details about the external representation. In the `table dietFoods`

Table 7.26. Reading and writing database tables for the diet problem.

```
model diet.mod;
table dietFoods IN "ODBC" "diet1.mdb" "Foods":
   FOOD <- [foods], cost, f_min, f_max;
table dietNutrs IN "ODBC" "diet1.mdb" "Nutrs":
   NUTR <- [nutrients], n_min, n_max;
table dietAmts IN "ODBC" "diet1.mdb" "Amounts":
   [nutrs, foods] amt;
read table dietFoods;
read table dietNutrs;
read table dietAmts;
solve;
table dietResults OUT "ODBC" "diet1.mdb" "Scen3":
   [foodlist], Buy, Buy.rc ~ BuyRC,
   {j in FOOD} Buy[j]/f_max[j] ~ BuyFrac;
write table dietResults;
```

declaration, for instance, "ODBC" identifies AMPL's ODBC table handler, which works on Microsoft systems with various representations, including Access and Excel. The next string, "diet1.mdb", is the name of the database file, and "Foods" is the name of a table within it. The syntax FOOD <- [foods] causes set FOOD to be assigned the set of names in external database column "foods"; the AMPL parameters cost, f_min, and f_max are also assigned from database columns having the corresponding names.

Sometimes the external (database) and internal (AMPL) names differ. For instance, the table dietResults declaration specifies a new table, with external column names "foodlist", "Buy", "BuyRC", and "BuyFrac", that is written after a solution is determined. As this declaration illustrates, specification of database columns can involve iteration over sets; in fact, columns can also be iterated, and tables can be subscripted. Many more details appear in Chapter 10 of [76].

7.6.2 Internet Optimization Services

Network services for optimization have grown in sophistication along with the Internet and the World Wide Web [77]. One successful experiment of this sort is the NEOS project, which makes various facilities related to mathematical programming freely available over the Internet. The NEOS solver facility offers many solvers for a growing list of problem areas, including those shown in Table 7.27. Many of these solvers accept AMPL input, as can be seen from the detailed listing at the NEOS web site, www-neos.mcs.anl.gov/neos/.

Table 7.27. Problem types accepted by the NEOS Server.

Linear programming
Linear network optimization
Linear integer programming
Nonlinear programming
Nonlinear integer programming
Nondifferentiable & global optimization
Stochastic optimization
Complementarity problems
Semidefinite programming

The central NEOS Server can accept problem submissions via web forms, e-mail, or a specialized client (the NEOS "submission tool") that provides a graphical user interface. The Server has various workstations, in other Internet locations, at its disposal; upon receiving a problem submission, it selects an available workstation and sends it the problem. While the problem is being worked on, NEOS may provide periodic progress reports. Once a problem is solved, NEOS returns the solution or information about it; the details depend on the solver and means of submission.

For submissions via web, e-mail, or submission tool, NEOS provides what is essentially batch processing. This is fine for some purposes, but for applications where one must manipulate the computed solution, a closer connection to AMPL's interactive environment is highly desirable. Such a connection is provided via NEOS's Kestrel client. Tables 7.28 and 7.29 show Kestrel in a typical use. To an AMPL session running locally, Kestrel appears to be a locally installed solver. When a problem is passed to Kestrel, it is not solved

Table 7.28. Kestrel example, part 1: sending a problem instance to a remote solver.

```
ampl: model gridneta.mod; data gridneta.dat;
ampl: option solver kestrel, kestrel_options 'solver=knitro';
ampl: option knitro_options 'hessopt=6 feastol=1.0e-5 iprint=2';
ampl: solve;

Job has been submitted to Kestrel
Kestrel/NEOS Job number    : 269646
Kestrel/NEOS Job password  : QAYkBqEW
Check the following URL for progress report :
    http://www-neos.mcs.anl.gov/neos/neos-cgi/
    check-status.cgi?job=269646&pass=QAYkBqEW
In case of problems, e-mail :
    neos-comments@mcs.anl.gov
```

Table 7.29. Kestrel example, part 2: receiving the solver listing and displaying results from the remote solver.

```
Intermediate Solver Output:
Checking the {\AMPL} files
Executing algorithm...

Nondefault Options
------------------
 hessopt      =  6
 feastol      =  1.00E-05

 Iter   Res   Objective     Feas err  Opt err   ||Step||  CG its     mu
------  ---  ------------  --------  --------  --------  ------  --------
     0       6.733333E-01  1.00E+01
     1  Acc  1.050234E+00  4.78E+00  1.05E-01  5.23E+00       2  1.00E-01
     2  Acc  2.512883E+00  1.78E+00  1.21E-01  6.70E+00       2
     3  Acc  5.126285E+00  5.77E-15  1.11E-01  5.87E+00       2
     4  Acc  5.072647E+00  3.11E-15  2.78E-02  7.87E-01       2  2.00E-02
     5  Acc  5.042755E+00  1.78E-15  6.83E-03  8.79E-01       3  2.00E-04
     6  Acc  5.038339E+00  6.66E-16  1.13E-03  3.95E-01       4
     7  Acc  5.038202E+00  4.44E-16  5.03E-04  7.04E-02       6
     8  Acc  5.038191E+00  8.88E-16  2.52E-04  1.31E-02       7
     9  Acc  5.038189E+00  6.66E-16  6.48E-05  7.40E-03       7  4.00E-05
    10  Acc  5.038188E+00  8.88E-16  1.48E-05  2.86E-03       7  4.00E-07
    11  Acc  5.038188E+00  4.44E-16  5.02E-06  6.20E-04       6
    12  Acc  5.038188E+00  1.33E-15  1.84E-06  3.93E-04       5
    13  Acc  5.038188E+00  4.44E-16  7.21E-07  9.79E-05       5

EXIT: OPTIMAL SOLUTION FOUND.
Results completed on remote station.

ampl: display Cost;
Cost = 5.03819

ampl: display {(i,j,0) in A} x[i,j,0];
x[i,j,0] [*,*] (tr)
:     0          1          2          3          4          :=
0   5.01906    3.57129    2.08151    1.48575    0.826271
1   2.0354     1.42124    1.44185    0.891306   0.336723
2   0.663005   1.39555    1.26844    1.06157    0.838679
3   0.761372   0.632623   0.991102   0.966883   0.673966
4   0.250817   0.96421    1.19955    1.41281    1.59119
5   1.27034    2.01509    3.01755    4.18168    5.73317
;
```

locally, however, but is instead sent by the Kestrel routines to the NEOS Server. Eventually the completed solution from the NEOS Server is passed back, and is accessible through the AMPL command environment as if it had been computed locally. Kestrel's calls to the remote server employ the CORBA protocols, as summarized in [58] and explained more fully in [57].

Kestrel allows for a simple sort of parallel processing. The idea is to use AMPL scripts `kestrelsub` to submit multiple solve requests and `kestrelret` to retrieve the results in the order of submission. The NEOS server distributes the requests to multiple workstations, queuing some of the requests if necessary. The `kestrelret` script does not return control to AMPL until the results of the corresponding `kestrelsubmit` are available. Table 7.30 shows an example from a decomposition script.

Table 7.30. Kestrel parallel processing.

```
for {p in PROD} {
        problem subI[p];
        commands kestrelsub;
}
for {p in PROD} {
        problem subI[p];
        include kestrelret;
        display Artif_Reduced_Cost[p];
        ...
}
```

Table 7.30 also illustrates both AMPL's `commands` command and its `include` facility. The former reads and processes the indicated file, `kestrelsub`, anew during each iteration of the loop. Rereading would be important if `kestrelsub` were updated elsewhere in the loop. Since neither `kestrelsub` nor `kestrelret` is modified by the loop in which it appears, it suffices to read them with AMPL's `include` facility, which simply interpolates text from the indicated file during parsing of the loop, so the file is read only once no matter how many times the loop body is executed.

Kestrel uses some powerful machinery (CORBA and NEOS) to run solvers on remote machines, but simpler approaches are possible where the AMPL user is also the owner of a copy of a solver on a different machine. The key is that AMPL runs solvers as separate processes that can do whatever they want, such as cause the problem to be solved remotely. On Unix and Linux systems, one can easily use a shell script as a solver, and such a solver script can use `ssh` or (when security is not an issue) `rsh` to invoke a remote solver, perhaps after copying the `.nl` file to the user's directory at the remote location.

7.6.3 Communication with Solvers via Suffixes

AMPL represents auxiliary information associated with model components by appending to a component's name a construction of the form *.suffix-name*. For a variable, the auxiliary information includes the variable's lower and upper bounds and, after a solve, its reduced cost. In Table 7.26, for instance, `Buy.rc` denotes the reduced cost of variable `Buy`.

Some suffixes are built-in, including `.sstatus` for exchanging basis information with simplex-based solvers, but AMPL also lets one declare new suffixes, both for arbitrary use in AMPL scripts and for exchanging auxiliary information with suitable solvers. Mixed-integer programming solvers, for example, often use priorities on integer variables to help decide which variable to branch on next. A properly written AMPL interface to such a solver would recognize suffix `.priority` on integer variables and act accordingly. After declaring a suffix, one can assign values to it with a `let` command, as in

```
model multmip1.mod; data multmip1.dat;
suffix priority;
let {i in ORIG, j in DEST} Use[i,j].priority
   := sum {p in PROD} demand[j,p];
```

One can also specify suffix values in a variable's declaration:

```
suffix priority;
var Use {ORIG, j in DEST} binary
   suffix priority sum {p in PROD} demand[j,p];
```

Solvers can return information in suffixes and even declare new suffixes if necessary. For instance, on request, when a problem is infeasible, some solvers return in suffix `.iis` an indication of which variables and constraints belong to an irreducible infeasible subset, a minimal set of mutually inconsistent variable bounds and constraints. An IIS can help pinpoint a cause of infeasibility.

Table 7.31 demonstrates finding an IIS and illustrates some details of suffixes and generic synonyms. In response to the first infeasible `solve`, this solver returns in suffix `.dunbdd` a direction of dual unboundedness. Since `.dunbdd` has not yet been declared as a suffix, AMPL creates and exhibits a `suffix` declaration statement automatically. The following `display` command involves the generic synonyms `_conname` for the names of all constraints and `_con` for the constraints themselves; in this case, `_con.dunbdd` shows the `.dunbdd` values of all the constraints. The output suggests an inconsistency between the bounds on sodium (NA) and vitamin B2.

For further confirmation, a suitable `option` command and second `solve` request an IIS. Following the `solve`, AMPL creates and exhibits another `suffix` declaration, for the suffix `.iis`. As this is a `symbolic` suffix, the solver has provided symbolic synonyms for the `.iis` values, shown in the `option iis_table`

Table 7.31. Retrieval of auxiliary solver information via solver-defined suffixes.

```
ampl: model diet.mod; data diet2.dat;
ampl: option solver cplex; solve;
CPLEX 8.0.0:  infeasible problem.
4 dual simplex iterations (0 in phase I)
constraint.dunbdd returned

suffix dunbdd OUT;
ampl:  display _conname, _con.dunbdd;
:      _conname        _con.dunbdd    :=
1   "Diet['A']"        0
2   "Diet['B1']"       0
3   "Diet['B2']"       0.322951
4   "Diet['C']"        0
5   "Diet['NA']"      -0.00409836
6   "Diet['CAL']"      0
;

ampl: option cplex_options 'iisfind=1'; solve;
CPLEX 8.0.0:  iisfind=1
CPLEX 8.0.0:  infeasible problem.
0 simplex iterations (0 in phase I)
Returning iis of 7 variables and 2 constraints.
constraint.dunbdd returned

suffix iis symbolic OUT;

option iis_table '\
0 non not in the iis\
1 low at lower bound\
2 fix fixed\
3 upp at upper bound\
';

ampl: display _varname, _var.iis, _conname, _con.iis;
:      _varname     _var.iis     _conname     _con.iis     :=
1   "Buy['BEEF']"    non      "Diet['A']"      non
2   "Buy['CHK']"     low      "Diet['B1']"     non
3   "Buy['FISH']"    low      "Diet['B2']"     low
4   "Buy['HAM']"     upp      "Diet['C']"      non
5   "Buy['MCH']"     low      "Diet['NA']"     upp
6   "Buy['MTL']"     upp      "Diet['CAL']"    non
7   "Buy['SPG']"     low            .             .
8   "Buy['TUR']"     low            .             .
;
```

command. (Solvers always deal with numeric suffix values; in the AMPL session, the numeric values are available in the .iis_num suffix.) Finally, the display output shows that, in this example, the IIS consists of the bounds on several Buy variables plus the lower limit on vitamin B2 and the upper limit on sodium. This can be interpreted as indicating that, for instance, if it is essential to keep all Buy amounts within their specified bounds, then the infeasibility can be alleviated only by relaxing either the lower limit on B2 or the upper limit on sodium.

More information about declaring and using suffixes appears in Chapter 14 of [76].

7.7 Updated AMPL Book

The first edition of the AMPL book appeared in late 1992, with a 1993 copyright. The second edition [76] appeared a decade later (late 2002, with a 2003 copyright). It describes many extensions made during that decade, with new chapters on modeling commands, database access, command scripts, complementarity problems, display commands, and interactions with solvers. The other chapters and the reference manual in the appendix are extensively revised. More details about the book appear at www.ampl.com/ampl/BOOK/.

7.8 Concluding Remarks

Since its inception in the mid-1980s, AMPL has evolved in many ways, but several goals and principles have guided its evolution from the beginning:

- Stay close to conventional mathematical notation, but use notation that is easy to enter on an ordinary keyboard.
- Use a consistent style in designing notation.
- Permit entities to be iterated whenever this makes sense.
- Make most indexing explicit, to avoid surprise interpretations and to allow complicated relationships to be expressed.
- Encourage separation of model and data.
- Automatically update derived entities as needed after fundamental data values have changed.
- Provide reasonable default values and behaviors and make specifying alternatives easy.
- Use open interfaces where possible, for example to solvers, databases, and user-defined functions.

Chapter 8

GENERAL ALGEBRAIC MODELING SYSTEM (GAMS)

Michael R. Bussieck & Alex Meeraus
GAMS Development Corporation
Washington D.C., U.S.A.
{MBussieck,AMeeraus}@gams.com

Abstract In this chapter we will categorize the development of mathematical programming tools into three major phases and will define three major categories of models and their use. This classification will be used to describe some of the features found in GAMS to support different application environments. Selected language features, and examples of external functions and privacy and security issues are discussed and separate annexes. Finally, we will list the inputs to a problem that has been provided to all the authors by the editor of this volume.

Introduction

Today, algebraic modeling languages are widely accepted as the best way to represent and solve mathematical programming problems. Their main distinguishing features are the use of relational algebra and the ability to provide partial derivatives on multidimensional, very large and sparse structures. In this chapter we will describe some of the origins of GAMS and provide background information that shaped early design decisions. Initial Research and Development of GAMS was funded by the International Bank for Reconstruction and Development, usually referred to as The World Bank, through the Bank's Research Committee (RPO 671-58, RPO 673-06) and carried out at the Development Research Center in Washington DC. Since 1987, R&D has been funded by *GAMS Development Corporation.*

The system was developed in close cooperation of mathematical economists who were and still are an important group of GAMS users. The synergy between economics, computer science and operations research was the most important success factor in the development of the system. Mathematical Programming and economics theory are closely intertwined. The Nobel Prize in Economics awarded to Leonid Kantorovich and Tjalling Koopmans in 1975 for their "contribution to the theory of optimal allocation of resources" was really a prize in mathematical programming. Other Nobel laureates like Kenneth Arrow in 1972, Wassily Leontief in 1973, and Harry Markowitz in 1990 are well known names in math programming. Another early example of this synergy is the use of LP in refining operations, which was started by Alan Manne, an economist, with his book on *Scheduling of Petroleum Refinery Operations* in 1956 [168].

The origins of linear programming algorithms all go back to George Dantzig's early work in the 1940s and 1950s [51, 52]. Computing technology and algorithmic theory had developed at a rapid pace. Thirty years later, we could solve problems of practical size and complexity that allowed us to test of the economic theory on real life problems. The research agenda at the World Bank in the 1970s and 1980s created the perfect environment to bring different disciplines together to apply mathematical programming to research and operational questions in Economic Development.

8.1 Background and Motivation

From the very beginning [23], the driving force behind the development of the General Algebraic Modeling System (GAMS) [31] has been the users of mathematical programming who believed in optimization as a the powerful and elegant framework for solving real life problems in the sciences and engineering. At the same time, these users were frustrated with the high cost, skill requirements, and overall low reliability of applying optimization tools. Most of our initiatives and support for new development came from the worlds of economics, finance, and chemical engineering. These disciplines find it natural to view and understand the world and its behavior as a mathematical program.

GAMS's impetus for development arose out of the frustrating experiences of a large economic modeling group at the World Bank. In hindsight, one may call it a historical accident that in the 1970's mathematical economists and statisticians were assembled to address problems of development. They used the best techniques available at the time to solve multisectoral economy-wide models and large simulation and optimization models in agriculture, steel, fertilizer, power, water use, and other sectors. Although the group produced impressive research, initial successes were difficult to reproduce outside their well functioning research environment. The existing techniques to construct, manipulate, and solve such models required several manual, time-consuming, and error-prone translations into the different, problem-specific representations required by each solution method. During seminar presentations, modelers had to defend the existing versions of their models, sometimes quite irrationally, because the time and money needed to make proposed changes were prohibitive.

Their models just could not be moved to other environments, because special programming knowledge was needed, and data formats and solution methods were not portable.

The idea of an algebraic approach to represent, manipulate, and solve large-scale mathematical models fused old and new paradigms into a consistent and computationally tractable system. Using matrix generators (see appendix *GAMS versus FORTRAN Matrix Generators*) for linear programs taught us the importance of naming rows and columns in a consistent manner. The connection to the emerging *relational data* model became evident. Painful experience using traditional programming languages to manage those name spaces naturally lead one to think in terms of sets and tuples, and this led to the relational data model. Combining multidimensional algebraic notation with the relational data model was the obvious answer. Compiler writing techniques were by now widespread, and languages like GAMS could be implemented relatively quickly. However, translating this rigorous mathematical representation into the algorithm specific format required the computation of partial derivatives on very large systems. In the 1970s, TRW developed a system called PROSE [231] that took the ideas of chemical engineers to compute *point derivatives* that were exact derivatives at a given point, and to embed them in a consistent, fortran-style calculus modeling language. The resulting system allowed the user to use automatically generated exact first and second order derivatives. This was a pioneering system and an important demonstration of a concept. However, in our opinion PROSE had a number of shortcomings: it could not handle large systems, problem representation was tied to an array-type data structure that required address calculations, and the system did not provide access to state-of-the art solution methods. From linear programming, we learned that exploitation of sparsity was the key to solve large problems. Thus, the final piece of the puzzle was the use of *sparse data structures*.

With all pieces in place, all we had to do was adopt the techniques to fit into one consistent framework and make it work for large problems.

8.2 Design Goals and Changing Focus

The original and still valid goal is to improve the model builder's productivity, reduce costs, and improve reliability and overall credibility of the modeling process. To achieve this, we established the following key principles to guide the GAMS development:

- The problem representation is independent of the solution method.
- The data representation follows the relational data model.
- The problem and data representations are independent of computing platforms.

- The problem and data representations are independent of user interfaces.

- Optimization methods will fail, and systems have to be designed to be fail-safe.

Another way to express these principles is to think in terms of layers of representations and capabilities that have clearly defined interfaces and functions. The oldest and most basic layer is the *solver layer* or implementation of a specific algorithm. Above the solver is the model layer, expressed in an algebraic modeling language. The *modeling layer* translates the mathematical representation into a computational structure required by a specific solution method and provides various services such as function and derivative evaluations and error recovery. Above the modeling layer is the *application or domain layer*, which is highly context sensitive and has knowledge about the problem to be solved and the kind of user interacting with the system.

It is instructive to put the development of modeling systems into some historic perspective and see how the focus and technical constraints have changed in the last 30 years. We can observe three major phases that shift the emphasis from computational issues to modeling issues and finally the application or the real problems. Each phase defined one of the main system layers discussed above. The dominant constraints in the first phase were the *computational limits* of our algorithms. Problem representation had to abide by algorithmic convenience, centralized expert groups managed large, expensive and long lasting projects and end users were effectively left out. The second phase has the *model* in focus. This volume is about languages and systems supporting this stage. Applications are limited by modeling skill, project groups are much smaller and decentralized, the computational cost are low and the users are involved in the design of the application. Applications are designed to be independent of computing platforms and frequently operate in a client-server environment.

We believe that we are entering a third phase which has the *application* as its focus and the optimization model is just one of many analytic tools that help making better decisions. The users are often completely unaware of any optimization model or use a mental model that is different from the actual model to solved by optimization techniques. User interfaces are build with off-the-shelf components and frequently change to adjust to evolving environments and new computing technologies. As with databases, modeling components have a much longer life than user interfaces. We have observed cases where the model has remained basically unchanged over many years, whereas the computing environments and user interfaces have changed several times. The solvers used to solve the models have changed, the computing platforms have changed, the user interfaces have changed and the overall performance of the model has changed without any change in the model representation.

8.3 A User's View of Modeling Languages

Today's modeling systems have achieved an interface between the solver and the model world. They also attempt to provide solutions for interfacing models and applications. These attempts focus mainly on data exchange capabilities, source code generation, and access to model object libraries. None of these suggested solutions has yet been widely accepted. This area will remain very active for the next decade, and may ultimately decide which optimization technology becomes the most widely used in tomorrow's applications. In this section we describe requirements for a modeling system for different classes of models and users. We segment the world of models into three categories:

- academic research models,
- domain expert models, and
- black box models.

We also illustrate various concepts, currently implemented in the GAMS system, that support the modeling process from each user's perspective. Each section starts by describing the typical environment and user supported by some actual examples. Furthermore, we discuss how particular GAMS features, that are not necessarily found in other systems, might help these users.

8.3.1 Academic Research Models

Academic research models implement mathematical programming problems often found in academic publications. The model source consists almost exclusively of highly complex algebra that defines the variables and equations. Frequently, these kinds of models, that are benchmarked using a given set of publicly available test instances, support statements made in research papers. An impressive library of test problems from Operations Research can be found in the OR Library [10]. In addition to obvious selection criterion such as personal taste for syntax and the development environment, the most important criteria for selecting a modeling system for these kind of models are the availability of the user's platform of choice and the availability of the model type and appropriate solver.

GAMS offers mixed integer linear and nonlinear programming model types, mixed complementarity problem (MCP) type, and mathematical programs with equilibrium constraints (MPEC). MCP and MPEC play an important role in economic modeling. In addition, GAMS can represent multi-level stochastic optimization problems and models with cone constraints. Furthermore, application languages like MPSGE for general equilibrium models and LOGMIP for disjunctive programming provide access to optimization techniques to researchers outside our narrow field of mathematical programming. To seamlessly support new model types or solvers, the GAMS language offers sufficient flexibility to represent new ideas within the current syntax (although certain constructs may look somewhat awkward). The idea is to first understand each new concept's model, algorithms, and most important application areas, before adding syntax from that concept to an increasingly cluttered language.

One of GAMS' early design criteria was to be both platform and solver independent. GAMS users immediately benefit from hardware and operating system improvements, and improvements on the solver side. Throughout its evolution, the GAMS system has seen the rise and fall of quite a few platforms and mathematical programming solvers. We anticipate a similar pattern in the future. The solver suite of the actual GAMS system includes about 20 supported solvers and a handful of contributed plug-and-play solvers. Each supported GAMS solver complies to a strictly defined solver interface, making the seamless exchange of solvers possible for a GAMS modeler. Common solver attributes (*e.g.*, maximum resource time) can be set through general GAMS options, and solver-specific options can be set through option files. Linking a solver to GAMS means complying with this strict interface, which can be a challenging task[1]. There are easier ways to connect to a modeling system, especially for research codes. For example, the AMPL (see Chapter 7) input/output library is geared towards convenience for the algorithm provider. We find many research codes hooked up to AMPL (*e.g.*, most NEOS solvers have an AMPL interface). Making these codes available for GAMS models would allow researchers to benchmark their GAMS models against a larger set of solvers, without reformulating their model in different modeling languages. Rather than compromising the strict GAMS solver interface, the GAMS system comes with a translation *solver* called CONVERT [34].

Modeling languages have a rich syntax that is usually based on sets and indexed variables, equations, and parameters. This syntax, the corresponding structure in the model, and the data is very useful to the model developer. However, this structure is not used by the solvers. Most solvers see the world as consisting of a list of variables (X_1 to X_n), a list of equations or constraints (E_1 to E_m), and the relationship between these variables and equations, as represented in some form. It is therefore acceptable for a translator to remove the structure, as long as the model as seen by the solver remains unchanged.

The translation solver CONVERT transforms models into a very simple internal scalar format. This internal format can then be written out in many different formats.

With GAMS as output format, the scalar model consists of

[1]The title of the corresponding documentation says it all: *Linking your Solver to GAMS. The Complete Notes. Don't Panic!!*

```
Set I Products          /P1*P2/
    J Cutting Patterns /C1*C2/;

Parameter c(J)     cost of raw material                    /C1 1, C2 1/
          cc(J)    cost of change-over of knives           /C1 0.1, C2 0.2/
          b(I)     width of product roll-type I            /P1 460, P2 570/
          nord(I)  number of orders of product type I      /P1 8, P2 7/
          Bmax     width of raw paper roll                 /1900/
          Delta    tolerance for width                     / 200/
          Nmax     max number of products in cut           /   5/
          bigM     max number of repeats of any pattern /  15/;

Variable  y(J)     cutting pattern
          m(J)     number of repeats of pattern j
          n(I,J)   number of products I produced in cut J
          obj      objective variable;

Binary Variable y; Integer Variable m, n;

Equation defobj, max_width(J), min_width(J), max_n_sum(J),
                 min_order(I), cut_exist(J), no_cut(J);

defobj..         sum(j, c[j]*m[j] + cc[j]*y[j]) =e= obj;
max_width(j)..   sum(i, b[i]*n[i,j])            =l= Bmax;
min_width(j)..   sum(i, b[i]*n[i,j]) + Delta    =g= Bmax;
max_n_sum(j)..   sum(i, n[i,j])                 =l= Nmax;
min_order(i)..   sum(j, m[j]*n[i,j])            =g= nord[i];
cut_exist(j)..   y[j]                           =l= m[j];
no_cut(j)..      m[j]                           =l= bigM*y[j];

m.up[j] =  bigM; n.up[i,j] = nmax;

model trimloss /all/;
solve trimloss minimize obj using minlp;
```

Figure 8.3.1. The orginial GAMS model for trim loss optimization

- declarations of the variables, with extra declarations for the subsets of positive, integer, or binary variables;
- declarations of the equations;
- the symbolic form of these equations; and
- assignment statements for non default bounds and initial values.

All operations involving sets are unrolled, and all expressions involving parameters are evaluated and replaced by their numerical values. Since there are no sets or indexed parameters in the scalar models, most of the differences between modeling systems have disappeared. Therefore, the GAMS format can be transformed easily into the format of another language. In AMPL, for example, the keyword *var* is used instead of *Variable*, bounds and initial values are writ-

```
*  MINLP written by GAMS Convert

Variables  b1,b2,i3,i4,i5,i6,i7,i8,x9;

Binary Variables b1,b2;
Integer Variables i3,i4,i5,i6,i7,i8;

Equations  e1,e2,e3,e4,e5,e6,e7,e8,e9,e10,
           e11,e12,e13;

e1..  0.1*b1 + 0.2*b2 + i3 + i4 - x9 =E= 0;
e2..  460*i5 + 570*i7 =L= 1900;
e3..  460*i6 + 570*i8 =L= 1900;
e4..  460*i5 + 570*i7 =G= 1700;
e5..  460*i6 + 570*i8 =G= 1700;
e6..  i5 + i7 =L= 5;
e7..  i6 + i8 =L= 5;
e8..  i3*i5 + i4*i6 =G= 8;
e9..  i3*i7 + i4*i8 =G= 7;
e10.. b1 - i3 =L= 0;
e11.. b2 - i4 =L= 0;
e12.. - 15*b1 + i3 =L= 0;
e13.. - 15*b2 + i4 =L= 0;

* set non default bounds
i3.up = 15; i4.up = 15; i5.up = 5;
i6.up = 5;  i7.up = 5;  i8.up = 5;

Model m / all /;
Solve m using MINLP minimizing x9;
```

```
#  MINLP written by GAMS Convert

var b1 binary;
var b2 binary;
var i3 integer >= 0, <= 15;
var i4 integer >= 0, <= 15;
var i5 integer >= 0, <= 5;
var i6 integer >= 0, <= 5;
var i7 integer >= 0, <= 5;
var i8 integer >= 0, <= 5;


minimize obj: 0.1*b1 + 0.2*b2 + i3 + i4;
subject to

e2:  460*i5 + 570*i7 <= 1900;
e3:  460*i6 + 570*i8 <= 1900;
e4:  460*i5 + 570*i7 >= 1700;
e5:  460*i6 + 570*i8 >= 1700;
e6:  i5 + i7 <= 5;
e7:  i6 + i8 <= 5;
e8:  i3*i5 + i4*i6 >= 8;
e9:  i3*i7 + i4*i8 >= 7;
e10: b1 - i3 <= 0;
e11: b2 - i4 <= 0;
e12: - 15*b1 + i3 <= 0;
e13: - 15*b2 + i4 <= 0;
```

Figure 8.3.2. The Scalar GAMS and AMPL versions of the trim loss model

ten using a different format, the equation declarations are missing, the equation definitions start with *subject to Ei* instead of just *Ei..*, and a few operators are named differently. In a few cases, a model cannot be translated into a particular language because special functions (*e.g.*, errorf) or variable types (*e.g.*, SemiCont) are not available in that language. GAMS models have an objective variable rather than an objective function. If there is one defining equation for that variable, CONVERT will eliminate the objective variable and will introduce a real objective for formats that support objective functions (*e.g.*, AMPL).

Fig. 8.3.1 represents a user-written GAMS model for trim loss minimization, and Fig. 8.3.2 represents the scalar models GAMS and AMPL versions.

Currently, CONVERT translates into AMPL [75], BARON [202], LINGO [213], LGO [188], MINOPT [218], and the usual MPS and LP formats. Details about CONVERT can found in the GAMS Solver Guide [79]. There is a translation server to support those who do not have access to a GAMS system. The interface to this server is email-based. An email to gms2xx@gamsworld.org with a GAMS model attached and the requested language name (*e.g.*, AMPL) in the subject line triggers a translation process on this server. Details are provided at the translation server web site at http://www.gamsworld.org/translate.htm.

8.3.2 Domain Expert Models

Domain expert models are often used as tools for problem analysis. To be a useful tool and to be able to support a user over a decade or more, these *models* need to be highly flexible. During its lifetime, a domain expert model grows and changes with every new project the domain expert takes on. The meaning of the term *model* also changes dramatically, from a collection of equations and variables to a compilation of sophisticated modules that prepare data, find a *solution* to a problem, and provide reporting capabilities. Mathematical programs are often included in the data preparation model (*e.g.*, calibration of data), and in the solution module, intertwined with other algorithmically steps. The part that represents mathematical programs (*i.e.*, equations and variables) does not, however, often exceed 10% of the overall code (the major part of GAMS source code is dedicated to data calculations and reporting).

The domain experts who use these models have entirely different needs than do the users of academic research models. A perfect example of such a domain expert user is the consulting firm Hill & Associates in Annapolis, Maryland, U.S.A. This firm specializes in providing sound advise and management services to the coal and electricity markets, based on supply, demand, and transportation data gathered over the past two decades. The models used in their *Electric Generation, Coal and Emission Forecasting System* include several extremely challenging linear and mixed-integer linear problems with millions of variables and hundreds of thousands of constraints. Since the requirements for this system change with nearly every project, the computing environment needs to be highly modular and to grow over time. Since the final product of a consultant is some written report, the models need to communicate with many data systems, including databases of different vintage and presentation systems such as MapInfo. Highly skilled experts, primarily economists and engineers, operate these models, and have deep insight into them and their application. Failures are the norm in such a constantly changing system, and handling failures is part of the routine operation.

The GAMS system fits nicely into such a diverse environment because it interfaces well with other systems. Data exchange with other modules is accomplished through the GAMS Data eXchange (GDX) facilities, a collection of ready-made programs that communicate with standard applications and an application programming interface (API) for custom-made data exchange. Furthermore, the GAMS execution system seamlessly integrates with external procedures, including a model's equations (see *External Functions in GAMS* in the appendix). This capability makes it very easy to integrate a GAMS model into a large application (a collection of such application systems was presented on the back cover of OR/MS Today issues from 1999 to 2001).

The GDX facilities complement the `GAMS` system's ASCII-based input and output. In addition to improving input/output time for bulk data movements, the data coming through the GDX interface is guaranteed to be syntactically correct. This inconspicuous feature, combined with a well-defined data contract (a mapping between the name-space in the `GAMS` model and the name-space in the external data source), allows the user to segment responsibility for interfacing the model with other systems. This segmentation is often reflected in the optimization project structures found at larger companies, where model analysts and developers ensure the correctness of the model, and the IT department guarantees the accurate transfer of data. Smooth operation is guaranteed if both groups adhere the data contract. If an error occurs, it can easily be traced to its source, to ensure that repair responsibility is properly assigned.

In modeling systems where the data exchange is part of the language, the data contract must reside inside the model. `GAMS` and GDX allow this contract to reside anywhere. This is especially helpful with spreadsheets and other cases where it is advantageous to place the data contract inside the data source.[2] In other applications where model and database are components of a larger system, it may be necessary to place the data contract between the model and the database inside the driving application.

The GDX interface can also be used to capture data in a database type file, called the GDX file. This file contains parameters, sets, and even scalar-like level and marginal values of variables and equations. All input data required for a model run can be represented in this platform-independent file (even across big and little endian computers). This ensures capture of the instance of a model failure. The problem can then be analyzed off-line together with the model source, and does not require otherwise vital connections with the rest of the application system.

Capturing the model source can frequently be as complex a process as obtaining the data itself, since the source can reside in different files located in different directories. Therefore, `GAMS` provides a source dumping mechanism (dumpopt) to collect all model sources in one file.

Independent of the original data source, the GDX file represents a highly structured database. Supplemented by a collection of GDX tools, the GDX facilities provide an environment that makes it easy to analyze both the data that goes into a model and the results that come out of each model run. For example, the tool *gdxdiff* compares two GDX files and reports differences similar to the UNIX *diff* text utility. To compare numeric data settings, an absolute tolerance allows the user to filter the differences. The utilities *gdxmerge* and *gdxsplit* combine structurally identical GDX files and add a scenario index to

[2]The GDX utility *gdxxrw*, which exchanges data with MS-Excel, supports this with its *param=index* option. For details see http://www.gams.com/dd/docs/gams/gdxutils.pdf.

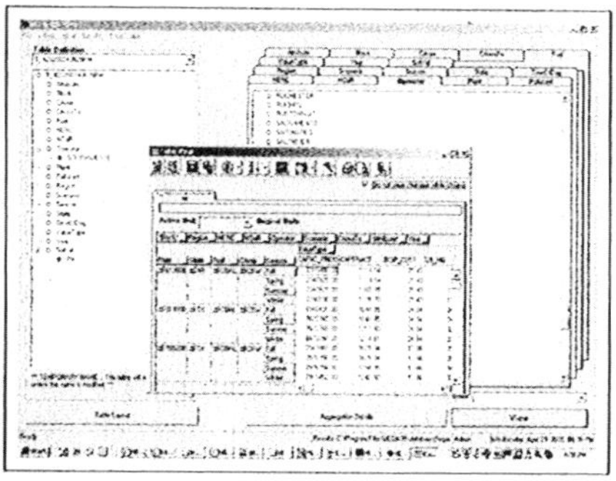

Figure 8.3.3. Scenario Management Tool: VEDA

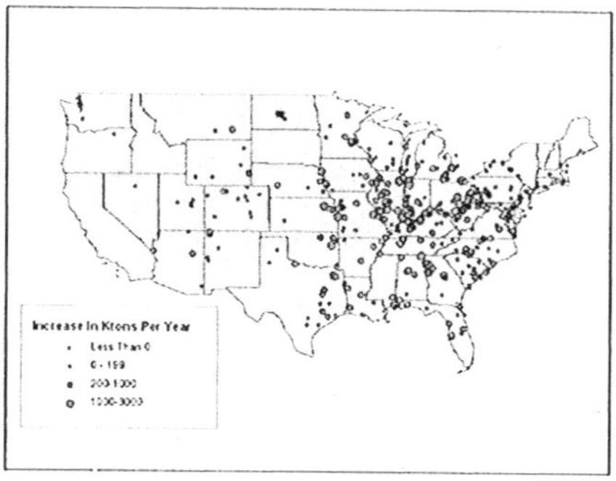

Figure 8.3.3. Geographical Information with MapInfo

every symbol for managing scenario analysis. *Gdxsplit* reverses the effect of *gdxmerge*. GDX files also represent a convenient way of storing model results in a compact format for deferred reporting of results. Using the GDX API data links to several non-standard reporting systems (including MapInfo, a geographical information system), MATLAB, and VEDA, a scenario management tool and data cube, have been successfully built. The screen shots in Figs. 8.3.3 and 8.3.3 show VEDA and MapInfo interfacing with `GAMS` models from Hill and Associates. The screen shot in Fig. 8.3.3 is a MATLAB visualization of a treatment plan calculated by a `GAMS` model for radiosurgery with the Gamma Knife [65].

8.3.3 Black Box Models

Developing the models used as black boxes in large applications is an extremely challenging task. The black box model user is often unaware that an optimization engine is involved when generating the application problem's so-

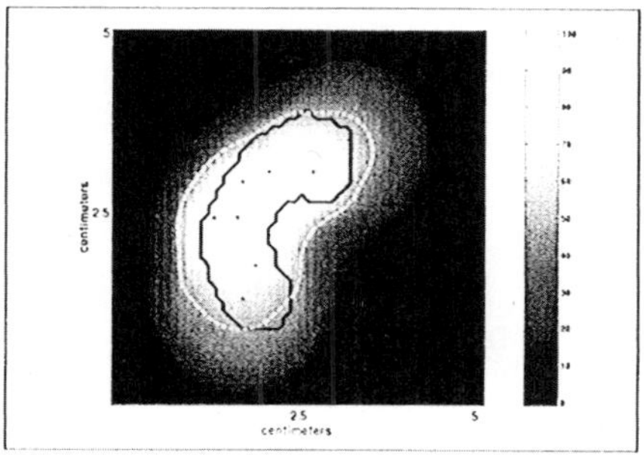

Figure 8.3.3. Visualization with MATLAB

lution. There are two certainties in optimization software that complicate the design of black box models: a) optimization will eventually fail, and b) optimization takes longer than the user is willing to wait. These nasty certainties contrast sharply with the overall system's requirements to a) always deliver a solution, and b) to do so in some measure of real time.

To prevent foreseeable model failures, black box models have an exhaustive and continuously growing list of data checks. Data checks are required to protect the model from delivering useless results, such as those infeasible or unbounded. The black box user does not care if a model is infeasible, but does require detailed, application-specific information on how to overcome the fact that the black box cannot generate a satisfactory solution to his or her problem. Data checks concentrate on particular problems in the data, and can report these problems in terms of data input. Most are simple record-by-record checks, such as ensuring that some numbers are in a particular range (*e.g.*, positive). When the data source is a modern database these simple integrity checks can be built into the database system, but over the model's lifetime such components can be replaced and external data checks can be short-circuited. A fail-safe black box model therefore needs all data checks. Some of these data checks are difficult, and cannot be easily implemented outside the model because they require a mathematical program's solution.

A scheduling project at the United States Military Academy (USMA), West Point, N.Y., U.S.A., is a good example of an effort primarily focused on detecting possible infeasibilities and providing application specific information to overcome them. Unlike other academic institutions, USMA schedules are built around the requirements of each of approximately 4,000 cadets. Each cadet's time is regimented, and USMA must insure that each cadet's schedule allows him and her to succeed in all academic, military, and physical programs at the academy. In some cases the activities of the cadet, the offered courses, and the strict rules of USMA collide, and no feasible schedule can be found for a cadet. Goal programming that penalizes any violation of a constraint while optimizing room assignments and other objectives does not help in this case. The human scheduler must acknowledge each violation of a constraint. Furthermore, assigning proper penalty weights depends on both the constraint and the individual cadet. The complexity of these interacting requirements was making data handling impossible. The solution was to solve a small mixed integer model for each cadet. Cadets with no feasible schedule resulted in an infeasible model. For these cadets, the model generated a set of infeasible schedules and so provided the human scheduler with several suggested ways to resolve each conflict on a cadet-by-cadet basis. This sophisticated data check ensured that the overall model that assigns schedules based on system-wide objectives was feasible, added value to the overall scheduling process, and helped the human scheduler become more productive at resolving scheduling conflicts.

A black box optimization model is often a central but small part of a large application system. The most prominent requirement is reliability in the model and the underlying modeling system. Data checks are one way to make the model reliable. Another way is to anticipate the solution module's failure and have a back-up solution ready. The `GAMS` execution system's flow control statements allow the user to build simple heuristics for most problems. The user gets a back-up solution, and the model can automatically capture the problematic instance (*e.g.*, in a GDX file) for off-line analysis by the model developer. The overall model cannot collapse just because a solver crashed with some hard exception error. The modeling system must be able to recover from such failures, and also needs watchdog capabilities to keep solver resources (*e.g.*, time and memory) in check. The `GAMS` system can run the solver completely detached from the overall model processing. This protects the model from unintentional interference from the solver or other external processes.

The world of platform independent development with Java and advances in internet-ready application shed new light on the importance of platform independence for modeling systems. For example, an optimization application's development might start on Windows PCs, undergo testing on Solaris workstations, deploy an optimization module on an HP workstation, and move to a 64-bit Linux PC after five years.

8.4 Summary and Conclusion

In this chapter we provided some background information to support and explain some of the early design decisions, which we believe are still valid today and guide us into the future. As a language and modeling system designers it is important to keep the focus on the existing and future user community. Mundane concepts as backward compatibility are an essential element in protecting the investments of our users. They demand that models built 10 years ago can be operated today without any change and at the same time are able to take advantage of the latest computer architecture and technology as well allow the use of any state-of-the-art solvers. The same holds for an unknown future modeling environments. The users of our systems are entitled to expect that their models will operate in 10 years from now with out any change. The incorporation of new technologies and theoretical advances will have to be done in a way to protect existing investments.

Finally we would like to recognize the contribution of the editor of this volume, Dr. Josef Kallrath, to bring together a group of highly individualistic and sometimes very competitive developers at the 2003 meeting in Bad Honnef and manage to make them contribute the chapters in a timely fashion. Congratulations.

Appendix

A Selected Language Features

The modeling systems described in this volume are closely related to each other and share many important design principles and in many cases look very similar. The following is a somewhat arbitrary list of features that are not shared by other systems:

Model Types. A model is a collection of symbolic equations which can be used to solve a specific model type with a statement like *solve mymodel using NLP minimizing myobjective*. GAMS requires the user to specify the general model type. This allows the system to verify that a suitable solver is available and that the symbolic algebra fits into the general problem class. Since a nonlinear mixed-integer problem is magnitudes more difficult to solve than a linear program, GAMS requires the user to be aware of the kind of model he or she wants to build and solve. The second feature is that GAMS does not have an objective function and uses instead a simple variable to express the maximand. This allows any variable or indirectly any constraint to serve as an objective at the point of model instantiation.

Attached Comments. Comment and explanatory texts are permanently attached to all symbols and data elements. Descriptive information will never be lost and can automatically be retrieved when producing reports or display intermediate results. This encourages good modeling practice and provides automatic documentation.

Domain Checking. Index positions of all data types are *domain checked*. This is very similar to the concept of referential integrity in data base systems and will guarantee that we will never attempt to reference an index position with an incorrect elopement, set or subset that does not belong there. For example, a variable declared to have the index context of the sets (I,J) can only be indexed by the *domain sets* I and J, their subsets or elements belonging to them. All common errors of inadvertently transposing index references like (J,I) will be flagged as errors.

No Dummy Index. Index references do not use *dummy indices*. In most cases, the use of *dummy indices* makes the algebra more difficult to read and leads to a loss of context. For example, a matrix multiplication is simply written as *c(i,j) = sum(k, a(i,k)*b(j,k))*. A good naming convention then leads to a very compact and direct representation. A consequence of this choice is need for 'aliases' to allow driving more than one index with the same set.

Relaxed Punctuation. A certain amount of context sensitivity and *relaxed* punctuation is allowed. Non-programmers become very impatient if every string has to be quoted, or a missing comma at the end of a line in some list causes a compilation error. Another example is a missing statement separator ';' not causing any ambiguity will not cause an error.

Common Syntax for Declarative and Procedural Components. The syntax for data transformations and symbolic equation definitions are alike and treated internally in the same way. Symbolic equation, model and data definitions are declarative, all other statements are procedural. As problems become more complex, the procedural aspects of the system become more important. Models of different types feed result into other models, all embedded in some complex control structures.

Simple Macros. Extensive compile time source preprocessing and recursive calling of GAMS itself allows easy tailoring of complex applications.

Computational Performance. Sparsity and parallelism are exploited to skip over "dead" indices and in most cases the order of computational burden is proportional to the number of no default (zero) entries in its operands. For example, the computational burden of *sum((i,j), a(i,j)*x(i,j))* will be proportional to the non-zero entries in 'a' and not the product of the cardinality of the sets i and j.

Save and Restart. The state of the system can be saved completely and allows to be restarted compilation and execution at any time, possibly on a different computing platform and even encrypted (see appendix *Secure Work Files*). A simple example is to do the report generation at different times from the solution of the model. All information belonging to different scenarios can easily be stored away and later reused to produce specific reports.

B GAMS External Functions

Although GAMS provides a powerful language for manipulating data and defining highly structured collections of variables and equations, there are times when one would like to define some parts of a model using a more traditional programming language such as FORTRAN or C.

GAMS External Functions connect code written in FORTRAN, C, Java, Delphi, or some other programming language to equations and variables in a GAMS model. We will refer to these GAMS equations as *external equations*, and the compiled version of the programming routines as the *external module* defining the *external functions*. The form of the external module depends on the operating system used. The external module under Windows is a Dynamic Link Library (.dll), and the external module under Unix is a shared object (.so). In principle, any language or system can be used to build the .dll or shared object defining the external module, as long as the interface conventions are not changed.

The basic mechanism is to declare all the equations and variables using the normal GAMS syntax. The interpretation of the external equations is done in a special way. Instead of the usual semantic content, the external equations specify the mapping between 1) the equation and variable names used in GAMS and the function, and 2) variable indices used in the external module which can be written in C, FORTRAN or most other programming languages.

The external equation interface is not intended as a way to bypass some of the very useful model checking done by GAMS when external equations are used with an NLP solver. External equations are still assumed to be continuous, and to have accurate and smooth first derivatives. The continuity assumption implies that the external functions must have very low noise levels, considerably below the feasibility tolerance used by the solver. The assumption about accurate derivatives implies that derivatives must be computed more accurately that can be done with standard finite differences. If these assumptions are not satisfied, there is no guarantee that the NLP solver can find a solution that has the mathematical properties of a local optimum (*i.e.*, that satisfies the Karush-Kuhn-Tucker conditions within the standard tolerances used by the solver). More information about GAMS External Functions can be found at http://www.gams.com/docs/extfunc.htm.

C Secure Work Files

Issues of privacy, security, data integrity and ownership arise when models are distributed to users other than the original developers, or are embedded in applications to be deployed by other developers. We may have to hide, protect, or purge some parts of the model before it can be released. The information to be protected can be of numeric or symbolic nature, and the protection requirements may be driven by:

Privacy. A Social Accounting Matrix supplied by a Statistical Office is required in a general equilibrium model to be used by the Ministry of Finance. The data from the statistical office needs to be protected for obvious privacy reasons and the model experiments are used to evaluate policy options that are highly confidential. Most of the model structure is public, most of the data however is private and model results need to be transformed in such a way as to prohibit the discovery of the original data.

Security. Components of a model contain proprietary information that describes mathematically a chemical reaction. The associated algebra and some of the data are considered of strategic importance and need to be hidden completely. However, the final model will be used at different locations around the world.

Integrity. Data integrity safeguards are needed to assure the proper functioning of a model. Certain data and symbolic information needs to be protected from accidental changes that would compromise the model's operation.

To address these secure work file issues, access control at a symbol level and secure restart files have been added to the GAMS system.

Access Control. The access to GAMS symbols, including sets, variables, parameters, and equations, can be changed once with the compile time commands $purge, $hide, $protect and $expose. $Purge will remove any information associated with this symbol. $Hide will make the symbol and all its information invisible. $Protect prevents changes to information. $Expose will revert the symbol to its original state.

Secure Restart Files. The GAMS licensing mechanism can be used to save a secure model in a secure work file (work files are used to save and restart the state of a GAMS program). A secure work file behaves like any other work file, but is locked to a specific users license file. A privacy license, the license file of the target users, is required to create a secure work file. The content of a secure work file is disguised and protected against unauthorized access via the GAMS license mechanism.

A special license is required to set the access controls and to create a corresponding secure work file. Reporting features have been added to allow audits and traces during generation and use of secure work files. More information about secure work files can be found at http://www.gams.com/docs/privacy.pdf

D GAMS versus FORTRAN Matrix Generators

In the late 70's and early 80's, the MPS file was the universally accepted format to describe a linear program. Those MPS files were either described using specialized Matrix Generator languages, like the very successful and durable MaGen/OMNI family of products from Haverly Systems [100], or general purpose programming languages, like FORTRAN and PL/I were used to write matrix generator programs. Considerable programming experience and a large stock of programs had been accumulated and novel approaches like algebraic modeling languages were viewed as irrelevant academic exercises. Computing power was still the limiting factor and the idea of replacing the MPS file with an algebraic model representation with the need to create a new instance of this model for each optimization step was considered totally impractical.

The first GAMS implementation was done on Control Data Corporation's (CDC) super computers and interfaced with the APEX [38] linear programming system which was the first high performance system utilizing super sparse matrix technology allowing all in-core solutions. The GAMS development team was determined to disprove the allegation of inherently poor performance of a mathematical approach. The implantation took full advantage of the CDC computing architecture and implemented novel dynamic data structures and just in time generation of machine code was employed. The system was blazingly fast, and attracted the hoped for attention in the traditional LP shops. The second implementation of GAMS focused on computing platform and solver independence with the main goal to operate on IBM's mainframes in conjunction with the MPSX [125] or MPSIII [144] linear programming systems, a computing environment that supported 90 percent of all commercial linear programming. A new product from IBM, the Optimization Subroutine Library (OSL) was used as the new performance benchmark. The performance target was to generate a GAMS models instance and load it into the OSL workspace in less time than it took to run the FORTRAN generator and read an MPS file into OSL. One of the benchmark models was a water resource model for Pakistan which was implemented using a FORTRAN generator and GAMS, one of the GAMS version called INDUS89 [251] can be found in the GAMS model library. The LP matrix has 2,726 row, 6,570 columns and 39,489 non-zero entries. Timings on an high end IBM 3090 costing some 3 million dollars were: OSL optimization time excluding input was 137 seconds, the FORTRAN MPS generation and OSL import time were 37 and 19 seconds an added 56 seconds, the GAMS generation and OSL import were 41 and 1 second added only 42 seconds. For comparison, the timings on an 1Ghz PC costing only 1,500 dollars with GAMS and OSL 3 are 4 seconds for the optimization phase and 0.5 seconds for generating the model instance.

Over past 15 years, the hardware cost has declined by a factor of 2000 and the speed has increased by a factor of 40. The computational cost of generating and solving one instance of this model has been reduced by a factor of 800,000 to essentially zero.

E Sample GAMS Problem

A portfolio optimization model is used to illustrate in more detail some of the GAMS language features. The same model has been used in other chapters to illustrate the representation and use of other modeling languages. Most of the presentation below will be self-evident and we will just add a few comments for clarification:

Dollar Control Statements. Lines starting with $ are compiler directives. Text between $ontext and $offtext are comments. $eolcom sets the end-of-line comment symbols. The $if statements is patterned after the Windows scripting language and in line 106 we set the local environment string scenario to the value of *s1* if it is still undefined at this point.

Dynamic Sets. Only static sets can be used to define the allowable index domain. Actual references can be done by any single element or set that is contained in the index domain. For example, the equation *TR(rR)* in lines 133 is defined over the domain set *rR*, in the definition at line 145, however, we use the dynamic set *r*. This is a convenient way to resize any model dimension at will.

Suffix Notation. Variables and equations have attributes like primal and dual values, lower and upper bounds, scale factors or priorities. Those attributes are spefied with a suffix notation. For example, in the equation definition in line 215 we reference the variables *f* as *f.lo* which is the lower bound of this variable at the time of model instantiation.

Maps and Filters. Set tuples can be used to limit the domain over which indexed operations are carried out. In line 214 we define an equation over a three-tuple *rpi* and reference the variable *pT2* with the tuple *rpi* insead of the three sets *r*, *p* and *i*. The use of n-tuples in domain definitions is a convenient way to manage domain restrictions which results in an easy to read, compact notation.

```
\index{GAMS Applications,product portfolio optimization}
  1 $Title  Product Portfolio Optimization
  2 $ontext
  3
  4 This problem computes minimal cost solutions satisfying the
  5 demand of pre-given product portfolios. It determines the number
  6 and size of reactors and gives a schedule of how may batches of
  7 each product run on each reactor. There are two scenarios (s1 20
  8 products. s2 40 products), add --scenario s2 as a GAMS
  9 parameter to specify the second scenario.
 10
 11 The global optimal reactor volumes are:
 12 data set s1
```

```
vr.fx('r1') = 132.5; vr.fx('r2') = 250;

data set s2
vr.fx('r1') = 20; vr.fx('r2') = 100; vr.fx('r3') = 250;

Two formulations are presented, a compact MINLP formulations
and a linearized MIP formulation using special ordered sets.

Problem sizes for data set s1

                              MINLP      MIP
variables                      102       548
equations                       28       418
Non-zeros                      190      1574
discrete variables              40       186


Kallrath, J. Exact Computation of Global Minima of a Nonconvex
Portfolio Optimization Problem. In Frontiers in Global
Optimization. Eds Floudas C A andd Pardalos P M.
Kluwer Academic Publishers, Dortrecht, 2003.

$offtext
$eolcom //

Sets
       s       scenario / s1,s2 /
       rR      reactors / R1*R3 /
       pP      products / L1*L37 /
       r(rR)   reactors considered in scenorio
       p(pP)   products considered in scenorio

Table RData(rR,s,*) Reactor data

         s1.VMIN  s1.VMAX  s2.VMIN  s2.VMAX
R1        102.14     250       20        50
R2        176.07     250      52.5      250
R3                           151.25     250


Table PData(pP,s,*) Product data

      s1.Dem s1.PTime s2.Dem s2.Ptime
L1     2600       6    2600       6
L2     2300       6    2300       6
L3     1700       6     450       6
L4      530       6    1200       6
L5      530       6     560       6
L6      280       6     530       6
L7      250       6     530       6
L8      230       6     140       6
L9      160       6     110       6
L10      90       6     110       6
L11      70       6      10       6
L12     390       6     110       6
L13     250       6      90       6
L14     160       6      90       6
```

```
L15      100      6        90      6
L16       70      6        70      6
L17       50      6        50      6
L18       50      6        30      6
L19       50      6        10      6
L20                        10      6
L21                        10      6
L22                       190      6
L23                       180      6
L24                        70      6
L25                        70      6
L26                        40      6
L27                        40      6
L28                        40      6
L29                        30      6
L30                        20      6
L31                        20      6
L32                        20      6
L33                        10      6
L34                        10      6
L35                        10      6
L36                        10      6
L37                        10      6


Parameters
      VMIN(rR)     volume flow of products in m^3 per week
      VMAX(rR)     volume flow of products in m^3 per week
      DEMAND(pP)   volume flow of products in m^3 per week
      PRODTIME(pP) production time in hours per batch
Scalars
      WHRS   hours in a week                               / 168  /
      CSTI   in kEuro depreciation per m^3 reactor and week / 0.97 /
      CSTF   in kEuro per week and reactor                  / 2.45 /
      ESF    economies of scale factor                      / 0.5  /;

$if not set scenario $set scenario s1
VMIN(rR)      = RDATA(rR,'%scenario%','VMIN');
VMAX(rR)      = RDATA(rR,'%scenario%','VMAX');
DEMAND(pP)    = PDATA(pP,'%scenario%','Dem');
PRODTIME(pP) = PDATA(pP,'%scenario%','PTime');

* Determine scenario sets
r(rR) = VMAX(rR)    > 0;
p(pP) = DEMAND(pP) > 0;

* definition of compact MINLP model

Variables  cTotal       total  costs
           cInvest      invest cost
           cFixed       fix    costs
           f(rR,pP)     utilization rate
           vR(rR)       reactor volume in m^3
           pS(pP)       surplus production
           bvr(rR)      indicating whether reactor r is active
           nB(rR,pP)    number of batches of product p in reactor r

```

```
Positive variables f,vR,pS; Integer variable nB; Binary Variables bvr;


Equations DEFcT         total costs
          DEFcF         fix costs
          DEFcI         invest cost
          TR(rR)        production time of reactor r
          SPP(pP)       compute surplus production p
          RVUB(rR)      maximal volume of reactor r
          RVLB(rR)      minimal volume reactor r;


* define the total cost
DEFcT..  cTotal  =e= cFixed + cInvest;
DEFcF..  cFixed  =e= sum (r, CSTF*bvr(r));
DEFcI..  cInvest =e= sum (r, CSTI**ESF*vR(r)**ESF);

* limit the total production time of reactor r
TR(r)..    sum(p, PRODTIME(p)*nB(r,p) ) =l= WHRS*bvr(r);

* compute the surplus production
SPP(p)..   pS(p) =e= SUM(r, nB(r,p)*f(r,p)*vR(r))/DEMAND(p) - 1 ;

* lower and upper bounds on reactor volume
RVLB(r)..  vR(r) =g= VMIN(r)*bvr(r);
RVUB(r)..  vR(r) =l= VMAX(r)*bvr(r);

Model     portfolioMINLP / DEFcT, DEFcF, DEFcI, TR, SPP, RVLB, RVUB / ;

f.lo(r,p) = 0.4; f.up(r,p)  = 1;  // bounds on the utilization rates
pS.lo(p)  =   0; pS.up(p)   = 1;  // bounds on the surplus production

* bounds on the number of batches
nB.lo(r,p) = 0;
nB.up(r,p) = min(WHRS/PRODTIME(p),floor(2*DEMAND(p)/(VMIN(r)*f.lo(r,p))));
nB.up(r,p)$(2*DEMAND(p) < f.lo(r,p)*VMIN(r)) = 0;

vR.l(rR) = 99;
vR.lo(r) = VMIN(r);

solve portfolioMINLP using minlp minimizing cTotal;

* additional variables and equations to define the MIP formulation
* first we need to linearize the product terms:

Sets i             dyadic represenation set        / 0*10 /
     j             discretization points for SOS2 / 0*10 /
     rpi(rR,pP,i) i required for representing np

Parameters
    vRj(rR,j)    x part of SOS2 construct
    ESFvRj(rR,j) y part of SOS2 construct

Positive Variable
      pT(rR,pP)     number of batches x reactor volume in m^3
      pT2(rR,pP,i) same for in dyadic representation
      ESFvR(rR)     economies of scale for vR
```

```
Binary Variables
      nBx(rR,pP,i) dyadic represenation of nB
SOS2 Variables
      lambda(rR,j) approximation of economies of scale function
Equations
      CNP(rR,pP)  compute the nonlinear products nB(rp)*f(rp)*vR(r)
      SPPx(pP)    compute surplus production p
      CNP10(rR,pP)   linearized version of CNP
      CNP11(rR,pP)   linearized version of CNP
      CNP12(rR,pP,i) linearized version of CNP
      CNP13(rR,pP,i) linearized version of CNP
      CNP14(rR,pP,i) linearized version of CNP
      DEFSOSx(rR)    SOS2 x construct
      DEFSOSy(rR)    SOS2 y construct
      DEFSOSone(rR)  SOS2 sum construct
      DEFcIlp        linearized version of DEFcI;
* new surplus production equation
SPPx(p)..   pS(p) =e= SUM(r, pT(r,p))/DEMAND(p) - 1 ;
* computes batches x volume
CNP(r,p).. pT(r,p) =e= nB(r,p)*f(r,p)* vR(r);
* Linearized version of CNP
CNP10(r,p)..          pT(r,p)  =e= sum(rpi(r,p,i),2**(ord(i)-1)*pT2(rpi));
CNP11(r,p)..          nB(r,p)  =e= sum(rpi(r,p,i),2**(ord(i)-1)*nBx(rpi));
CNP12(rpi(r,p,i))..   pT2(rpi) =l= VMAX(r)*nBx(rpi);
CNP13(rpi(r,p,i))..   pT2(rpi) =l= vR(r);
CNP14(rpi(r,p,i))..   pT2(rpi) =g= f.lo(r,p)*(vR(r)-VMAX(r)*(1-nbx(rpi)));
* SOS2 approximation of economies of scale function
DEFSOSx(r)..  vR(r) =e= sum(j, vRj(r,j)*lambda(r,j));
DEFSOSy(r)..  ESFvR(r) =e= sum(j, ESFvRj(r,j)*lambda(r,j));
DEFSOSone(r).. sum(j, lambda(r,j)) =e= 1;
DEFcIlp..  cInvest =e= sum (r, CSTI**ESF*ESFvR(r));
rpi(r,p,i) = ord(i) <= ceil(log(max(1,nB.up(r,p)))/log(2)) + 1$nB.up(r,p);
vRj(r,j) = (VMAX(r)-VMIN(r))*(ord(j)-1)/(card(j)-1) + VMIN(r);
ESFvRj(r,j) = vRj(r,j)**ESF;
Model PortfolioMIP  /TR, SPPx, RVLB, RVUB, DEFcF, DEFcIlp, DEFcT,
CNP10, CNP11, CNP12, CNP13, CNP14, DEFSOSx, DEFSOSy, DEFSOSone/;
portfolioMIP.iterlim = 1e8; portfolioMIP.optcr = .05;
solve portfolioMIP using mip minimizing cTotal;
```

Chapter 9

THE LINGO ALGEBRAIC MODELING LANGUAGE

Kevin Cunningham
LINDO Systems
Chicago, IL, USA
http://www.lindo.com
kevin@lindo.com

Linus Schrage
LINDO Systems
Chicago, IL, USA
http://www.lindo.com
linus.schrage@gsb.uchicago.edu

Abstract We give an overview of the LINGO modeling system, developed by LINDO Systems. LINGO can solve a wide variety of optimization problems. It has solvers for: linear, integer, quadratic, general nonlinear, and global optimization. Although not specifically customized for complementarity problems, the global solver capability allows LINGO to solve complementarity problems, as encountered in market equilibrium computations. LINGO has interfaces to the Excel spreadsheet program, as well as to standard database systems.

Keywords: LINGO, optimization, global optimization, modeling language, optimization, Linear Programming, Integer Programming, Nonlinear Programming, Mathematical Programming, spreadsheets, complementarity problems

9.1 History

Before 1987, the main modeling system available from LINDO Systems was the LINDO program. LINDO is a scalar-based modeling language that does not have the set based or subscripted variable features that are common to today's modeling languages. In 1987, the Intel 80386 chip became widely available. This chip's 32-bit architecture, coupled with the DOS extenders of the time, made it possible to build programs that far exceeded the previous 640Kb DOS memory limit. In addition, software designed for the new 32-bit architecture executed an order of magnitude faster than its 16-bit brethren. Suddenly, the PC had become a viable machine for solving large, real-world optimization problems.

To exploit these new capabilities we began the development of the `LINGO` algebraic modeling system in 1987 and shipped the first versions in late 1988. At the same time, spreadsheets were becoming popular, and we introduced a spreadsheet based optimizer, What'sBest, that had essentially the same algorithmic engine as `LINGO`. The design of the `LINGO` language owes much to the structured modeling concepts Art Geoffrion was beginning to develop around that time (Geoffrion, 1987).

The early, rudimentary versions of `LINGO` were command driven, solved only linear and integer models, and had no external links for importing data and exporting solutions. Current versions of `LINGO` incorporate linear, integer, barrier, quadratic, nonlinear, and global solvers, along with interfaces to flat files, databases, spreadsheets and programming languages. Current Windows versions also include an easy-to-use graphical user interface with a "syntax aware" editor and graphical display of solutions and generated model matrices.

9.2 Design Philosophy

Our primary goal, like all optimization software vendors, is to provide state-of-the-art optimization tools. However, we feel our additional goal of providing easy-to-use software serves to distinguish our products. Some `LINGO` features that illustrate this concept are:

- Simplified syntax for small models
- Closely coupled solvers
- Interface to Excel
- Automatic identification of appropriate solvers (linear, quadratic, nonlinear)
- Automatic linearization and global optimization
- Model debugging
- Programming interface

9.2.1 Simplified Syntax for Small Models

Many users of optimization software have a need to solve small, simple models, and may not want to invest heavily in familiarizing themselves with the intricacies of a full-blown modeling language. In fact, for a casual modeler, the power of an algebraic modeling language (AML) might be considered overkill. As an example, suppose we'd like to solve a simple two product, three resource product mix model. We could draw upon the set-based modeling features of LINGO and formulate such a model as follows:

```
MODEL:

SETS:
  Products:  Profit, Production;
  Resources:  Supply;
  RxP( Resources, Products):  Usage;
ENDSETS

DATA:
  Products, Profit =
   Regular,  20
   Super ,   30
  ;

  Resources, Supply =
   Metal,    50
   Plastic,   60
   Glass    120
  ;

     RxP, Usage =
   Metal,Regular, 1
   Plastic,Super, 1
   Glass,Regular, 1
   Glass,Super   ,2
  ;
ENDDATA

!Maximize profit;
MAX = @SUM( Products( p):  Profit( p) * Production( p));

!Subject to constraints on resources;
@FOR( Resources( r):
 @SUM( RxP( r, p):  Usage( r, p) * Production( p))
  <= Supply( r)
);

END
```

Product mix model, set-based formulation

Although, this set-based version of our model would appear trivial to a modeler experienced in algebraic modeling systems, it might prove intimidating to a casual or novice modeler. For this reason, LINGO allows for scalar-based models, which, at least for small models, can be more comforting for a number of users. A scalar-based equivalent of our sample model could be entered into LINGO simply as:

```
MODEL:
  MAX = 20 * REGULAR + 30 * SUPER;
  REGULAR <= 50;
  SUPER <= 60;
  REGULAR + 2 * SUPER <= 120;
END
```

Product mix model, scalar-based formulation

9.2.2 Close Coupled Solvers

Traditionally, AMLs have existed separately from their solvers. The AML takes the user's model formulation and constructs a file written in a well defined format containing the generated math program. The solver of choice is then invoked as a separate process to read the generated model file, optimize the math program, and pass the solution back to the AML in another file, also written in a specific format.

This open coupling of an AML and its solvers has its advantage in that any solver that conforms to the agreed upon file interface may be accessed from the AML. The ability to select from an array of solvers can be a big plus for the technically trained optimization professional. However, it's our contention that for every user capable of choosing amongst all the various solvers now available, there are a multiple of users that don't know/don't care about the engine under the hood.

Over the years, modeling systems have become considerably easier to use. This has made it possible for many non-technically trained users to try their hand at modeling. It's our experience that there are many users capable of formulating valid and relevant models that have had little, if any, formal training in math programming. Many such users are unaware of even the most fundamental concepts in math programming, *e.g.*, linearity vs. nonlinearity. In our opinion, the majority of users would simply prefer for the AML to intelligently select a reliable and fast solver that is most appropriate for their particular model.

The above considerations are the primary reasons we opted to follow a less traditional approach of closely coupling a suite of solvers to the LINGO AML front-end. By closely coupled, we mean that all the solvers are bundled with the LINGO system and execute as part of the same process as the AML. The LINGO AML performs extensive symbolic analysis and reduction of the model to determine if it is linear, quadratic or nonlinear, and then picks the most appropriate solver from its bundled selection.

In addition to freeing the user from having to select a solver, close coupling offers additional advantages of:

Faster execution—the generated model and solution are passed directly through memory rather than through disk files.

Less prone to interface errors—open systems are flexible but more prone to error when one or more vendors misinterpret interface specifications or mismatch releases of their systems. When such errors do occur, the user may need to contact both the AML vendor and the solver vendor to determine where the fault lies. Solver links can be more readily tested and verified before shipping when and AML is closely coupled with the solvers.

9.2.3 Interface to Excel

Almost all math programming models need to incorporate data pertaining to the system being modeled. Data could certainly be embedded in the mathematical model and managed within the AML; however, there are much more appropriate vehicles for data maintenance. An excellent, and certainly the most popular, way to manage small to moderate amounts of data is within spreadsheets. For this reason, we've always made it a priority to provide data links to spreadsheets.

Prior to 1998, `LINGO` supplied a link to read data from and write solutions to Lotus WK1 format files. This feature was accepted well enough by users, but it was somewhat cumbersome in that the spreadsheet file needed to be saved before running the model, and it needed to be reloaded to view the solution once the model was optimized.

With the subsequent rise in popularity of Excel and the development of Object Linking and Embedding (OLE) technology in Windows, `LINGO` added real-time OLE links to Excel in 1998. OLE data transfers were somewhat slow at first, but current implementations can export 100,000 values to Excel in a fraction of a second, while importing 100,000 values from Excel requires about three seconds. The real-time links eliminated the burden of saving and reloading the spreadsheet file each time the model was run, and it was at this point that `LINGO` users started to make extensive use of the spreadsheet interface.

Below is a `LINGO` example of a simple transportation model that draws all model data from an external workbook. The solution is also exported back to the same workbook.

```
MODEL:

!  A 3 Warehouse, 4 Customer Transportation model;
!  Set members and data are imported from Excel
  and the solution is exported back to Excel using @OLE();

SETS:
 WAREHOUSE : CAPACITY;
 CUSTOMER  : DEMAND;
```

```
 ROUTES( WAREHOUSE, CUSTOMER) : COST, VOLUME;
ENDSETS

!  The objective;
   MIN = @SUM( ROUTES: COST * VOLUME);

!  The demand constraints;
   @FOR( CUSTOMER( J):
    @SUM( WAREHOUSE( I): VOLUME( I, J)) >= DEMAND( J));

!  The supply constraints;
   @FOR( WAREHOUSE( I):
    @SUM( CUSTOMER( J): VOLUME( I, J)) <= CAPACITY( I));

!  Here are the parameters;
   DATA:
     !Import the set members and their data attributes;
     WAREHOUSE, CUSTOMER, CAPACITY, DEMAND, COST = @OLE();
     !Export the solution back to the workbook;
     @OLE() = VOLUME;
   ENDDATA

END
```

A transportation model linked to Excel

This model imports the WAREHOUSE and CUSTOMER set members along with the data on capacities, demand and shipping costs from an Excel workbook using the @OLE() function in the DATA section of the model. The VOLUME array, containing optimal shipping quantities, is exported back to the same workbook using the @OLE() function, as well. The workbook used resembles the following once optimization is completed and the solution values have been returned:

Supporting Excel workbook for a transportation model

The @OLE() function will accept a workbook name and a list of ranges for retrieving or exporting data. If the workbook name is omitted, the current open workbook is used. If the range names are omitted, the names of the arrays in the `LINGO` model are used instead.

This example illustrates the concept of *data independence*, whereby the details of the model are totally isolated from the data of a particular instance of the model. In this case, the model resides in `LINGO`, while all data is managed within an Excel workbook. Data independence is vital for creating models that are easy to maintain and scale up or down in size.

There has been much debate regarding the relative merits of spreadsheet solvers vs. AMLs. On the one hand, given that data management can take up a major portion of one's modeling efforts, a spreadsheet solver is a natural way of combining the solver engine with the data source. Professor Sam Savage,

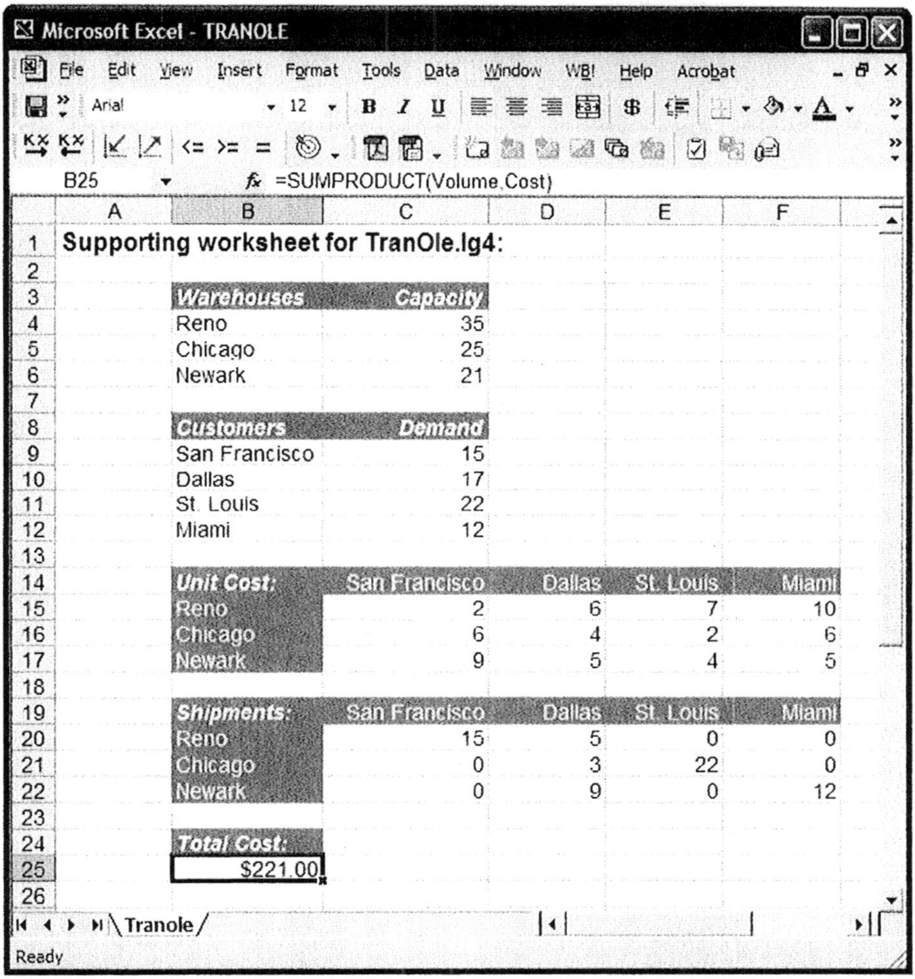

Figure 9.2.1. OLE in LINGO

a long-time advocate of spreadsheet modeling, jokingly advances the Willie Sutton theory behind the popularity of spreadsheet modeling. Namely, one models in spreadsheets because "that's where the *data* are." On the other hand, the concise, structured nature of an AML's modeling language makes reading and maintaining even very large models relatively easy. This contrasts with a spreadsheet model where it can often be a challenge to discern the underlying mathematical relationships of large models with formulaespread throughout numerous cells on multiple tabbed sheets. Another advantage of modeling in AMLs over spreadsheets is that models are generally much easier to scale up or down in size. As the dimensions of a problem change, spreadsheet models typically require the insertion or deletion of rows or columns and the editing of cell formulas. Whereas, if an AML model is constructed so as to be data independent, it will generally require little or no modification when re-dimensioning. It could be said that combining the power of model expression of an AML with the ease of data handling in spreadsheets gives the modeler the best of both worlds.

9.2.4 Model Class Identification

A LINGO modeler need not be aware of whether their model is linear, quadratic or nonlinear. LINGO automatically determines the model's class as part of a two phase process during the matrix generation phase. Initially, any variables that are determined to be fixed in value are substituted out of the math program, which frequently converts what would otherwise be a nonlinear model into a linear one. LINGO then symbolically executes the Reverse Polish Notation (RPN) of the reduced model. The symbolic execution phase identifies all quadratic and nonlinear terms in the model. These terms are identified on a (row,variable) coordinate basis to allow the solvers to exploit any sparsity in the model's nonlinear terms. Once the model's class has been identified, LINGO selects an appropriate solver and commences optimization.

As an example, consider a variation on the small product mix model presented above:

```
MODEL:
  MAX = PROFIT_R * REGULAR + PROFIT_S * SUPER;
  REGULAR <= LABOR;
  SUPER <= LAND;
  CAP_R * REGULAR + CAP_S * SUPER <= CAPITAL;
  PROFIT_R = 20;
  PROFIT_S = 1.5 * PROFIT_R;
  LABOR = 50;
  LAND = 60;
  CAPITAL = 120;
  CAP_R = 1;
  CAP_S = CAP_R + 1;
END
```

The majority of "variables" in this model are actually fixed in value. LINGO determines this as part of the model generation phase and actually only passes the following reduced linear program to the solver:

```
MODEL:
  MAX= 20 * REGULAR + 30 * SUPER;
  REGULAR <=   50;
  SUPER <=  60;
  REGULAR + 2 * SUPER <=   120;
END
```

9.2.5 Automatic Linearization and Global Optimization

Functions and operators that are nonsmooth, discontinuous, or nonconvex, may cause difficulties for many solvers. First, the introduction of such functions force linear models to become nonlinear. Secondly, most large-scale nonlinear solvers assume that all functions are continuous and differentiable and tend to perform quite poorly on models that do not conform to this assumption.

Examples of problematic functions include MIN(), MAX(), ABS(), IF() and all the logical operators (*e.g.*, equals-to). On the other hand, these functions and operators can be used to succinctly express certain relations within a model and are frequently used by novice modelers. An experienced modeler knows to avoid discontinuous functions. Experienced modelers also know that any model containing piecewise-linear, discontinuous functions can be transformed, through the use of integer variables, so that the result is linear and mathematically equivalent to the original model (although we believe most would prefer not to go through such gyrations were they not necessary). In reality, the percentage of practicing modelers that are trained in linearizing discontinuous functions is quite small. In fact, most practicing modelers are unaware that something as innocuous as an absolute value function can convert an otherwise well-crafted model into a nonlinear, non-smooth nightmare for the solver.

For years, our support staff assisted users with re-formulating discontinuous functions in order to convert their models to being continuous and, hopefully, linear. This motivated us to begin experimenting with ways to automate the linearization process by incorporating it as a feature in the software. This led to the introduction of automatic linearization in our spreadsheet solver, What'sBest!, in 2002 and in `LINGO` in 2003. As an option, `LINGO` will reformulate models in order to linearize the following: ABS(), MAX(), MIN(), IF(), products of binary and continuous variables, and all the logical operators.

To illustrate linearization, consider the trivial model:

$Minimize : |x - 3|$
Subject to:
$x = 2$

The model, as written, is nonlinear due to the absolute value term in the objective. An equivalent, linear version of this model would be:

$Minimize : a$
Subject to:
$a = p + n$
$p = M * z$
$n = M * (z - 1)$
$x - 3 = p - n$
$x = 2$
p, n nonnegative
$z = 0$ or 1

where M is a large constant term, and is traditionally referred to as the Big M coefficient. The second model simply partitions the absolute value term, $|x-3|$, into two parts around the breakpoint at $x = 3$. The variable a is forced to take on the value of $|x - 3|$.

This same transformation can also be viewed in LINGO. Here again is the original model using LINGO syntax:

```
MODEL:
 MIN = @ABS( X - 3);
 X <= 2;
END
```

Enabling the linearization option in LINGO and generating the model yields the following formulation:

```
MIN    _C3
SUBJECT TO
  X <=   2
  - _C1 - _C2 + _C3 =    0
 _C1 - 100000 _C4 <=   0
 _C2 + 100000 _C4 <=   100000
  X - _C1 + _C2 =   3
END
INTE _C4
```

The variables *_C1* through *_C4* were added as part of linearization and correspond to the variables *p*, *n*, *a* and *z*, respectively. Note that LINGO set the value for Big M at 100,000. The user can control the Big M value as an option setting in LINGO.

Linearization can be a powerful feature for certain classes of models. However, there are several caveats. First, linearization works best if a model can be completely linearized. If a model can only be partially linearized, then the resulting model tends to be a large, mixed-integer, nonlinear program (MINLP). A large MINLP can actually be more difficult to solve than the un-linearized, original NLP. Second, the choice of the Big M coefficient is critical. If Big M is too small, the linearized model will be infeasible. On the other hand, setting Big M too high can lead to numeric stability problems in the solver. Recently, in light of these considerations, we have introduced general global optimization algorithms. Global solution algorithms partition non-convex models into a number of smaller convex models and then use a branch-and-bound manager to find the global optimal point over the entire variable domain.

9.2.6 Debugging Models

In an ideal world, all models would return an optimal solution. Unfortunately, this is not the case. Sooner or later, one is bound to run across either an infeasible or unbounded model. This is particularly true in the development phase of a project when the model will tend to suffer from typographical errors. To this end, LINGO supports a debugging facility to find the irreducible infeasible subset (IIS) of constraints in infeasible models and the irreducible unbounded subset (IUS) of variables on unbounded models (Chinneck, 1996,[39]).

Consider a model that is infeasible because of a formulation error. The debug feature identifies two types of constraints: *sufficient* and *necessary*. Removing any sufficient constraint from the model is sufficient to fix the entire model. Not all models will have a sufficient set. A "necessary" set has the feature that is is necessary to delete at least one constraint from the necessary set to get a feasible set, although this may not be sufficient, *i.e.*, there may be other constraints not in the necessary that also need to be deleted or repaired.

The following example illustrates; the coefficient .55 in row 4 should have been 5.5:

```
Model:
  [Row1] Max = 3*X + 7*Y;
  [Row2] X + 2*Y <= 3;
  [Row3] 2*X + Y <= 2;
  [Row4] 0.55*X + Y >=4;
End
```

Infeasible linear program

When we attempt to solve this formulation, LINGO returns an infeasible model error message. At this point, we invoke the model debugger and are presented with the following analysis:

```
Sufficient Rows:
ROW4]  .55 X + Y >=   4

Necessary Rows:
ROW2] X + 2 Y <=   3

Necessary Variable Bounds:
Y >=   0
```

The debugger has correctly identified that the erroneous Row4, when eliminated, is sufficient to make the entire model feasible. Row2 is necessary in the sense that as long as Row2 and Row4 are present, the model is infeasible. Removal of Row2 is not sufficient, however. If Row2 is dropped, Row3 and Row4 still make the model infeasible.

Typically, the debugger helps to substantially reduce the search effort for model bugs. The first version of this feature was implemented in response to a user who had an infeasible model. The user had spent a day searching for a bug in a model with 400 constraints. The debug feature quickly found a necessary set with 55 constraints, as well as one sufficient set constraint. The user immediately noticed that the right-hand side of the sufficient set constraint was incorrect.

9.2.7 Programming Interface

LINGO provides an Application Programming Interface (API) that allows programmers to incorporate LINGO's features into their own software applications. Once a LINGO model is developed, it can be can be loaded and solved from a calling application with a few simple calls to the LINGO API. Here is some sample C++ code demonstrating the LINGO API link (with calls to the LINGO API in bold type):

```
// create a LINGO environment
pLSenvLINGO pLINGO;
pLINGO = LScreateEnvLng();
if ( !pLINGO) goto ErrorExit;

// Open LINGO's log file
nError = LSopenLogFileLng( pLINGO, "\\LINGO8\\LINGO.log");
if ( nError) goto ErrorExit;

// Build the LINGO script
CString csScript;
csScript = "SET ECHOIN 1\n"; // Echo input to log file
csScript = csScript +      // Import model from a file
 "TAKE \\LINGO8\\SAMPLES\\STAFFPTR.LNG\n";
csScript = csScript +      // Solve the model
 "GO\n";
csScript = csScript +      // Exit
 "QUIT\n";

// Run the script
nError = LSexecuteScriptLng( pLINGO, (LPCTSTR) csScript);

// Close the log file
LScloseLogFileLng( pLINGO);
if ( nError) goto ErrorExit;

// Delete the LINGO environment
LSdeleteEnvLng( pLINGO);
```

Models may also make use of LINGO's @POINTER() function for directly loading data from or returning solutions to the calling application. These data transfers occur via direct memory access, bypassing the need for intermediary text files.

The user may also supply the address of a callback routine for the LINGO API to call periodically during optimization. The calling application can query the LINGO API from the callback function regarding the current state of the solver, and, if desired, interrupt the solver at any point. An additional callback function can be established for notification of any errors that may occur during processing of the `LINGO` command script.

9.3 Future Directions

Twenty years ago the typical mathematic modeler was well trained in modeling techniques and optimization theory. Only specialized researchers had the mainframe hardware and the technical wherewithal to operate the optimization applications of the time. The introduction of the IBM PC in 1983 and, in particular, the Compaq 386 in late 1986 changed all that. Powerful desktop hardware became readily available at a relatively inexpensive price.

Readily available hardware brought many new optimization software developers onto the scene. These vendors focused on developing a new generation of easier to use optimization software, primarily in the areas of spreadsheet solvers and AMLs. Gradually, users untrained in traditional mathematic modeling began to experiment with this friendlier optimization software. Spreadsheet solvers, in particular, have served to introduce many novice modelers to optimization. The sheer number of untrained modelers began to outnumber the more traditional, technical users.

The trend toward less technically trained modelers has continued over the years to the point where the typical modeler now has had little or no formal training in mathematical modeling. In order to respond to this trend, our hope is to incorporate more modeling "horse sense" into our applications, with the goal of making life easier for the less experienced modeler. Areas of interest include constraint programming, model reformulation and extensions to global optimization.

Chapter 10

THE LPL MODELING LANGUAGE

Tony Hürlimann
Department of Informatics
University of Fribourg, Switzerland
tony.huerlimann@unifr.ch

Abstract LPL is a structured mathematical, logical modeling and programming language with a powerful index mechanism, which allows one to build, maintain, modify, and document large linear, non-linear, and other mathematical models. It is also a modeling system with a point-and-click user interface to browse, edit, and run models. LPL includes some special features: Integration of algorithmic and declarative knowledge, logical modeling, automatic generation of model documentation, and others.

Keywords: LPL, modeling, logical modeling, declarative knowledge, algorithmic knowledge

10.1 History

For more than twenty years, several larger linear models (LPs and MIPs) are maintained and used at the Department of Informatics (formerly Institute of Automation and Operations Research) at the University of Fribourg, Switzerland. At the beginning of the eighties these models were stored and maintained in a MPSX-standard format. Ad-hoc tools were written to make specific modifications to the files. The first general and successful model manipulation tool on the IBM-PC was a generator written in dBASEII (from Ashton-Tate) that generated – from some constraint specifications – several database tables that represented the ROWS, COLUMNS, BOUNDS, and RHS section of a MPSX-file [185]. In 1986, we started a research project to store the models in an algebraic form, where the tables would be in a database and the mathematical structure in a text file. The idea was that in such a way we only needed to modify the data in the tables to generate a new model instance.

The result of the author's dissertation [117] was a small specialized algebraic language – called LPL (Linear Programming Language) – which was able to formulate all our models in a very readable and short way. The language even had some interesting features still not found in present languages: (1) It was possible to define hierarchical indexed sets, (2) One could automatically generate databases in dBASE format, the algebraic structure guided the generation, and (3) it had some limited way of generating the bipartite activity-constraint graph, an extremely valuable tool for certain models to browse and understand the mathematical structure of a model instance. Hierarchical sets were especially appealing since many constraints were organized in a way which were best viewed as trees: Constraints were generated recursively from the top down the tree. LPL was also the first language to integrate units of measurement (see [118], for example).

With the completion of the dissertation the research project ended. But the program was used on a regular base at the department and on need more and more features were added to the language, where the design principle always was (and still is): "Make it as simple as possible, but not simpler" (A. Einstein) which in turn is Occam's razor dating back to the medivial ages. (That means, I always had be very reluctant to all modeler requirements, because users tend to ask you to add thousands of features they would not use thereafter.)

LPL was always meant to be a language to maintain our models at the Department. Although a free copy was put on the Web already in 1992, LPL was regarded as a productive internal product, and – more and more – as a test-bed for developing new ideas in modeling generation and manipulation: report generation, logical modeling, other graphical views of a large model, etc. The aspect of test-bed became more important when a new research project started in 1996, financed by the Swiss Federal Research Foundation. This project ended in 2001 and the result was basically, what LPL is today: A full-fetched declarative and algorithmic modeling language.[1]

Hence, LPL is an old language created in the eighties with the knowledge of the last twenty years in modeling language design, but it is a modern language and has changed considerably over time. The radical changes were possible due to the absence of an "LPL-community", although quite a lot of users downloaded it and apparently used it over the years. But nothing was undertaken to make it widely known. At the present time, LPL is firmly supported at the Department of Informatics as a permanent ongoing research project and LPL also has its own home page on the Web [124]:

[1] LPL started as a **L**inear **P**rogramming **L**anguage, but soon it became a **L**ogical **P**rogramming **L**anguage, because it is possible to use logical operators in constraints. A recent extension of automatic model documentation generation suggested still another meaning: **L**iterate **P**rogramming **L**anguage. But the true and hidden meaning of LPL is: "**L**'art **P**our **L**'art".

10.2 Some Basic Ideas

The concept of *Modeling Language* has a standard meaning in the community of Operations Research where it comes from: an algebraic language. Although recent new languages add the paradigm of constraint programming, it is still meant basically as a *declarative* language, in contrast to an *algorithmic* language, under which paradigm most present *programming* languages could be subsumed.

Most programming languages could be classified into three paradigms: imperative, functional, and logic programming. Imperative languages (*e.g.*, PASCAL, C, Java) are closely related to the way a computer works: given a set of memory locations a program is a sequence of well defined instructions on retrieving, storing and transforming the content of these locations. A functional language (Lisp) implements Church's λ-calculus, where everything is considered as a function. In a logical language (*e.g.*, PROLOG), the computation can be viewed as a kind of (constructive) proof. Common to all three paradigms is that they implement a Turing machine, which allows them to implement any possible algorithm. In this sense they are all *algorithmic* languages, although the way they implement and execute a computation is very different from each other. Implementing a problem in such a language, means *to model* the way the problem must proceed. Hence, a (computer) program is a special *model*, a model which implements an algorithm.

However, modeling is a much broader idea than programming. We use the concept of *model* in a large number of contexts. We use modeling clay to form small replicas of physical objects; children – and sometimes also adults – play with a model railway or model aeroplane; architects build model houses in order to show them to new clients; some people work as photo models, others take someone for a model, many would like to have a model friend. Models can be much smaller than the original (an orrery, a mechanical model of the solar system) or much bigger (Bohr's model of the atom).

But besides these less formal meanings, we also *model* a feasible (or sometimes infeasible!) space as a conjunction (or disjunction) of mathematical constraints. This is very different from a computer program: A mathematical formula models a space in a purely *declarative* way. There is no hint or instruction on how the model should be "processed". It is there just waiting for somebody to extract its hidden properties. While the run-time system of a compiled algorithmic program is eager to execute its instructions step by step, a declarative model waits to be transformed into another form. The model $\{x+y = 10, x-2y = 4\}$, for example, could be transformed into the model $\{x = 8, y = 2\}$, which is called its "solution" – but it is still the same model. No notion of "computation" or "processing" is implied when we have a mathematical model. We merely want to *represent* this model viewed as a two-dimensional x-y-diagram, for example, or to generate an explanation on its behavior (what-if-queries). To carry out these specific "transformations", a particular algorithm is required.

Hence, the broad concept of modeling implies that the concept of modeling language must be viewed in a much broader context too. Most actual so called "modeling languages" are algebraic (or declarative) languages with scripting capability. `Mosel`, `AIMMS`, and also `LPL` take a different approach: They try to integrate the two important paradigms into a single language: the algorithmic and declarative paradigm to create a new kind of language, which at the end is a generalized programming language, a true *modeling language*. The discussion here – seemingly on terms – is not merely theoretical. The reason why we would need languages containing both paradigms is real: Some problems are "best" formulated in an algorithmic way, others in purely declarative way and still others – and this is very important – as a blend of both. Below I give an example of such a problem.

All this means that modeling must become a discipline which must follow certain criteria of quality such as software engineering, the discipline of good programming. These criteria are *reliability* to ensure correctness and robustness of models, and *transparency* to enhance model extendability, reusability and compatibility. Reliability can be achieved by a unique notation, in which models are coded, by a formal specification of its semantics, and by introducing types, units and other checking mechanisms, in order to enforce domain consistencies. Transparency can be obtained by flexible decomposition techniques which fractionates the whole model into easily maintainable software modules and components. These are the requirements not only for present programming languages, but also for future modeling languages: (1) A unique notation for all kind of models; (2) Error checking mechanisms; (3) Decomposition techniques to break down a complex problem into manageable components. (For more arguments and information on these theoretical aspects see [120].)

10.3 Highlights

The most difficult thing in designing a computer language is to decide what feature to add and what to leave out. `LPL` is no exception. Nothing is easier than to add all kinds of nice features needed once in a while, with the result of having an inconsistent and unmanageable giant language. The history of programming languages is full of instructive anecdotes with fights and angriness, hopes and deceptions. The success of Wirth's Pascal in contrast to the goliath Algol68 was a striking event, and was hopefully a lesson learnt. Creating a language in which all problems can be coded has a strange fascination. And what else does the author propose when recommending a combined language of algorithmic and declarative knowledge? Yes, we are in quest of the language that allows to formulate all kind of models, but not in ten different ways!

LPL has been designed in order to fulfill very different tasks in modeling which we judge to be important:

1 Orthogonality: Each element introduced into the language should have a general meaning if possible. For example: introducing the power operator in expression implies using it in constraints too – which leads us to non-linear models; other example: introducing the logical OR operator in expression implies using it again in constraints – which leads use to logical based models.

2 Unique notation: Every task should be written using a unique notation, for example: declarative and algorithmic knowledge formulation, data modeling and structure modeling.

3 Decomposition: It should be possible to break down a complex model into different pieces of code.

Using these design principles, LPL was implemented with the following features:

1 As a language which combines declarative and algorithmic knowledge in the same way. One can formulate complete mathematical linear, non-linear, and logical models in a purely declarative way. The data can be stored within the model or outside in databases. The interface to various solvers is configurable. It also allows the user to write algorithms. Define data structures and it has all the control structures of another programming language. The language permits one to break down the whole into logical modules – which are themselves models or model-parts.

2 As a language that allows the user to define and manipulate the data. It can be used to generate a database (DDL: data definition language). Like a data manipulation language (DML) in databases, LPL can also manipulate (join, select, project, etc.) multi-dimensional data (data cubes). The data are read from or written to databases or text files in various ways. Tools exist to generate various pivot tables from a multi-dimensional data cube.

3 As a modeling environment with an easy-to-use interface. It allows the user to view and traverse the model in different ways and to find and modify all elements with a single mouse click.

4 As a model documentation tool: From an LPL model specification enriched with inline text blocks, one can automatically generate a LaTeX-file or a PDF-file as a printable or publishable model documentation.

5 As a library ready to be used by other applications: LPL comes as a dynamic link library and its whole functionality can be used (and hidden

from the user) in another application. One can create other modeling systems based on the library. LPL's own model browser is only one that was put on top of the basic design.

6 As a client/server architecture: The same basic architecture of LPL can be used as a client/server application. Installing an LPL-Client on a machine and an LPL-Server on another, the two can communicate: The client can send the model to the server where the model can be manipulated, calculated or solved by an LPL-server in the same way as locally. It is completely transparent to the user (true ASP: Application Service Providing). On the server machine, one only needs to install a WWW-Server, a FTP-Server, Tomcat (a free Java-based distributed software), and an LPL-Server.

Three of the mentioned features will be illustrated in more detail in the next three sections: (1) combination of algorithmic and declarative knowledge, (2) logical models, (3) documentation generation using a model. The complete specification of the LPL syntax and semantic can be found in the Reference Manual of LPL [123].

10.4 The Cutting Stock Problem

A paper machine produces a standard width of 100 inches. This paper has to be cut into four widths: 97 rolls of width 45 inches, 610 rolls of width 36 inches, 395 rolls of width 31 inches, and 211 rolls of width 14 inches. The cut should be such that the total waste is minimal, [for more information on the problem see Chvátal (1983,[42])].

From the model implementation point of view we can say that this problem can be solved efficiently using the following three modules: (1) We need a LP formulation for a cutting stock problem with a limited number of patterns. (2) We need to solve an integer knapsack problem, which decides which pattern must be entered into the basic matrix of (1), in order to improve the objective. (3) We need an algorithm that solves these two declarative models in an iterated way, until some criteria are fulfilled. Hence, we can formulate this problem by composing it from three "pieces of knowledge": two of which are declarative representable and one that is a purely algorithmic piece.

Hence in LPL, the complete problem would be formulated as following:

```
MODEL CuttingStock "The Cutting Stock Problem";
  SET
    widths ALIAS w      "the different ordered smaller rolls";
    patterns ALIAS p    "possible cutting patterns";
  PARAMETER
    B                   "the width of the initial roll";
    width{w}            "the width of smaller roll w";
    demand{w}           "the demand for smaller roll w";
```

```
    a{w,p}  ::= IF(w=p,FLOOR(B/width)) "number of w rolls in p";
    du{w}                "cuts duals";

  MODEL cutting_stock ALIAS cs;
    VARIABLE X{p}       "number of initial rolls cut according to p";
      yy[1,1]; zz[1,1];  --asserts that the duals of 'cuts' are available
    CONSTRAINT cuts{w}: SUM{p} a*X + yy-zz >= demand;
    MINIMIZE obj: SUM{p} X;
  END

  MODEL find_pattern ALIAS fp;
    INTEGER VARIABLE y{w} [0,7]  "the number of rolls of size w in new patt";
    CONSTRAINT patt: SUM{w} width*y <= B;
    MAXIMIZE  cont: SUM{w} du*y;
  END

  MODEL DATA myData3 "Chvatal p.196";
  BEGIN
    w   := /1:4/;
    p   := /1:4/;
    B   := 100;
    width{w}  := [45  36  31  14];
    demand{w} := [97 610 395 211];
  END

BEGIN
  cutting_stock;
  du{w}:=cs.cuts.dual;
  find_pattern;
  WHILE (fp.cont > 1.0001) DO
    p:=ADDM('1');
    a{w,p}:= IF(p=#p,fp.y,a);
    cutting_stock;
    du{w}:=cs.cuts.dual;
    find_pattern;
  END;
  cutting_stock;
  WRITE "obj: ####.#\n####.# of pattern [ ## ]\n"
        : cs.obj, ROW{p|cs.X} (cs.X, COL{w} a);
END
```

The basic syntax of is:

```
  MODEL ModelName;
    <declarative part of the model>
  [ BEGIN
    <algorithmic part of the model> ]
  END
```

Each model begins with the keyword `MODEL` and ends with the keyword `END`. In the declarative part the user declares the sets (`SET`), the parameters (`PARAMETER`), the variables (`VARIABLE`), the constraints (`CONSTRAINT`), and recursively other sub-models beginning with another `MODEL` keyword. The optional algorithmic part begins with `BEGIN` and contains a sequence of instructions.

Each declaration begins with a keyword `SET`, `PARAMETER`, etc. Then comes a name and several attributes. For example the declaration of the set $widths$

contains only one attribute: an alias, which is simply another (in general a shorter) name. LPL does not distinguish between index names and set names, so these aliases can be used as indexes in expressions. Parameters can be declared over sets: `a{w,p}`, for example, declares a matrix over the sets w and p. It also contains two other attributes: an expression and a comment. The expression is: `IF(w=p,FLOOR(B/width))`, which implements the assignment:

$$a_{w,p} = \begin{cases} \lceil B/width_w \rceil, & w = p \\ 0, & w \neq p \end{cases}$$

The indexes can be omitted if the context permits it. This makes the whole expression shorter and sometimes more readable. The second attribute is a comment in `"..."`. It allows the user to explain the object. This text is part of an automatically generated model documentation.

The next model part is the submodel `cutting_stock` which contains the declarations of three variables, a constraint, and a minimizing function. The submodel implements the following linear program:

$$\{\min \sum_p X_p \quad | \quad \sum_p a_{w,p} \cdot y_w \geq demand_w, x_p \geq 0, \quad \forall w\}$$

that is, minimize the waste (the number of patterns), subject to the constraints that all demands must be fulfilled. The two singleton variables y and z are technical variables to assert that the reduced prices of the constraints exist. Their values are fixed at 1. The attribute `[1,1]` says, that the lower and upper bounds of these variables `(y,z)` are 1. (The lower and upper bounds could be arbitrary expressions.)

Next comes the submodel `find_pattern` which consists of the declarations of an integer variable, a constraint, and a maximization function. This submodel implements the following integer linear knapsack problem:

$$\{\max \sum_w du_w \cdot y_w \quad | \quad \sum_w width_w \cdot y_w \leq B, \quad y_w \in \{0 \ldots 7\}$$

The last declaration is another submodel, in a very special model: it is a `MODEL DATA`, where all data are collected. This model only has an algorithmic part and assigns the data defined in Chvátal's booki [42]. It defines four different widths and four different patterns, and assigns the values to the numerical vectors `width` and `demand`, as well as to the singleton parameter `B`.

This ends the declaration of the main model and an algorithmic part is started with the keyword `BEGIN`. This contains a sequence of instructions. The first instruction (`cutting_stock;`) says to proceed with (actually solve, since it contains a minimizing function) the first submodel `cutting_stock`. The values of the dual cuts are now assigned to the vector `du`. All objects are accessible anywhere within the model using the dot-notation in LPL. The expression `cs.cuts.dual`, for example, refers to an object `cuts` within model `cs` (note that `cs` is an alias for `cutting_stock`). Hence `cs.cuts` refers to the constraints within the submodel `cutting_stock`. Each constraint contains several values, one of which is the reduced prices calculated by a simplex solver, and LPL refers to them with the word `dual`. The third instruction is `find_pattern;`, hence the solving of the knapsack problem.

As long as the objective value of the knapsack model is larger than one, the instructions within the `WHILE` sequence are executed. That is: first a new element is added to the set `pattern` using 's function `ADDM()`. Then the values of the variables `y` of the knapsack problem are assigned to the newly created column within the matrix `a`. Finally the two submodels are solved again.

At the end, the result is printed out with a `WRITE` instruction. The output of this `WRITE`-statement (which is not explained here) is:

```
obj:  452.3
  48.5 of pattern [  2 0 0 0 ]
 100.7 of pattern [  0 2 0 0 ]
 105.5 of pattern [  0 2 0 2 ]
 197.5 of pattern [  0 1 2 0 ]
```

The model was decomposed into four model components, which follows the logic of the problem: two declarative submodels and an algorithmic part that guides their solution process. Furthermore, the data are clearly separated from the model specification and collected in another specialized submodel. To treat another model instance (with different data), one only need to exchange it with another `MODEL DATA` or to load another one.

10.5 Liquid Container

A company produces two liquids at unknown quantities x and y. The liquids are stored in two containers of capacity a and b. The two liquids can be stored in both containers, but cannot be mixed in the same container. We wish to formulate the capacity constraint.

Three cases can be distinguished: Either (1) we produce only the first liquid, or (2) only the second, or (3) both. In the first case we have: $y = 0$ and $x \leq a+b$. In the second case we have: $x = 0$ and $y \leq a + b$. In the third case we have: $x \leq a$ and $y \leq b$. At least one of the three cases must be true. Hence, this constraint could be expressed as follows:

$$((y = 0) \wedge (x \leq a + b)) \vee ((x = 0) \wedge (y \leq a + b)) \vee ((x \leq a) \wedge (y \leq b)).$$

In LPL it is easy to verify that this formula corresponds to

$$((x > a) \rightarrow (y \leq 0)) \wedge ((y > b) \rightarrow (x \leq 0))$$

which is much shorter. Note that we suppose that $x, y \geq 0$. In LPL, this could be coded as the following model:

```
MODEL Logic4 "Two Liquid Container";
  PARAMETER a:=10; b:=12;
  VARIABLE x [0,a+b]; y [0,a+b];
  CONSTRAINT capa: (x>a -> y<=0) AND (y>b -> x<=0);
  MAXIMIZE obj:x+y;
  WRITE x, y;
END
```

We can see that LPL allows one to mix logical and mathematical operators within a constraint. LPL automatically translates this specification into a linear MIP model – adding new 0-1 variables – which then can be solved by a standard MIP solver. Considerable effort has been invested in this translation. It is difficult in the worst case to find the sharpest MIP for a given logical formulation. The number of variables or constraint may grow exponentially with the number of existing propositions or real variables in the logical formulation. To avoid this, LPL uses some heuristic to limit the generation, with the effect that the resulting MIP may be far from the sharpest formulation. Nevertheless, even in such a case, the translation proposed by LPL can be useful, because a user can then manually manipulate and improve LPL's proposition and reformulate the problem on this basis.

10.6 Model Documentation

Documenting a model is extremely important. However, this is still badly neglected by the model language designers as well as by the users. Documenting a model takes time and is the least rewarding job, because "the boss" is only interested in the results! The same is true in programming. The only language that now supports documentation extensively is *Java* with its tool, the *javadoc*. There exist tools created a long time ago by Knuth (1984,[147];1992,[148]) that allow a programmer to generate a complex documentation of the source code, but they seem not to be widely used. However, a well documented model is an invaluable investment (supposing that the model is not just used once). Many models take months to build – and documenting this process makes it much more transparent and accessible to the decision makers. Model documentation can also be an excellent means to prepare case-studies in learning modeling building. Building models is a creative skill and must be learnt. One of the best ways to learn it is to study existing models. However, having only the simple mathematical formulation is sometimes not very helpful in understanding the formulas. Hence, a model enriched with explanation is much more useful. In the Internet, thousands of mathematical models – in various language formulations – can be downloaded. How much more useful they would be, if an explanatory text would accompany them!

Well then, let's code the model here and write a documentation separate from it. Why bother? Why integrate the documentation into the already heavily overloaded code? Because this is the only way to keep the code and the documentation consistent. How often is the user willing to write the documentation – at the end of the project, and then the model changes again... . Both should be done at the same time and in the same environment. Integrating the documentation into the model itself has another benefit: It can be regenerated (semi-)automatically at any time.

LPL includes a LaTeX generation tool. The mathematical and logical formulas are typeset with the free software TeX. Qualified comments and additional text as well as figures and graphics can be included into the formal model specification. From TeX then one can generate a PDF-file, HTML-files, or even XML-files using free software.

As an example, let's take a queen position problem on a chessboard. Suppose the problem is as follows: How many queens are needed, and in what position, so that every unoccupied square of the chessboard is under direct attack from some queen? We introduce a binary variable X for each (i, j) position on the chessboard, which is 1 if the position is occupied by some queen, otherwise it is 0. The objective function now is to minimize the number of queens:

$$\min \sum_{i,j} X_{i,j}$$

The constraint can be formulated as follows: "For each (i, j) position on the chessboard one must require that it is under attack from at least one queen." "At least" in this context means OR. Therefore, one can formulate this in LPL as following (where S is also an index over row/column like i or j):

```
CONSTRAINT
  ATT{i,j}: OR{S|S<>i and S-i+j>=1 and S-i+j<=#S} X[S,S-i+j]
    OR OR{S|S<>i and i+j-S>=1 and i+j-S<=#S} X[S,i+j-S]
    OR OR{S|S<>i} X[S,j] OR OR{S|S<>j} X[i,S];
```

The LaTeX generation tool of LPL translates this constraint automatically into the following LaTeX typeset formula:

$$\begin{aligned} ATT_{i,j} : \exists_{S|S\neq i \wedge S-i+j\geq 1 \wedge S-i+j\leq \#S} X_{S,S-i+j} \\ \vee \exists_{S|S\neq i \wedge i+j-S\geq 1 \wedge i+j-S\leq \#S} X_{S,i+j-S} \\ \vee \exists_{S|S\neq i} X_{S,j} \vee \exists_{S|S\neq j} X_{i,S} \end{aligned} \qquad (?)$$

10.7 Conclusion

A short and necessarily limited introduction to LPL was given in this chapter. LPL is still an ongoing research project at the Department of Fribourg at the University of Fribourg, Switzerland and keeps it status as a test-bed for new and exciting ideas in modeling representation and manipulation. At the same time, there exists a stable version ready to be used in productive applications.

Chapter 11

THE MINOPT MODELING LANGUAGE

Carl A. Schweiger
Pavilion Technologies, Inc.
11100 Metric Blvd. #700
Austin, TX 78758-4018
cschweiger@pav.com

Christodoulos A. Floudas
Department of Chemical Engineering
Princeton University,
Princeton, NJ, 08544
floudas@titan.princeton.edu

Abstract MINOPT is a modeling language for solving a broad range of optimization problems formulated as mathematical programs. The motivation for developing MINOPT was the need for a framework capable of addressing mixed-integer nonlinear programming problems as well as problems involving dynamic problems described by differential and algebraic equations. The result is a comprehensive system that can address linear programs, mixed-integer linear programs, nonlinear programs, nonlinear programs with differential and algebraic constraints, mixed-integer nonlinear programs with differential and algebraic constraints, optimal control problems, and mixed-integer optimal control problems. MINOPT features an expressive modeling language, implementations of Generalized Benders Decomposition and Outer Approximation algorithms, and interfaces to commercial optimization software. Examples are presented to illustrate the features of the modeling language.

Keywords: MINOPT, Mixed Integer Nonlinear Programming, dynamic optimization, optimal control, differential and algebraic equations

11.1 Introduction

MINOPT is a tool for modeling and solving mathematical programming problems that arise in many areas of mathematics and science. MINOPT has two key components: a modeling language and an algorithmic framework. The modeling language is used to write the model for the mathematical program in a clear and concise form and provides an interface to the computer implementation of numerous solution algorithms. These solvers constitute the algorithmic framework of MINOPT.

The primary goal of the design of MINOPT was to create a framework capable of addressing two areas of problem formulations. The first was the ability to handle Mixed Integer Nonlinear Programming (MINLP) problems through the direct application of suitable algorithms. The second was the ability to address dynamic problems described by systems of differential and algebraic equations (DAEs). To achieve these goals, a new modeling language and algorithmic framework were developed.

As a consequence of the broad design goals of MINOPT, the resulting software is capable of addressing a broad range of problem types. These include Linear Programs (LP), Mixed-Integer Linear Programs (MILP), and Nonlinear Programs (NLP) as well as the MINLPs. The dynamic extension of MINOPT allows it to solve problems that are be classified as Nonlinear Programs with Differential and Algebraic constraints (NLP/DAE), Optimal Control Problems (OCP), Mixed-Integer Nonlinear Programs with Differential and Algebraic constraints (MINLP/DAE), and Mixed-Integer Optimal Control Problems, (MIOCP).

In addition to the enhanced capabilities of the MINOPT modeling language, the algorithmic framework of MINOPT is also highly developed. MINOPT interfaces to state-of-the-art solvers for LP, MILP, and NLP problems and includes implementations of several algorithms for the solution of MINLP problems. MINOPT implements robust numerical integration techniques for solving problems involving differential equations.

Through the connection of the modeling language to the algorithmic framework, the user can switch among the various solution algorithms and specify the options and parameters for the solution algorithms.

11.1.1 Motivation

Modeling languages have been developed to aid in the formulation and solution of mathematical programming problems. They allow the model to be written in a way that looks close to standard mathematical notation and serves as direct input to the computer implementation of the solution algorithm. The translation of the model to input that the algorithm understands is handled automatically by the computer. This facilitates the process of applying computer implementations of algorithms and helps reduce errors that may be introduced when the translation is done by the user. The use of the modeling language facilitates writing the model, entering the data, modifying the model, changing algorithms, adjusting algorithm parameters, and on the whole makes mathematical programming more reliable

The key features of an effective modeling language are the following:

- high level language for compact representation of large and complex models,
- ability to change models simply and safely,
- unambiguous representation of mathematical expressions, and
- model descriptions that are independent of solution algorithms.

Probably the most challenging feature to uphold is that of maintaining the separation between the model and the solution algorithm. For the model description to be truly independent of the solution algorithm, it should be possible to write the model without knowing what the solution algorithm is. This does not mean that the modeling should be done without concern for the solution algorithm since the modeling can affect the efficiency of the solution algorithm. This means that the description of the model should not contain any references to a particular solution algorithm.

Many of the existing modeling languages such as `GAMS` (citegBKM88,[32]) and `AMPL` [75] have the features of an effective modeling language and are excellent tools for solving problem types such as LP, NLP, and MILP. Recently, these tools have become better equipped at addressing MINLP problems, but still have limited solution resources. For example, the implementation of Generalized Benders Decomposition requires that the steps of the solution algorithm be explicitly defined employing the scripting capabilities of the modeling language.

Although there have been advances in the ability to address MINLP problems, these modeling languages focus on algebraic models. They do not have provisions to address dynamic models described by differential and algebraic equations. In order to use these modeling languages to solve these types of problems, the dynamic system needs to be discretized explicitly to create an algebraic formulation. In effect, this is an implementation of a particular solution algorithm for the dynamic problem. Since the model formulation encodes the algorithmic approach, the model description is no longer independent of the solution algorithm, which violates the separation of the model and the solver.

The main motivation for the development of `MINOPT` was the lack of existing tools capable of solving MINLP problems using the different MINLP solution algorithms and the lack of tools capable of modeling and directly solving problems described by systems of DAEs. The development of `MINOPT` modeling language has maintained the key features of a modeling language as listed above.

MINLP problems can be defined and solved using several algorithms without having to describe the steps of the algorithm. Systems involving differential equations can be modeled and solved using a clear, concise representation of the DAE system.

11.1.2 MINOPT Overview

MINOPT is a general purpose package for the solution of a various types of mathematical programming problems. MINOPT consists of two components: a modeling language for the representation of the mathematical programming problem and an algorithmic framework for the solution of the mathematical problem. MINOPT reads the mathematical representation of the problem from an input file, transforms the input, and applies the appropriate algorithm in an efficient manner. The following are some of the key features of MINOPT:

Modeling Language: MINOPT features a modeling language that is clear and concise. It allows for explicit variable naming, clear mathematical representation, set notation, etc. It handles different variables types, recognizes different constraint types and determines the problem type. All model information is provided in the form of a single input file that MINOPT reads using a front-end parser. The parser is user-friendly and helps point out problems in the input file. It also has internal checks to ensure that the input is consistent and well-posed.

Automatic Differentiation: Many of the solution algorithms require the derivatives of the constraint expressions with respect to the variables. MINOPT employs automatic differentiation to determine the derivative expressions and calculations.

Algorithmic Framework: MINOPT employs a number of different solution algorithms for the various types of optimization problems. It includes implementations of a number of different algorithms for the solution of these problems and uses existing software packages for the solutions of various problems and subproblems.

Solver Packages: MINOPT includes several packages for the solution of mixed integer nonlinear programs and has interfaces to various solver packages to solve the various subproblems.

Options: MINOPT features an extensive list of options that can be used to select various algorithms, algorithm parameters, and algorithmic procedures.

Expandable Platform: Although not directly visible to the user, MINOPT has been developed with a flexible platform for the addition of algorithms when they become available.

11.2 Model Types and Solution Algorithms

The key elements of a mathematical model are the variables, parameters, constraints, and mathematical relationships. As a modeling language, `MINOPT` is capable of handling a wide variety of models described by the types of variables and mathematical relationships employed in the model. `MINOPT` supports the following variable and constraint types

continuous time invariant	linear
continuous dynamic	nonlinear
control	dynamic
integer	dynamic path
	dynamic point

and the following different problem classifications:

- Linear Program (LP)
- Nonlinear Program (NLP)
- Mixed Integer Linear Program (MILP)
- Mixed Integer Nonlinear Program (MINLP)
- Nonlinear Program with Differential and Algebraic Constraints (NLP/DAE)
- MINLP with Differential and Algebraic Constraints (MINLP/DAE)
- Optimal Control Program (OCP)
- Mixed Integer Optimal Control Program (MIOCP)

The problem formulations for the LP, MILP, and NLP problem types are well known and will not be discussed here. The following sections focus on the MINLP, NLP/DAE, MINLP/DAE, OCP, and MICOP formulations. The variable and constraint notation used throughout this section is shown in Table 11.1.

Table 11.1. Notation for variable and constraint used in the algorithm descriptions.

Variables			
x	continuous time invariant variables		
y	integer variables		
$z(t)$	dynamic state variables		
$u(t)$	control variables		
w	parameters		
t	independent variable (time)		
t_i	time instant		
	Constraints		**Dual variables**
J	objective function		
f	differential and algebraic	ν	adjoint variables (for f)
g	inequality constraints	μ	Lagrange multipliers (for g)
h	equality constraints	λ	Lagrange multipliers (for h)

11.2.1 Mixed-Integer Nonlinear Program (MINLP)

The MINLP problems we consider have the following general form [see also (1.2.1):

$$\begin{aligned} \min_{\boldsymbol{x},\boldsymbol{y}} \quad & f(\boldsymbol{x}, \boldsymbol{y}) \\ \text{s.t.} \quad & \boldsymbol{h}(\boldsymbol{x}, \boldsymbol{y}) = \mathbf{0} \\ & \boldsymbol{g}(\boldsymbol{x}, \boldsymbol{y}) \leq \mathbf{0} \\ & \boldsymbol{x} \in \mathbb{R}^n \\ & \boldsymbol{y} \in \{0, 1\}^m \end{aligned}$$

A number of algorithms have been developed to address problems with the above form. Some deal with the formulation as stated while others deal with a restricted class of the problem. The following is a listing of these algorithms.

1 Generalized Benders Decomposition, *GBD* ([83, 186, 69])

2 Branch and Bound, *B&B* ([8, 96, 183, 29])

3 Outer Approximation, *OA* ([61])

4 Feasibility Approach, *FA* ([170])

5 Outer Approximation with Equality Relaxation, *OA/ER* ([149])

6 Outer Approximation with Equality Relaxation and Augmented Penalty, *OA/ER/AP* ([247])

7 Generalized Outer Approximation, *GOA* ([66])

8 Generalized Cross Decomposition, *GCD* ([105])

An overview of MINLP algorithms and extensive theoretical, algorithmic, and applications-oriented description of *GBD*, *OA*, *OA/ER*, *OA/ER/AP*, *GOA*, and *GCD* algorithms is found in [67]. `MINOPT` has implementations of *GBD* and the *OA*, *OA/ER*, and *OA/ER/AP* algorithms.

11.2.1.1 Generalized Benders Decomposition (*GBD*). The basic idea behind *GBD* is the generation of upper and lower bounds on the solution of the MINLP model through the iterative solution of subproblems formulated from the original problem. The upper bound results from the solution of the *primal* problem while the lower bound is the result of the solution of the *master* problem. The primal problem corresponds to the solution of the original problem with the values of the $\boldsymbol{y}$ variables fixed. The solution of this problem is in the $\boldsymbol{x}$ space only and it provides information about the Lagrange multipliers for the constraints. The master problem is formulated by making use of the Lagrange multipliers and nonlinear duality theory. Its solution provides a lower bound as well as a new set of $\boldsymbol{y}$ variables for the next primal iteration. The algorithm iterates between the primal and master problems generating upper and lower bounds that converge in a finite number of iterations. `MINOPT` automatically generates the primal and master subproblems and iterates through their solutions to solve the MINLP problem. The *GBD* algorithm as implemented in `MINOPT` is outlined as follows:

$k \Leftarrow 1, \boldsymbol{y}^1 \Leftarrow \boldsymbol{y}^{\text{init}}$ {Initialize the y variables}
$LBD \Leftarrow -\infty, UBD \Leftarrow +\infty$
$\mathrm{F} \Leftarrow \emptyset, \bar{\mathrm{F}} \Leftarrow \emptyset$ {Initialize the feasibility and infeasibility sets}
$\epsilon \Leftarrow$ tolerance {Set the convergence tolerance}
repeat
$\boldsymbol{y} \Leftarrow \boldsymbol{y}^k$ {Fix the values of the binary variables}
Solve the Primal Problem

$$\begin{aligned} \{\boldsymbol{x}^*, \boldsymbol{\lambda}^*, \boldsymbol{\mu}^*\} \Leftarrow \arg \min_{\boldsymbol{x} \in \mathbb{R}^n} \quad & f(\boldsymbol{x}, \boldsymbol{y}^k) \\ \text{s.t.} \quad & \boldsymbol{g}(\boldsymbol{x}, \boldsymbol{y}^k) \le \boldsymbol{0} \\ & \boldsymbol{h}(\boldsymbol{x}, \boldsymbol{y}^k) = \boldsymbol{0} \end{aligned}$$

if Primal is feasible **then**
$\boldsymbol{x}^k \Leftarrow \boldsymbol{x}^*$ {Obtain the optimal variable values}
$\boldsymbol{\lambda}^k \Leftarrow \boldsymbol{\lambda}^*, \boldsymbol{\mu}^k \Leftarrow \boldsymbol{\mu}^*$ {Obtain the optimal Lagrange multipliers}
$\mathrm{F} \Leftarrow \mathrm{F} \cup k$ {Update the feasibility set}
$UBD \Leftarrow \min(UBD, f(\boldsymbol{x}^*, \boldsymbol{y}^k))$
else
Solve the Infeasibility Problem

$$\begin{aligned} \{\boldsymbol{x}^*, \boldsymbol{\lambda}^*, \boldsymbol{\mu}^*\} \Leftarrow \arg \min_{\substack{\boldsymbol{x} \in \mathbb{R}^n \\ \boldsymbol{\alpha}, \boldsymbol{\beta}^+, \boldsymbol{\beta}^- \in \mathbb{R}^+}} \quad & \sum_{i \in I^{\text{ineq}}} \alpha_i + \sum_{i \in I^{\text{eq}}} (\beta_i^+ + \beta_i^-) \\ \text{s.t.} \quad & \boldsymbol{g}(\boldsymbol{x}, \boldsymbol{y}^k) - \boldsymbol{\alpha} \le \boldsymbol{0} \\ & \boldsymbol{h}(\boldsymbol{x}, \boldsymbol{y}^k) + \boldsymbol{\alpha}^+ - \boldsymbol{\beta}^- = \boldsymbol{0} \end{aligned}$$

$\boldsymbol{x}^k \Leftarrow \boldsymbol{x}^*$ {Obtain the optimal variable values}
$\boldsymbol{\lambda}^k \Leftarrow \boldsymbol{\lambda}^*, \boldsymbol{\mu}^k \Leftarrow \boldsymbol{\mu}^*$ {Obtain the optimal Lagrange multipliers}
$\bar{\mathrm{F}} = \bar{\mathrm{F}} \cup k$ {Update the infeasibility set}
end if

Solve the Master Problem

$$
\begin{aligned}
&\{\boldsymbol{y}^*, \mu_b^*\} \Leftarrow \\
&\quad \arg\min_{\boldsymbol{y},\mu_b} \quad \mu_b \\
&\qquad \text{s.t.} \quad \mu_b \geq f(\boldsymbol{x}^l, \boldsymbol{y}) + (\boldsymbol{\lambda}^l)^T \boldsymbol{g}(\boldsymbol{x}^l, \boldsymbol{y}) + (\boldsymbol{\mu}^l)^T \boldsymbol{h}(\boldsymbol{x}^l, \boldsymbol{y}) \quad l \in \mathrm{F} \\
&\qquad\qquad 0 \geq (\boldsymbol{\lambda}^l)^T \boldsymbol{g}(\boldsymbol{x}^l, \boldsymbol{y}) + (\boldsymbol{\mu}^l)^T \boldsymbol{h}(\boldsymbol{x}^l, \boldsymbol{y}) \quad l \in \bar{\mathrm{F}}
\end{aligned}
$$

$\boldsymbol{y}^{k+1} \Leftarrow \boldsymbol{y}^*$
$LBD \Leftarrow \mu_b^*$
$k \Leftarrow k + 1$
until $UBD - LBD \leq \epsilon$

Note that the above formulation and algorithm is for a subclass of problems to which *GBD* can be applied. The more general class is one that has the set of $\boldsymbol{y}$ variables containing both continuous and integer variables. Although the algorithmic procedure remains the same when $\boldsymbol{y}$ contains continuous variables, the theoretical developments are altered somewhat.

11.2.1.2 Outer Approximation/Equality Relaxation/ Augmented Penalty (*OA/ER/AP*).

The outer approximation methods [see also Section 6.2] deal with a subclass of MINLP problems in which the functions $f(\boldsymbol{x}, \boldsymbol{y})$, and $\boldsymbol{g}(\boldsymbol{x}, \boldsymbol{y})$ are linear in the $\boldsymbol{y}$ variables and separable in $\boldsymbol{x}$ and $\boldsymbol{y}$. The set of $\boldsymbol{y}$ variables is also strictly binary variables. The class of MINLP problems the *OA/ER* algorithm addresses has the following formulation:

$$
\begin{aligned}
\min_{\boldsymbol{x},\boldsymbol{y}} \quad & f(\boldsymbol{x}) + \boldsymbol{c}^T \boldsymbol{y} \\
\text{s.t.} \quad & \boldsymbol{g}(\boldsymbol{x}) + \boldsymbol{B}\boldsymbol{y} \leq \boldsymbol{0} \\
& \boldsymbol{h}(\boldsymbol{x}) + \boldsymbol{C}\boldsymbol{y} = \boldsymbol{0} \\
& \boldsymbol{x} \in \mathbb{R}^n \\
& \boldsymbol{y} \in \{0, 1\}^m
\end{aligned}
$$

`MINOPT` automatically detects whether or not the problem fits this form.

The *OA* algorithm is similar to the *GBD* algorithm in that it iterates between upper and lower bounding primal and master subproblems. Both algorithms have the same primal problem formulation, but the *OA* algorithm has a different master problem formulation. The master problem for *OA* is formed by a projection onto the $\boldsymbol{y}$ space and an outer approximation of the objective function and feasible region. Again, `MINOPT` automatically generates this master problem and iterates through the subproblem solutions.

The *OA/ER* algorithm is a generalization of the Outer Approximation algorithm to handle nonlinear equality constraints. It relaxes the equality constraints to inequalities to generate the outer approximation of the feasible region. The *OA/ER/AP* algorithm is a modification of the *OA/ER* algorithm to avoid convexity assumptions necessary for finding the global solution using the *OA/ER* algorithm. The *OA/ER/AP* algorithm as implemented in `MINOPT` is stated as follows:

$k \Leftarrow 1$, $\boldsymbol{y}^1 \Leftarrow \boldsymbol{y}^{\text{init}}$ {Initialize the y variables}
$LBD \Leftarrow -\infty, UBD \Leftarrow +\infty$
$\mathrm{F} \Leftarrow \emptyset$, $\bar{\mathrm{F}} \Leftarrow \emptyset$ {Initialize the feasibility and infeasibility sets}
$\epsilon \Leftarrow$ tolerance {Set the convergence tolerance}
repeat
$\boldsymbol{y} \Leftarrow \boldsymbol{y}^k$ {Fix the values of the binary variables}
Solve the Primal Problem

$$\{\boldsymbol{x}^*, \boldsymbol{\lambda}^*, \boldsymbol{\mu}^*\} \Leftarrow \arg \min_{\boldsymbol{x} \in \mathbb{R}^n} \quad f(\boldsymbol{x}) + \boldsymbol{c}^T \boldsymbol{y}^k$$
$$\text{s.t.} \qquad \boldsymbol{g}(\boldsymbol{x}) \leq -\boldsymbol{B}\boldsymbol{y}^k$$
$$\boldsymbol{h}(\boldsymbol{x}) = -\boldsymbol{C}\boldsymbol{y}^k$$

if Primal is feasible **then**
$\boldsymbol{x}^k \Leftarrow \boldsymbol{x}^*$ {Obtain the optimal variable values}
$\boldsymbol{\lambda}^k \Leftarrow \boldsymbol{\lambda}^*$, $\boldsymbol{\mu}^k \Leftarrow \boldsymbol{\mu}^*$ {Obtain the optimal Lagrange multipliers}
$F \Leftarrow F \cup k$ {Update the feasibility set}
$UBD \Leftarrow \min(UBD, f(\boldsymbol{x}^*, \boldsymbol{y}^k))$
else
Solve the Infeasibility Problem

$$\{\boldsymbol{x}^*, \boldsymbol{\lambda}^*, \boldsymbol{\mu}^*\} \Leftarrow \arg \min_{\substack{\boldsymbol{x} \in \mathbb{R}^n \\ \boldsymbol{\alpha}, \boldsymbol{\beta}^+, \boldsymbol{\beta}^- \in \mathbb{R}^+}} \quad \sum_{i \in I^{\text{ineq}}} \alpha_i + \sum_{i \in I^{\text{eq}}} (\beta_i^+ + \beta_i^-)$$
$$\text{s.t.} \qquad \boldsymbol{g}(\boldsymbol{x}) - \boldsymbol{\alpha} \leq -\boldsymbol{B}\boldsymbol{y}^k$$
$$\boldsymbol{h}(\boldsymbol{x}) + \boldsymbol{\beta}^+ - \boldsymbol{\beta}^- = -\boldsymbol{C}\boldsymbol{y}^k$$

$\boldsymbol{x}^k \Leftarrow \boldsymbol{x}^*$ {Obtain the optimal variable values}
$\boldsymbol{\lambda}^k \Leftarrow \boldsymbol{\lambda}^*$, $\boldsymbol{\mu}^k \Leftarrow \boldsymbol{\mu}^*$ {Obtain the optimal Lagrange multipliers}
$\bar{\mathrm{F}} = \bar{\mathrm{F}} \cup k$ {Update the infeasibility set}
end if
$\boldsymbol{T}^k \Leftarrow \text{diag}(t_{ii})$ where $t_{ii} = \text{sign}(\mu_i^k)$
$\boldsymbol{v}_k \Leftarrow \sigma \boldsymbol{\mu}^k$, $\boldsymbol{w}_k \Leftarrow \sigma \boldsymbol{\lambda}^k$

Solve the Master Problem

$$\{\boldsymbol{y}^*, \mu_b^*\} \Leftarrow$$

$$\begin{array}{lll}
\arg\min\limits_{\substack{\mu_b,\boldsymbol{x},\boldsymbol{y}\\ \boldsymbol{p},\boldsymbol{q}}} & j = \mu_b + \boldsymbol{c}^T\boldsymbol{y} + \sum\limits_{l\in \mathrm{F}\cup\bar{\mathrm{F}}} (\boldsymbol{v}_l^T\boldsymbol{p}_l + \boldsymbol{w}_l^T\boldsymbol{q}_l) & \\
\text{s.t.} & \mu_b + \geq f(\boldsymbol{x}^l) + \nabla_x f(\boldsymbol{x}^l)(\boldsymbol{x} - \boldsymbol{x}^l) & \forall l \in \mathrm{F} \\
& \boldsymbol{p}_l \geq \boldsymbol{g}(\boldsymbol{x}^l) + \nabla_x \boldsymbol{g}(\boldsymbol{x}^l)(\boldsymbol{x} - \boldsymbol{x}^l) + \boldsymbol{B}\boldsymbol{y} & \forall l \in \mathrm{F} \\
& \boldsymbol{q}_l \geq \boldsymbol{T}^l(\boldsymbol{h}(\boldsymbol{x}^l) + \nabla_x \boldsymbol{h}(\boldsymbol{x}^l)(\boldsymbol{x} - \boldsymbol{x}^l) + \boldsymbol{C}\boldsymbol{y}) & \forall l \in \mathrm{F} \\
& \boldsymbol{p}_l \geq \boldsymbol{g}(\boldsymbol{x}^l) + \nabla_x \boldsymbol{g}(\boldsymbol{x}^l)(\boldsymbol{x} - \boldsymbol{x}^l) + \boldsymbol{B}\boldsymbol{y} & \forall l \in \bar{\mathrm{F}} \\
& \boldsymbol{q}_l \geq \boldsymbol{T}^l(\boldsymbol{h}(\boldsymbol{x}^l) + \nabla_x \boldsymbol{h}(\boldsymbol{x}^l)(\boldsymbol{x} - \boldsymbol{x}^l) + \boldsymbol{C}\boldsymbol{y}) & \forall l \in \bar{\mathrm{F}} \\
& \sum\limits_{i\in B^l} y_i - \sum\limits_{i\in N^l} y_i \leq |B^l| - 1 & \forall l \in \mathrm{F}
\end{array}$$

$\boldsymbol{y}^{k+1} \Leftarrow \boldsymbol{y}^*$
$LBD \Leftarrow \max(LBD, j^*)$
$k \Leftarrow k + 1$
until $UBD - LBD \leq \epsilon$

A square diagonal matrix T is used for relaxing the equality constraints. The rows of the matrix correspond to the equality constraints, and the values of the diagonal elements depend on the signs of the corresponding multipliers obtained from the primal problem. These values are 1 if the corresponding multiplier is positive, -1 if the corresponding multiplier is negative, and zero if the multiplier is zero. If there are no equality constraints in the problem, the relaxation is not required and the algorithm reduces to *OA*.

The weights for the slack variables are determined from the values of the Lagrange multipliers from the solution of the primal problem. The value of σ is a parameter for the algorithm that determines the weights. It can be a scalar used for all of the weights, or it can be a vector of values corresponding to the constraints in the problem. According to [247], the value of σ should be 1000.

The integer cut constraint has the form

$$\sum_{j\in B} y_j - \sum_{j\in N} y_j \leq |B| - 1$$

where $|B|$ is the cardinality of the set of binary variables that are 1, B, in an iteration. This constraint is added to eliminate binary combinations that have been considered in previous primal problems. This constraint can be added either only for iterations where the primal where feasible, or for all of the iterations. (This is an option in `MINOPT`.)

11.2.2 Nonlinear Program with Differential and Algebraic Constraints (NLP/DAE)

This problem has a form similar to that of the NLP with the addition of dynamic and point constraints. The dynamic constraints include both differential and algebraic equations and are referred to as the DAE system. The problem has the form:

$$
\begin{aligned}
\min \quad & J(\dot{\boldsymbol{z}}_1(t_N), \boldsymbol{z}_1(t_N), \boldsymbol{z}_2(t_N), \boldsymbol{x}) \\
\text{s.t.} \quad & \boldsymbol{f}_1(\dot{\boldsymbol{z}}_1(t), \boldsymbol{z}_1(t), \boldsymbol{z}_2(t), \boldsymbol{x}, t) = \mathbf{0} \\
& \boldsymbol{f}_2(\boldsymbol{z}_1(t), \boldsymbol{z}_2(t), \boldsymbol{x}, t) = \mathbf{0} \\
& \boldsymbol{z}_1(t_0) = \boldsymbol{z}_1^0 \\
& \boldsymbol{z}_2(t_0) = \boldsymbol{z}_2^0 \\
& \boldsymbol{h}_i'(\dot{\boldsymbol{z}}_1(t_i), \boldsymbol{z}_1(t_i), \boldsymbol{z}_2(t_i), \boldsymbol{x}) = \mathbf{0} \qquad \forall i \in 0 \dots N \\
& \boldsymbol{g}_i'(\dot{\boldsymbol{z}}_1(t_i), \boldsymbol{z}_1(t_i), \boldsymbol{z}_2(t_i), \boldsymbol{x}) \le \mathbf{0} \qquad \forall i \in 0 \dots N \\
& \boldsymbol{h}_i''(\boldsymbol{x}) = \mathbf{0} \\
& \boldsymbol{g}''(\boldsymbol{x}) \le \mathbf{0} \\
& \boldsymbol{x} \in \mathbb{R}^n
\end{aligned}
$$

In this formulation, $\boldsymbol{f}_1(\cdot)$ represents the p differential equations, $\boldsymbol{f}_2(\cdot)$ represents the q dynamic algebraic equations, $\boldsymbol{z}_1(t)$ is a vector of p dynamic variables whose time derivatives, $\dot{\boldsymbol{z}}_1(t)$, appear explicitly, and $\boldsymbol{z}_2(t)$ is a vector of q dynamic variables whose time derivatives do not appear explicitly. The vector $\boldsymbol{x}$ represents the vector time invariant continuous variables that are parameters for the DAE system and variables for the optimization. Time, t, is the independent variable for the DAE system and t_0 is the initial time. The initial condition for the above system is determined by specifying p of the $2p+q$ variables $\boldsymbol{z}_1(t_0), \dot{\boldsymbol{z}}_1(t_0), \boldsymbol{z}_2(t_0)$. For DAE systems with index 0 or 1, the remaining $n+m$ values are determined. The algorithms in `MINOPT` can only handle DAE systems of index 0 or 1. The initial conditions for $\boldsymbol{z}_1(t)$ and $\boldsymbol{z}_2(t)$ are $\boldsymbol{z}_1^0$ and $\boldsymbol{z}_2^0$ respectively.

The constraints $\boldsymbol{h}_i'(\cdot)$, and $\boldsymbol{g}_i'(\cdot)$ are point constraints involving the dynamic variables at a specified time instances, t_i. The index i can have values from 0 to N where N is the number of time instances necessary in the problem and t_N is the final time. The constraints $\boldsymbol{h}''(\boldsymbol{x})$ and $\boldsymbol{g}''(\boldsymbol{x})$ are general constraints that involve only the $\boldsymbol{x}$ variables.

`MINOPT` solves this problem using a parametric method where the DAE system is solved as a function of the $\boldsymbol{x}$ variables. The solution of the DAE system is achieved through an integration routine that returns the values of the $\boldsymbol{z}$ variables at the time instances, $\boldsymbol{z}(t_i)$, along with their sensitivities with respect to the parameters, $\frac{dz}{dx}(t_i)$. The resulting problem is an NLP optimization over the space of the x variables, which has constraints that are implicit functions of the x variables through the integration. The NLP formulation of the problem has the form:

$$\begin{aligned}
\min \quad & J(\dot{\boldsymbol{z}}_1(t_N), \boldsymbol{z}_1(t_N), \boldsymbol{z}_2(t_N), \boldsymbol{x}) & \\
\text{s.t.} \quad & \boldsymbol{h}'_i(\dot{\boldsymbol{z}}_1(t_i), \boldsymbol{z}_1(t_i), \boldsymbol{z}_2(t_i), \boldsymbol{x}) = \mathbf{0} & \forall i \in 0 \ldots N \\
& \boldsymbol{g}'_i(\dot{\boldsymbol{z}}_1(t_i), \boldsymbol{z}_1(t_i), \boldsymbol{z}_2(t_i), \boldsymbol{x}) \leq \mathbf{0} & \forall i \in 0 \ldots N \\
& \boldsymbol{h}''(\boldsymbol{x}) = \mathbf{0} & \\
& \boldsymbol{g}''(\boldsymbol{x}) \leq \mathbf{0} & \\
& \boldsymbol{x} \in \mathbb{R}^n &
\end{aligned}$$

where the variables $\dot{\boldsymbol{z}}_1(t_i)$, $\boldsymbol{z}_1(t_i)$, and $\boldsymbol{z}_2(t_i)$ are determined by solving the DAE system through integration:

$$\begin{aligned}
\boldsymbol{f}_1(\dot{\boldsymbol{z}}_1(t), \boldsymbol{z}_1(t), \boldsymbol{z}_2(t), \boldsymbol{x}, t) &= \mathbf{0} \\
\boldsymbol{f}_2(\boldsymbol{z}_1(t), \boldsymbol{z}_2(t), \boldsymbol{x}, t) &= \mathbf{0} \\
\boldsymbol{z}_1(t_0) &= \boldsymbol{z}_1^0 \\
\boldsymbol{z}_2(t_0) &= \boldsymbol{z}_2^0
\end{aligned}$$

The values of the dynamic variables are sampled at specific times over the integration horizon to obtain the values of $z(t_i)$.

This reformulated problem is an optimization over the space of x variables where the variables $\boldsymbol{z}(t_i)$ are implicit functions of the x variables through the integration of the DAE system. Since the functions $J(\cdot)$, $\boldsymbol{g}'(\cdot)$, and $\boldsymbol{h}'(\cdot)$ include $\boldsymbol{z}(t_i)$, they are also implicit functions of x. The solution algorithms for the NLP require the evaluation of the objective and constraints and their gradients with respect to x. These are evaluated directly for the constraints $\boldsymbol{g}''(\boldsymbol{x})$ and $\boldsymbol{h}''(\boldsymbol{x})$. However, for the functions $J(\cdot)$, $\boldsymbol{g}'(\cdot)$, and $\boldsymbol{h}'(\cdot)$, the values of $\boldsymbol{z}(t_i)$, and the gradients $\frac{dz}{dx}(t_i)$ returned from the integration are used. The functions $J(\cdot)$, $\boldsymbol{g}'(\cdot)$, and $\boldsymbol{h}'(\cdot)$ are evaluated directly and the gradients $\frac{dJ}{dx}$, $\frac{d\boldsymbol{g}'_i}{dx}$, and $\frac{d\boldsymbol{h}'}{dx}$ are evaluated by using the chain rule:

$$\begin{aligned}
\frac{dJ}{d\boldsymbol{x}} &= \left(\frac{\partial J}{\partial \boldsymbol{z}}\right)\left(\frac{\partial \boldsymbol{z}}{\partial \boldsymbol{x}}\right) + \left(\frac{\partial J}{\partial \boldsymbol{x}}\right) \\
\frac{d\boldsymbol{h}_i'}{d\boldsymbol{x}} &= \left(\frac{\partial \boldsymbol{h}_i'}{\partial \boldsymbol{z}}\right)\left(\frac{\partial \boldsymbol{z}}{\partial \boldsymbol{x}}\right) + \left(\frac{\partial \boldsymbol{h}_i'}{\partial \boldsymbol{x}}\right) \\
\frac{d\boldsymbol{g}_i'}{d\boldsymbol{x}} &= \left(\frac{\partial \boldsymbol{g}_i'}{\partial \boldsymbol{z}}\right)\left(\frac{\partial \boldsymbol{z}}{\partial \boldsymbol{x}}\right) + \left(\frac{\partial \boldsymbol{g}_i'}{\partial \boldsymbol{x}}\right)
\end{aligned}$$

`MINOPT` determines the derivatives of the all of the constraints with respect to the different variable types using automatic differentiation. It manages these derivative expressions and automatically applies the chain rule to get the necessary derivative terms for the nonlinear optimizer.

With the function and gradient evaluations available, standard gradient based optimization techniques can be applied to solve this problem as an NLP. The computational expense of this method arises in the function evaluations for $J(\cdot)$, $\boldsymbol{g}_i'(\cdot)$, and $\boldsymbol{h}_i'(\cdot)$ and their gradients with respect to $\boldsymbol{x}$. Each time these functions must be evaluated, the entire DAE system must be integrated.

11.2.3 Mixed-Integer Nonlinear Program with Differential and Algebraic Constraints (MINLP/DAE)

The NLP/DAE problem becomes an MINLP/DAE when binary variables are added to the problem formulation. This problem has the form:

$$\begin{aligned}
\min \quad & J(\dot{\boldsymbol{z}}_1(t_N), \boldsymbol{z}_1(t_N), \boldsymbol{z}_2(t_N), \boldsymbol{x}, \boldsymbol{y}) \\
\text{s.t.} \quad & \boldsymbol{f}_1(\dot{\boldsymbol{z}}_1(t), \boldsymbol{z}_1(t), \boldsymbol{z}_2(t), \boldsymbol{x}, \boldsymbol{y}, t) = \boldsymbol{0} \\
& \boldsymbol{f}_2(\boldsymbol{z}_1(t), \boldsymbol{z}_2(t), \boldsymbol{x}, \boldsymbol{y}, t) = \boldsymbol{0} \\
& \boldsymbol{z}_1(t_0) = \boldsymbol{z}_1^0 \\
& \boldsymbol{z}_2(t_0) = \boldsymbol{z}_2^0 \\
& \boldsymbol{h}_i'(\dot{\boldsymbol{z}}_1(t_i), \boldsymbol{z}_1(t_i), \boldsymbol{z}_2(t_i), \boldsymbol{x}, \boldsymbol{y}) = \boldsymbol{0} \qquad \forall i \in 0 \ldots N \\
& \boldsymbol{g}_i'(\dot{\boldsymbol{z}}_1(t_i), \boldsymbol{z}_1(t_i), \boldsymbol{z}_2(t_i), \boldsymbol{x}, \boldsymbol{y}) \leq \boldsymbol{0} \qquad \forall i \in 0 \ldots N \\
& \boldsymbol{h}''(\boldsymbol{x}, \boldsymbol{y}) = \boldsymbol{0} \\
& \boldsymbol{g}''(\boldsymbol{x}, \boldsymbol{y}) \leq \boldsymbol{0} \\
& \boldsymbol{x} \in \mathbb{R}^n \\
& \boldsymbol{y} \in \{0, 1\}^m
\end{aligned}$$

The solution for this problem in MINOPT is handled by solving the problem as an MINLP with appropriate extensions for the DAE system. The DAE system in this formulation may or may not contain y variables and their presence leads to two distinct cases.

For the case where DAE system has no y variables, the problem is solved by directly applying the described MINLP algorithms where the NLP problem is solved using the NLP/DAE technique described above. For fixed values of the y variables, the primal problem is an NLP/DAE whose solution was described earlier. The point constraints are implicit functions of the x variables and can be viewed as general nonlinear function in x. Since the dual variables and gradient information for these constraints is known from the solution of the primal, the master problems can be formulated without considering the DAE system.

For the case where the DAE system contains y variables, the solution procedure becomes more complicated. For fixed values of the y variables, the primal problem is once again an NLP/DAE and can be solved as described above. However, since the DAE system has y variables, the master problem must include information from the DAE system. The formulation of the master problem is not possible for the *OA* algorithms. The algorithms for for *GBD* and *GCD* can be applied using dual information from the primal problem as well as the DAE system to formulate the master problem. The solution procedure for this type of problem is described in [217].

The *GBD* master problem is formulated using dual information and the solution of the primal problem. Provided that the y variables participate linearly, the problem is an MILP whose solution provides a lower bound and y variables for the next primal problem. Dual information is required from all of the constraints including the DAEs whose dual variables, or adjoint variables, are dynamic.

The dual variables μ', λ', μ'', and λ'' are generally obtained from the solution technique for the primal problem. Dual information from the DAE system is obtained by solving the adjoint problem for the DAE system, which has the following formulation:

$$
\begin{aligned}
p &= \left(\frac{df_1}{d\dot{z}_1}\right)^T \nu_1(t) \\
\dot{p} &= \left(\frac{df_1}{dz_1}\right)^T \nu_1(t) + \left(\frac{df_2}{dz_1}\right)^T \nu_2(t) \\
0 &= \left(\frac{df_1}{dz_2}\right)^T \nu_1(t) + \left(\frac{df_2}{dz_2}\right)^T \nu_2(t)
\end{aligned}
$$

This is a set of DAEs where the solutions for $\frac{df_1}{d\dot{z}_1}$, $\frac{df_1}{dz_1}$, $\frac{df_2}{dz_1}$, $\frac{df_1}{dz_2}$, and $\frac{df_2}{dz_2}$ are known functions of time obtained from the solution of the primal problem.

The variables $\boldsymbol{\nu}_1(t)$ and $\boldsymbol{\nu}_2(t)$ are the adjoint variables and the solution of this problem is a backward integration in time with the following final time conditions:

$$\left(\frac{d\boldsymbol{f}_1}{d\dot{\boldsymbol{z}}_1}\right)^T\Bigg|_{t_N} \boldsymbol{\nu}_1(t_N) = -\left(\frac{d\boldsymbol{J}}{d\boldsymbol{z}_1}\right) - \left(\frac{d\boldsymbol{g}'}{d\boldsymbol{z}_1}\right)^T \boldsymbol{\mu}' - \left(\frac{d\boldsymbol{h}'}{d\boldsymbol{z}_1}\right)^T \boldsymbol{\lambda}'$$

Thus, the Lagrange multipliers for the end-time constraints are used as the final time conditions for the adjoint problem and are not included in the master problem formulation.

The master problem is formulated using the solution of the primal problem, $\boldsymbol{x}^k$ and $\boldsymbol{z}^k(t)$ along with the dual information, $\boldsymbol{\mu}''^k$, $\boldsymbol{\lambda}''^k$, and $\boldsymbol{\nu}^k(t)$. The master problem has the following form:

$$\begin{aligned}
\min_{\boldsymbol{y},\mu_b} \quad & \mu_b \\
\text{s.t.} \quad & \mu_b \geq J(\boldsymbol{x}^k, \boldsymbol{y}) \\
& \quad + \int_{t_0}^{t_N} \boldsymbol{\nu}_1^k(t)\boldsymbol{f}_1(\dot{\boldsymbol{z}}_1^k(t), \boldsymbol{z}_1^k(t), \boldsymbol{z}_2^k(t), \boldsymbol{x}^k, \boldsymbol{y}, t)dt \\
& \quad + \int_{t_0}^{t_N} \boldsymbol{\nu}_2^k(t)\boldsymbol{f}_2(\boldsymbol{z}_1^k(t), \boldsymbol{z}_2^k(t), \boldsymbol{x}^k, \boldsymbol{y}, t)dt \\
& \quad + \boldsymbol{\mu}''^k\boldsymbol{g}''(\boldsymbol{x}^k, \boldsymbol{y}) + \boldsymbol{\lambda}''^k\boldsymbol{h}''(\boldsymbol{x}^k, \boldsymbol{y}) \qquad k \in K_{\text{feas}} \\
& \boldsymbol{0} \geq \int_{t_0}^{t_N} \boldsymbol{\nu}_1^k(t)\boldsymbol{f}_1(\dot{\boldsymbol{z}}_1^k(t), \boldsymbol{z}_1^k(t), \boldsymbol{z}_2^k(t), \boldsymbol{x}^k, \boldsymbol{y}, t)dt \\
& \quad + \int_{t_0}^{t_N} \boldsymbol{\nu}_2^k(t)\boldsymbol{f}_2(\boldsymbol{z}_1^k(t), \boldsymbol{z}_2^k(t), \boldsymbol{x}^k, \boldsymbol{y}, t)dt \\
& \quad + \boldsymbol{\mu}''^k\boldsymbol{g}''(\boldsymbol{x}^k, \boldsymbol{y}) + \boldsymbol{\lambda}''^k\boldsymbol{h}''(\boldsymbol{x}^k, \boldsymbol{y}) \qquad k \in K_{\text{infeas}} \\
& \boldsymbol{y} \in \{0,1\}^m
\end{aligned}$$

The integral term can be evaluated since the profiles for $\boldsymbol{z}^k(t)$ and $\boldsymbol{\nu}^k(t)$ both are fixed and known. Note that this formulation has no restrictions on whether or not $\boldsymbol{y}$ variables participate in the DAE system.

A schematic of the MINLP/DAE solution algorithm is given in Fig. 11.2.3, and the algorithmic is stated as follows:

$k \Leftarrow 1$, $\boldsymbol{y}^1 \Leftarrow \boldsymbol{y}^{\text{init}}$ {Initialize the y variables}
$LBD \Leftarrow -\infty$, $UBD \Leftarrow +\infty$
$\mathrm{F} \Leftarrow \emptyset$, $\bar{\mathrm{F}} \Leftarrow \emptyset$ {Initialize the feasibility and infeasibility sets}
$\epsilon \Leftarrow$ tolerance {Set the convergence tolerance}
repeat

$y \Leftarrow y^k$ {Fix the values of the binary variables}
Obtain the optimal solution, optimal x^k, optimal state profiles,$z(t)$, optimal Lagrange multipliers λ^k and μ^k, and gradients
Solve the Primal Problem
if Primal is feasible **then**
  $x^k \Leftarrow x^*$ {Obtain the optimal variable values}
  $\lambda^k \Leftarrow \lambda^*$, $\mu^k \Leftarrow \mu^*$ {Obtain the optimal Lagrange multipliers}
  $\mathrm{F} \Leftarrow \mathrm{F} \cup k$ {Update the feasibility set}
  $UBD \Leftarrow \min(UBD, f(x^*, y^k)$
else
  Solve the Infeasibility Problem
  $x^k \Leftarrow x^*$ {Obtain the optimal variable values}
  $\lambda^k \Leftarrow \lambda^*$, $\mu^k \Leftarrow \mu^*$ {Obtain the optimal Lagrange multipliers}
  $\bar{\mathrm{F}} = \bar{\mathrm{F}} \cup k$ {Update the infeasibility set}
end if
Solve the Adjoint Problem {If necessary}
$\nu^k(t) = \nu^*(t)$ {Obtain the adjoint variables for the DAE system}
Solve the Master Problem
$y^{k+1} \Leftarrow y^*$
$LBD \Leftarrow \max(LBS, \mu_b^*)$
$k \Leftarrow k + 1$
until $UBD - LBD \leq \epsilon$

11.2.4 Optimal Control Problem (OCP) and Mixed Integer Optimal Control

The formulations for the optimal control problem (OCP) and mixed-integer optimal control problem (MIOCP) involve time varying optimization variables called control variables, $u(t)$. The difference between the OCP and the MIOCP is the presence of binary variables in the problem formulation. Since their presence does not affect the solution strategy for the of the optimal control problem, they are omitted in the discussion of the OCP solution algorithm. The

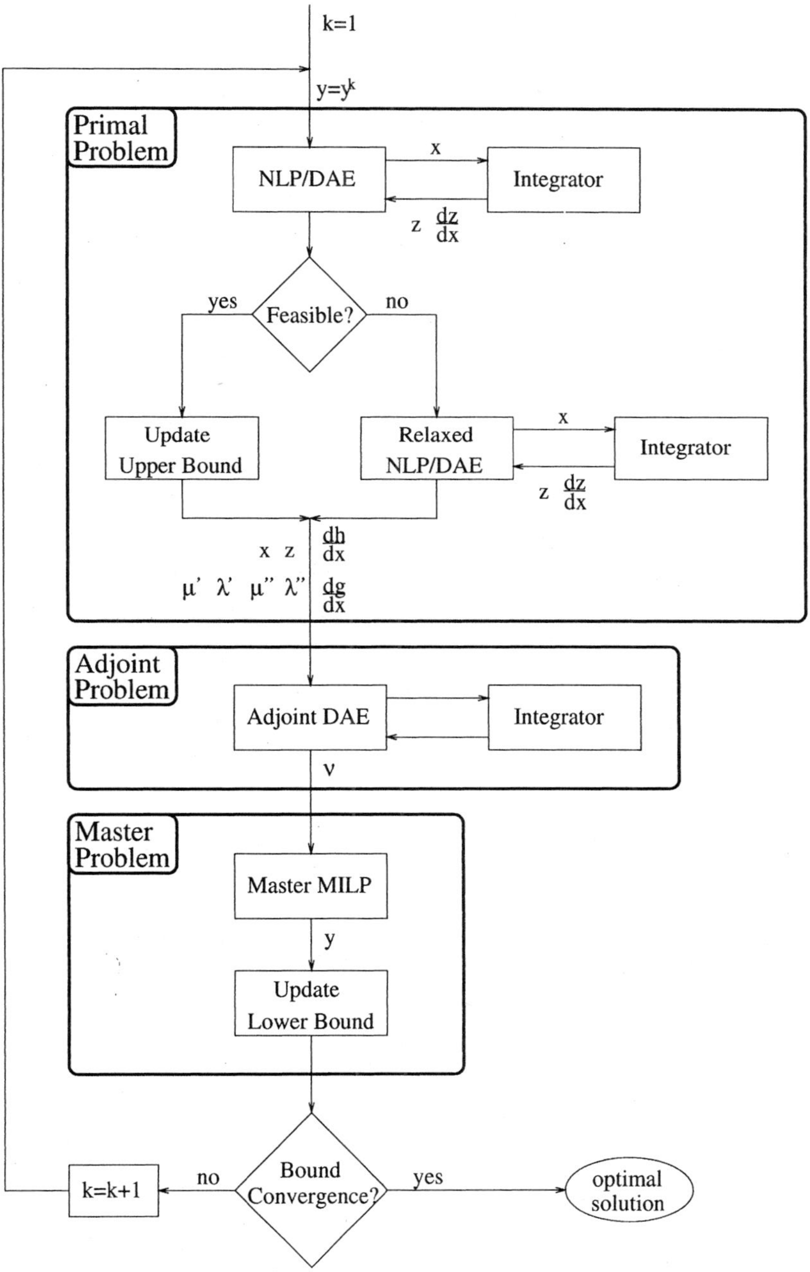

Figure 11.2.1. Schematic flowsheet for the MINLP/DAE algorithm

OCP has the following form:

$$
\begin{aligned}
\min \quad & J(\dot{\boldsymbol{z}}_1(t_N), \boldsymbol{z}_1(t_N), \boldsymbol{z}_2(t_N), \boldsymbol{u}(t_N), \boldsymbol{x}) \\
\text{s.t.} \quad & \boldsymbol{f}_1(\dot{\boldsymbol{z}}_1(t), \boldsymbol{z}_1(t), \boldsymbol{z}_2(t), \boldsymbol{u}(t), \boldsymbol{x}, t) = \boldsymbol{0} \\
& \boldsymbol{f}_2(\boldsymbol{z}_1(t), \boldsymbol{z}_2(t), \boldsymbol{u}(t), \boldsymbol{x}, t) = \boldsymbol{0} \\
& \boldsymbol{z}_1(t_0) = \boldsymbol{z}_1^0 \\
& \boldsymbol{z}_2(t_0) = \boldsymbol{z}_2^0 \\
& \boldsymbol{h}_i'(\dot{\boldsymbol{z}}_1(t_i), \boldsymbol{z}_1(t_i), \boldsymbol{z}_2(t_i), \boldsymbol{u}(t_i), \boldsymbol{x}) = \boldsymbol{0} \qquad \forall i \in 0 \dots N \\
& \boldsymbol{g}_i'(\dot{\boldsymbol{z}}_1(t_i), \boldsymbol{z}_1(t_i), \boldsymbol{z}_2(t_i), \boldsymbol{u}(t_i), \boldsymbol{x}) \leq \boldsymbol{0} \qquad \forall i \in 0 \dots N \\
& \boldsymbol{h}''(\boldsymbol{x}) = \boldsymbol{0} \\
& \boldsymbol{g}''(\boldsymbol{x}) \leq \boldsymbol{0} \\
& \boldsymbol{x} \in \mathbb{R}^n \\
& \boldsymbol{y} \in \{0, 1\}^m
\end{aligned}
$$

Since the control variables are functions of time, problems in this class are infinite dimensional optimization problems. (They effectively have an infinite number of decision variables.) Optimal control problems can be solved using a number of different solution techniques, but most of them convert the problem to one that has a finite dimension. These techniques differ in the way that this conversion is handled. A discussion of various optimal control techniques can be found in [33], [156], and [227]. The solution strategies for optimal control problems can be classified into one of four categories:

1 Dynamic programming ([11, 128, 163, 162, 164, 165])

2 Solution of necessary conditions based on the solution of a two point boundary value problem through quasi-linearization techniques ([176, 175, 174]), shooting algorithms ([54])

3 Complete discretization and conversion to a finite dimensional NLP ([35, 234, 18, 248, 196, 49, 158, 243])

4 Infinite nonlinear programming approximation–optimization in the space of the decision variables using control parameterization ([189, 204, 205, 179, 87, 244, 245])

The solution method that `MINOPT` employs is the control parameterization technique as described in [244] and [245].[1] In the control parameterization approach, the control variables, $\boldsymbol{u}(t)$, are expressed in terms of time invariant

[1] `MINOPT` could be used to implement a full collocation technique, but this would require that the user provide the input in the collocated form. MINOPT does not employ an automatic collocation strategy.

parameters. This can be done either open loop where the control variables are parameterized in terms of the independent variable t:

$$\boldsymbol{u}(t) = \phi(\boldsymbol{w}, t) \quad .$$

Alternatively, the parameterization can be closed loop where the control variables are parameterized in terms of the state variables $\boldsymbol{z}(t)$:

$$\boldsymbol{u}(t) = \psi(\boldsymbol{w}, \boldsymbol{z}(t)) \quad .$$

In either case, the control parameters $\boldsymbol{w}$ are introduced as a new set of time invariant decision variables. The set of time invariant parameters is now expanded to include the control parameters:

$$\boldsymbol{x} = \{\boldsymbol{x}, \boldsymbol{w}\}$$

Through the control parameterization, the control variable are converted to state variables:

$$\boldsymbol{z} = \{\boldsymbol{z}, \boldsymbol{u}\} \quad .$$

The set of DAEs is expanded by an equivalent amount to include the parameterization functions:

$$\boldsymbol{f}(\cdot) = \{\boldsymbol{f}(\cdot), \phi(\cdot), \psi(\cdot)\}$$

As a result of the control parameterization, the optimal control problem is expressed as an NLP/DAE.

The task of preparing an optimal control problem for `MINOPT` is to determine the control variables for the problem and the control parameterization functions ϕ and ψ. For open-loop control, piecewise continuous Lagrange polynomials are often used for the collocation. This collocation is performed over each of the time intervals defined by the time instances t_i. Different parameters are used in each interval for each control variable. Although the time instances are fixed, a scaling factor can be applied to each interval allowing the effective size of the interval to vary.

`MINOPT` employs an indexed function called interval() that activates parameters over a control parameterization interval. This allows the user to create piecewise functions that define the control trajectory. The function has the form

$$u(t) = \text{interval}_{i \in I}(f(\omega_i), t) \quad .$$

For example, a piecewise linear function could be expressed as

$$u(t) = \text{interval}_{i \in I}(\omega_i + (\omega_{i+1} - \omega_i)\frac{t - \hat{t}_i}{\hat{t}_{i+1} - \hat{t}_i}) \quad .$$

The values of ω_i are the values of the control variable $u(t)$ at the control parameterization points $\hat{t}_i$, and t is the time over the horizon. When t is in the control parameterization i, the values of the $u(t)$ is computed using the appropriate control parameters. Using this function, the OCP is converted to an NLP/DAE and solve as described above.

The solution of the MIOCP problem is handled by using the control parameterization to reduce the problem to an MINLP/DAE problem and then applying MINLP/DAE algorithmic approach to solve the problem.

11.2.5 External Solvers

`MINOPT` has implementations of the MINLP algorithms and dynamic solution approaches described previously. However, `MINOPT` does not directly solve the LP, MILP, NLP, and dynamic problems and subproblems that arise in these algorithms. It automatically creates the subproblems and interfaces to external packages in order to solve these problems. `MINOPT` leverages existing, robust computational implementations of optimization algorithms.

Since multiple solution algorithms exist for a given class of problem formulation and multiple implementations exist for a given algorithm, `MINOPT` was designed with an extensible solver interface. It interfaces to solvers capable of solving the LP, MILP, NLP, and dynamic problems and subproblems that `MINOPT` creates. A summary of the solvers currently interfaced with `MINOPT` are shown in Table 11.2. More solvers can be added as they become available.

Table 11.2. External solvers used by MINOPT.

Problem Type	Algorithm	Solver
LP	Simplex method	CPLEX
		LPSOLVE
		MINOS
		NPSOL
MILP	Branch and Bound	CPLEX
		LPSOLVE
NLP	Augmented Lagrangian/Reduced Gradient	MINOS
	Sequential Quadratic Programming	NPSOL
	Sequential Quadratic Programming	SNOPT
Dynamic	Integration (Backwards Difference Formulae)	DASOLV
		DAESSA

11.3 Example Problems

In order to illustrate the features of the `MINOPT` modeling language, two example problems are presented. The first is an MINLP problem that shows some of the modeling features of the language such as the set notation, parameter definitions, summations, and enumerated constraints.

The second example is and optimal control problem that illustrates how MINOPT is used to implement a control parameterization solution. The problem shows how to describe a DAE model within MINOPT along with the point constraints and control parameterization function.

11.3.1 Language Overview

A MINOPT model is written as a text file. The text file contains three sections. The first is the OPTIONS section where the options for the algorithms and subproblem solvers are defined. The second section is the DECLARATIONS section where the parameters, variables, and bounds are defined. The final section is the MODEL section where the objective and constraints are defined.

In the declarations section MINOPT requires the partitioning of the variables be explicitly defined. The primal variables are classified as x-variables and the complicating variables or binary variables in the MINLP algorithms are classified as y-variables. This coincides with the definitions used in the algorithm descriptions. The dynamic variables are classified as z-variables. MINOPT uses the keywords XVARIABLES, YVARIABLES, and ZVARIABLES to declare the variable partitioning. Parameters in the problem are declared using the keyword PARAMETER.

MINOPT allows for the declaration of sets (as well as sets of sets) to be used in the definition of the variables, parameters, and constraints. Summations over sets are represented using the notation << *index* E *set* | *expression(index)* > to represent the summation of the expressions over the set. Constraints that are enumerated over a set are expressed by using the notation *constraint* (*index* E *set*) : *equation* to denote that the equation is to be enumerated over the set.

11.3.2 MINLP Problem–Nonconvex Portfolio Optimization Problem

This problem is taken from [139] and is used to illustrate the modeling of an MINLP problem. The problem involves a set of 4 reactors, R, and 40 products, P, over which the parameters, variables, and constraints are defined. The MINOPT input is the following:

```
# Nonconvex Portfolio Optimization Problem %
# From "Exact Computation of Global Minima of a %
#      Nonconvex Portfolio Optimization Problem" %
# Josef Kallrath %
# Frontiers in Global Optimization, %
# C.A. Floudas and P.M. Pardalos, Editors %
# 2003 Kluwer Academic Publishers

OPTIONS {{ }}

DECLARATIONS {{
  INDEX {r, p};
  SET R = |1:4|;
```

```
  SET P = |1:40|;

#Parameters
  #Investment Cost [kEuro]
  PARAMETER CI(R) = {0.97, 0.97, 0.97, 0.97};
  #Fixed Cost [kEuro/week]
  PARAMETER CF(R) = {2.45, 2.45, 2.45, 2.45};

  #Time capacity  [hr]
  PARAMETER CT(R) = { 168,  168,  168,  168};
  #Utilization lower limit
  PARAMETER FL(R) = { 0.4,  0.4,  0.4,  0.4};
  #Reactor volume lower limit [m^3]
  PARAMETER VL(R) = {  20,   20,   20,   20};
  #Reactor volume upper limit [m^3]
  PARAMETER VU(R) = { 250,  250,  250,  250};

  #Batch time [hr]
  PARAMETER TP(P) = <p E P| 6 >;

  #Demand [m^3]
  PARAMETER D(P)  = {2600, 2300, 450, 1200, 560,
                 530,  530, 140,  110, 110,
                  10,  110,  90,   90,  90,
                  70,   50,  30,   10,  10,
                  10,  190, 180,   70,  70,
                  40,   40,  40,   30,  20,
                  20,   20,  10,   10,  10,
                  10,    4,   2,    2,   2};

  XVAR {v(R), f(R,P), pS(P), q(R,P) };

  YVAR {y(R), nB(R,P)};
  BINARY {y(R)};
  INTEGER {nB(R,P)};

  LBDS v(R) = <r E R| 0     >;
  STP  v(R) = <r E R| VL(r) >;
  UBDS v(R) = <r E R| VU(r) >;

  LBDS f(R,P) = <r E R| <p E P| FL(r) > >;
  STP  f(R,P) = <r E R| <p E P| FL(r) > >;
  UBDS f(R,P) = <r E R| <p E P|   1.0 > >;

  LBDS pS(P) = <p E P| 0.0 >;
  STP  pS(P) = <p E P| 1.0 >;
  UBDS pS(P) = <p E P| 1.0 >;

  LBDS q(R,P) = <r E R| <p E P|             0 > >;
  STP  q(R,P) = <r E R| <p E P| 28*1.0*VL(r) > >;
  UBDS q(R,P) = <r E R| <p E P| 28*1.0*VU(r) > >;

  LBDS nB(R,P) = <r E R| <p E P|  0 > >;
  STP  nB(R,P) = <r E R| <p E P|  D(p)/100/4 > >;
  UBDS nB(R,P) = <r E R| <p E P| 28 > >;

  STP y(R) = <r E R| 1 >;
```

```
}}

MODEL {{
  MIN: <<r E R| sqrt[CI(r)*v(r)] >>
     + <<r E R| CF(r)*y(r) >>;

#Total time
  time(r E R): <<p E P| TP(p)*nB(r,p)>> =l= CT(r)*y(r);

#Production amount
  production(r E R, p E P):
    q(r,p) =e= nB(r,p)*f(r,p)*v(r);

#Demand
  demand(p E P):  <<r E R| q(r,p)>> =g= D(p);

#Relative surplus production
  surplus(p E P): pS(p) =e= (<<r E R|q(r,p)>> - D(p))/D(p);

#Reactor volume Bounds
  RVUpper(r E R): v(r) =l= VU(r)*y(r);
  RVLower(r E R): v(r) =g= VL(r)*y(r);

}}
```

11.3.3 Optimal Control Problem–Dow Batch Reactor

To illustrate the ability of MINOPT to solve Optimal Control Problems, a batch reactor problem described in [154] is solved. The problem is based on the kinetic model of a batch reactor system presented by [19].

This problem is solved using a control parameterization over two stages of six intervals. The control variables are parameterized as piecewise constant over the horizon. The MINOPT input file is as follows:

```
# Batch Reactor System %
# From "A Boundary Value Problem Approach to the%
# Optimization of Chemical Processes Described by %
# DAE Models"%
# Daniel Leineweber, Hans Georg Bock, Johannes %
# P. Schloder, Johanna V. Gallitzendorfer, Andreas %
# Schafer, Peter Jansohn%

OPTIONS {{
  EVTOL  = 1e-9;
  FTOL   = 1e-6;
  INTTOL = 1e-7;
  OTOL   = 1e-6;
  SSTOL = 1e-10;
}}

DECLARATIONS {{
  INDEX {i};
  SET I = |1:12|;  SET I1 = |1:6|; SET I2 = |6:12|;
  SET P = |1:13|;

#Parameters
```

```
  PARAMETER a1  = 1.3708e12;  #kg/(gmol hr)
  PARAMETER b1  = 9.2984e3;   #K (E1/R)
  PARAMETER am1 = 1.6215e20;  #1/hr
  PARAMETER bm1 = 1.3108e4;   #K (Em1/R)
  PARAMETER a2  = 5.2282e12;  #kg/(gmol hr)
  PARAMETER b2  = 9.5999e3;   #K (E2/R)
  PARAMETER K1  = 2.575e-16;  #gmol/kg
  PARAMETER K2  = 4.876e-14;  #gmol/kg
  PARAMETER K3  = 1.7884e-16; #gmol/kg

  XVAR {p, tau,
        u(P)};

  LBDS p = {0.0};
  STP  p = {0.0131};
  UBDS p = {0.0262};

  LBDS tau = {0.01};
  STP  tau = {5};
  UBDS tau = {20};

  LBDS u(P) = {293.15};
  STP  u(P) = {373.15};
  UBDS u(P) = {393.15};

  ZVAR {x0, x1, x2, x3, x4, x5,
        z0, z1, z2, z3,
        k1, km1, k2,
        temp};

  ISPE {0,1,2,3,4,5};
  ICS  {5};
  ICP  {0};
  ZINC {1.5776, 8.32, 0, 0, 0, 0,
             7, 1e-8, 0, 0,
             20.651, 89957, 35.108,
             373.15};

  PARA ti(P) = {0,0.04167,0.0833,0.125,0.1667,0.2083,0.25,
                   0.37500,0.5000,0.625,0.7500,0.8750,1.00};
  TIME         {0,0.04167,0.0833,0.125,0.1667,0.2083,0.25,
                   0.37500,0.5000,0.625,0.7500,0.8750,1.00};
}}

MODEL {{
  MIN: tau + 100*p;

#kinetic model
  dae0:  x0' =e= tau*(-k2*x1*z1);
  dae1:  x1' =e= tau*(-k1*x1*x5 + km1*z3 - k2*x1*z1);
  dae2:  x2' =e= tau*( k2*x1*z1 + k1*x3*x5 - 0.5*km1*z2);
  dae3:  x3' =e= tau*(-k1*x3*x5 + 0.5*km1*z2);
  dae4:  x4' =e= tau*( k1*x1*x5 - km1*z3);
  dae5:  x5' =e= tau*(-k1*x1*x5 + km1*z3 - k1*x3*x5
                     + 0.5*km1*z2);

#electroneutrality condition
```

```
  dae6:   p - x5 + exp[-z0] - z1 - z2 - z3 =e= 0;

#equilibrium conditions
  dae7:   z1 =e= K2*x0/(K2 + exp[-z0]);
  dae8:   z2 =e= K3*x2/(K3 + exp[-z0]);
  dae9:   z3 =e= K1*x4/(K1 + exp[-z0]);

#Arrhenius expressions
  dae10: k1  =e= a1 *exp[-b1 /temp];
  dae11: km1 =e= am1*exp[-bm1/temp];
  dae12: k2  =e= a2 *exp[-b2 /temp];

#Control parameterization
  dae13: temp =e= interval[i E I|u(i) +
         (u(i+1) - u(i))*(t-ti(i))/(ti(i+1)-ti(i))];

#Point constraints
  pcon0[12]: x3 =g= 1;
  pcon1[i E I1]: x3 =l= 2*(tau*ti(i))^2;
}}
```

The dynamic constraints are defined using the same syntax as for the algebraic modeling. The main difference is that the character ' is used to denote the time derivative of a dynamic variable. This model employs the interval function `interval[` *index* E *set* | *expression(i)* `]` where the set corresponds to the set of control parameterization intervals defined by the `TIME` declaration. The point constraints are defined using the notation *constraint* `[` *index* `]`: where the index corresponds to the time instant where the constraint is to be enforced.

11.4 Summary

The `MINOPT` modeling language has been introduced as a tool for modeling and solving a variety of classes of mathematical programming problems. It is capable of handling a broad range of model types and is a particularly attractive tool for solving MINLP problems and problems involving DAEs. The two example problems illustrate the compact, concise notation for describing an MINLP problem and an optimal control problem. More information about `MINOPT` including many example problems can be found at the `MINOPT` website (`http://titan.princeton.edu/MINOPT/minopt.html`).

Chapter 12

MOSEL: A MODULAR ENVIRONMENT FOR MODELING AND SOLVING OPTIMIZATION PROBLEMS

Yves Colombani, Bob Daniel & Susanne Heipcke
Dash Optimization
Blisworth House, Blisworth, Northants NN7 3BX, U.K.
{yves.colombani,bob.daniel,susanne.heipcke}@dashoptimization.com

Abstract Xpress-Mosel, a commercial product from Dash Optimization since 2001, provides a complete environment for developing, testing and deploying optimization applications. This chapter explains the basics of the Mosel language that are required to use the software as a modeling and solution reporting interface to standard matrix-based solvers. It also shows how Mosel can be used to implement more complex solution algorithms, including graphical solution output.
A major novelty of Mosel is its concept of *modularity* which makes it possible to extend the Mosel language according to one's needs, in particular to deal with other types of problems and solvers. Besides giving examples of the use of modules alraedy provided with the Mosel distribution, the chapter explains how the user can extend the existing Mosel language to provide new functionality that may be required, for instance, to access other solvers.

Keywords: Mosel,Mathematical Programming, Linear Programming, Integer Programming, Nonlinear Programming, optimization, heuristics, Mathematical Programming software, modeling language, programming language, multiple solvers

12.1 Introduction

`Mosel` is a new environment for modeling and solving problems that is provided either in the form of libraries or as a standalone program. `Mosel` includes a language that is both a *modeling* and a *programming language* combining the strengths of these two concepts. As opposed to "traditional" modeling environments like `AMPL` [75] for which the problem is described using a "modeling language" and algorithmic operations are written with a "scripting language" (similarly for `OPL Studio` [235] with OPL-script), in `Mosel` there is no separation between a modeling statement (*e.g.*, declaring a decision variable or expressing a constraint) and a procedure that actually solves a problem (*e.g.*, call to an optimizing command). Thanks to this synergy, one can program a complex solution algorithm by interlacing modeling and solving statements.

12.1.1 Solver Modules

Each category of problem comes with its own particular types of variables and constraints and a single kind of solver cannot be efficient in all cases. To take this into account, `Mosel` does not integrate any solver by default but offers a dynamic interface to external solvers provided as *modules*. Each solver module comes with its own set of procedures and functions that directly extends the vocabulary and capabilities of the language. This architecture guarantees an efficient link between `Mosel` and the solver(s) being used. Similar connections are also provided by other systems (*e.g.*, `MPL` [171]) but due to the concept of modules, `Mosel` is not restricted to any particular type of solver and each solver may provide its specifics at the user level. For instance, an LP solver may define the procedure 'setcoeff(ctr,var,coeff)' to set the matrix coefficient of the constraint 'ctr' for the variable 'var' whilst such a procedure may make no sense for other types of solvers. Similarly, the module *mmquad* extends the syntax of the language with quadratic expressions that can be handled by a suitable solver (like Xpress-Optimizer).

A major advantage of the modular architecture is that there is no need to modify `Mosel` to provide access to a new solution technology. Recently Dash Optimization introduced a new SLP ("Sequential Linear Programming") solver for which a `Mosel` module is available: this module enables the user to define and solve problems with non-linear constraints. Another example is the interface to the CHIP system that is being developed by Cosytec in the context of the EU project LISCOS. This latter development transforms `Mosel` into a Constraint Programming language and opens the way for hybrid solution methods.

12.1.2 Other Modules

The notion of a module is not restricted to interfaces to solvers: more generally it is a means to open the `Mosel` environment to software and technology other than solvers. For example, one `Mosel` module (*mmodbc*) allows the user to access data sources for which an ODBC interface is defined (typically databases and certain spreadsheets) using standard SQL commands. With the help of the module *mmive* the user can define his own graphics in the graphical environment Xpress-IVE. Other modules could be written to provide the functionalities required to communicate with other applications for which no generic interface is available.

12.1.3 User Modules

Besides the provided modules, `Mosel` is open to any kind of addition by its users. The communication between `Mosel` and its modules uses specific protocols and libraries. This *Native Interface* is public and allows the user to implement his own modules. Any programming task that can be cast into the form of a C program may be turned into a module to be used from the `Mosel` language.

Possible uses of user-written modules include, but are not limited to,

- application specific data handling (*e.g.*, definition of composite data types, data input and output in memory),
- access to external programs,
- access from the model to solution algorithms and heuristics implemented in a programming language (possibly re-using existing code),
- access to efficient implementations of specific, time-critical programming tasks (such as a sorting algorithm that needs to be called frequently by some heuristic).

12.1.4 Contents of this Chapter

This chapter gives an introduction to the `Mosel` environment: its language, described in the first section, and extensions to it provided in the form of modules. After the description of the language is a short note about the deployment of models with the `Mosel` libraries. Using the simple example that is introduced in the first section to show the main features of the language, we then show how to use some of the modules provided with the `Mosel` distribution. The last section describes how a user can implement his own modules via the `Mosel` Native Interface.

`Mosel` is Dash's second "modeling language", the first being `mp-model` [5] which first appeared in 1983. Though `Mosel` bears little obvious resemblance to `mp-model`, the experience which Dash's users had with `mp-model` contributed to the factors influencing `Mosel`'s design. In particular, the tendency to avoid adding features for their own sake – thus avoiding an "epicycles-on-epicycles" phenomenon – has been carried through into `Mosel`.

12.2 The Mosel Language

We start this section with a small example that gives a first impression of how a mathematical model may be implemented and solved with `Mosel`. The following subsections then give a more systematic introduction to the `Mosel` language.

12.2.1 Example Problem

An investor wishes to invest a certain amount of money. He is evaluating ten different securities (the set $SHARES$) for his investment. He estimates the return on investment for a period of one year (RET_s). To spread the risk he wishes to invest at most 30% of the capital into any share. He further wishes to invest at least half of his capital in North-American shares and at most a third in high-risk shares. How should the capital be divided among the shares (with $frac_s$ the fraction invested into share s) to obtain the highest expected return on investment?

The following `Mosel` model implements and solves this problem as a Linear Programming (LP) problem.

```
model "Portfolio optimization with LP"
 uses "mmxprs"                        ! Use Xpress-Optimizer

 declarations
  SHARES = 1..10                      ! Set of shares
  RISK = {2,3,4,9,10}                 ! Set of high-risk values among shares
  NA = {1,2,3,4}                      ! Set of shares issued in N.-America
  RET: array(SHARES) of real          ! Estimated return in investment

  frac: array(SHARES) of mpvar        ! Fraction of capital used per share
 end-declarations

 RET:= [5,17,26,12,8,9,7,6,31,21]

! Objective: total return
 Return:= sum(s in SHARES) RET(s)*frac(s)

! Limit the percentage of high-risk values
 sum(s in RISK) frac(s) <= 1/3

! Minimum amount of North-American values
 sum(s in NA) frac(s) >= 0.5

! Spend all the capital
 sum(s in SHARES) frac(s) = 1

! Upper bounds on the investment per share
 forall(s in SHARES) frac(s) <= 0.3

! Solve the problem
 maximize(Return)

! Solution printing
 writeln("Total return: ", getobjval)
 forall(s in SHARES) writeln(s, ": ", getsol(frac(s))*100, "%")
end-model
```

General structure: Every `Mosel` program starts with the keyword `model`, followed by a model name chosen by the user. The `Mosel` program is terminated with the keyword `end-model`. All objects must be declared in a `declarations` section, unless they are defined unambiguously through an assignment (*e.g.*, `Return:= sum(s in SHARES) RET(s)*frac(s)` defines `Return` as a linear constraint and assigns to it the expression `sum(s in SHARES) RET(s)* frac(s)`). There may be several such `declarations` sections at different places in a model.

In the present case, we define three *sets*, and two arrays:
`SHARES` is a so-called *range set*, *i.e.* a set of consecutive integers (here: from 1 to 10).
`RISK` and `NA` are simply *sets of integers*.
`RET` is an array of real values indexed by the set `SHARES`, its values are assigned after the declarations.
`frac` is an array of decision variables of type `mpvar`, also indexed by the set `SHARES`. These are the decision variables in our model.

The model then defines the objective function, two linear inequality constraints and one equality constraint and sets upper bounds on the variables. Constraints may be named (as it is the case here for the objective function `Return`) if they need to be referred to later, for instance, to access solution information (dual, slack, activity).
Solving: With the procedure `maximize`, we call the Xpress-Optimizer to maximize the linear expression `Return`. Since there is no "default" solver in `Mosel`, we specify that the Xpress-Optimizer is to be used with the statement `uses "mmxprs"` at the begin of the program.
Output printing: The last two lines print out the value of the optimal solution and the solution values for all variables.
Line breaks: It is possible to place several statements on a single line, separating them by semicolons (like `RISK = {2,3,4,9,10}; NA = {1,2,3,4}`). But since there are no special "line end" or continuation characters, every line of a statement that continues over several lines must end with an operator (+, >= *etc.*) or characters like , that make it obvious that the statement is not terminated.
Comments: As shown in the example, single line comments in `Mosel` are preceded by `!`. Comments over multiple lines start with `(!` and terminate with `!)`.

Fig. 12.2.1 shows the result of an execution of our model in the development environment Xpress-IVE.

12.2.2 Types and Data Structures

`Mosel` provides the *basic types* that may be expected of any programming language: `integer`, `real` (double precision floating point numbers), `boolean` (symbols `true` and `false`), `string` (single character and any text). Together with the *MP types* `mpvar` (decision variables) and `linctr` (linear constraints)

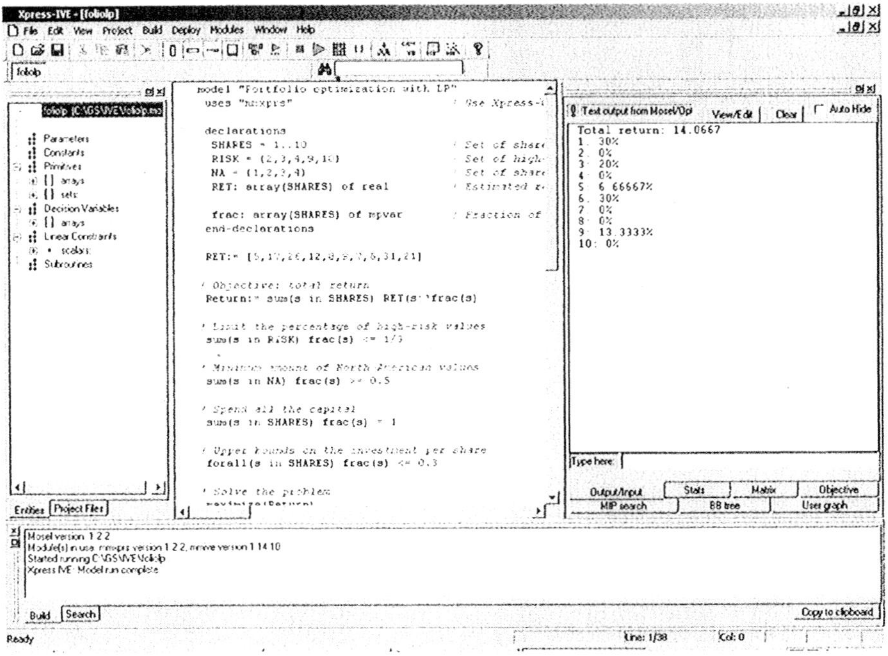

Figure 12.2.1. IVE display after model execution

that are provided specifically for mathematical programming, these form the *elementary types* of Mosel.

In addition to the elementary types, Mosel defines the types set (collection of elements of a given type) and array (collection of labeled objects of a given type). The data structures (sets and arrays) in Mosel are *dynamic* if at their creation their size and/or contents is not known. The following definitions some of which are used in the model of Section 12.2.1 result in *constant* sets (SHARES, SHARES2) and *static* arrays (RET, x):

```
declarations
 SHARES = 1..10
 RET: array(SHARES) of real
 SHARES2 = {"hardware", "telecom", "cars"}
 x: array(SHARES2, -2..1) of mpvar
end-declarations
```

Dynamic sets (R, S) and arrays (A, B) are created with definitions like

```
declarations
 S: set of string
 A: array(S,S) of linctr
 B: array(R:range) of boolean
end-declarations
```

The examples use a special type of set that we have already seen in the initial example: SHARES and R are *ranges* (= ordered sets of integers). 1..10 indicates that SHARES is the set of all integers from 1 to 10.

12.2.3 Initialization of Data/Data File Access

Dynamic arrays and sets are often used when data are read from a file, as is the case in this modification to our initial example that illustrates the use of the initializations section in Mosel.

```
declarations
 SHARES: set of string                ! Set of shares
 RISK: set of string                  ! Set of high-risk values among shares
 NA: set of string                    ! Set of shares issued in N.-America
 RET: array(SHARES) of real           ! Estimated return in investment
end-declarations

initializations from "folio.dat"
 RISK RET NA
end-initializations

writeln("SHARES: ", SHARES, "\nRISK: ", RISK, "\nNA: ", NA, "\nRET: ", RET)
```

If the data file folio.dat read by this program fragment has this contents

```
RET: [("treasury") 5 ("hardware") 17 ("theater") 26 ("telecom") 12
      ("brewery") 8 ("highways") 9 ("cars") 7 ("bank") 6
      ("software") 31 ("electronics") 21 ]
RISK: ["hardware" "theater" "telecom" "software" "electronics"]
NA: ["treasury" "hardware" "theater" "telecom"]
```

then the program prints the following output:

```
SHARES: {'treasury','hardware','theater','telecom','brewery','highways','cars',
'bank','software','electronics'}
RISK: {'hardware','theater','telecom','software','electronics'}
NA: {'treasury','hardware','theater','telecom'}
RET: [('treasury',5),('hardware',17),('theater',26),('telecom',12),('brewery',8),
('highways',9),('cars',7),('bank',6),('software',31),('electronics',21)]
```

Note that it is not required to initialize the set SHARES explicitly, it is filled when the array RET is read that is indexed by this set. The contents of dynamic arrays like RET is printed as a list of n-tuples where the first n-1 elements are the indices and the last the value of the array entry.

If the set SHARES is defined as a constant set (and hence RET defined as a static array) as in the original version of the example, then there is no need to specify the index values in the data file:

```
RET: [5 17 26 12 8 9 7 6 31 21]
```

and the array will be printed in the following form (that is, also without the index values since all its entries are defined):

```
[5,17,26,12,8,9,7,6,31,21]
```

For more flexibility, it is also possible to use the procedures `read/readln`, or `write/writeln` to read from or write to text files. To read the data file `readdata.dat` with the following contents

```
# Data:
A(1) = 5.2
A(2) = 3.4
```

we may write the following code in a `Mosel` program:

```
declarations
 j: integer
 B: array(range) of real
end-declarations

fopen("readdata.dat", F_INPUT)
fskipline("#")                   ! Skip lines starting with a '#'
repeat
 readln('A(',j,') =',B(j))      ! Read the indices and the value on a line
until (getparam("nbread")<4)     ! until a line is not properly constructed
fclose(F_INPUT)
```

12.2.4 Language Constructs

Besides data types, `Mosel` also provides the typical flow controls (selections and loops) that one will expect from a programming language.

12.2.4.1 Selections. The simplest form of a selection is the `if-then` statement which may be extended to `if-then-else` or even `if-then-elif-then-else` to test two conditions consecutively and execute an alternative if both fail as in the following example:

```
declarations
 A : integer
 x: mpvar
end-declarations

if A >= 20 then
 x <= 7
elif A <= 10 then
 x >= 35
else
 x = 0
end-if
```

The upper bound 7 is applied to the variable x if the value of A is greater or equal 20, and if the value of A is less or equal 10 then the lower bound 35 is applied to x. In all other cases (that is, A is greater than 10 and smaller than 20), x is fixed to 0.

If several mutually exclusive conditions (here: values of A) are tested, the `case` statement should preferably be used as in

```
declarations
 A : integer
 x: mpvar
end-declarations

case A of
 -MAX_INT..10 : x >= 35
 20..MAX_INT :  x <= 7
 12, 15 :        x = 1
 else            x = 0
end-if
```

12.2.4.2 Loops. Loops regroup actions that need to be repeated a certain number of times, either for all values of some index or counter (`forall`) or depending on whether a condition is fulfilled or not (`while`, `repeat-until`). The `forall` and `while` loops in `Mosel` exist in two versions: an inline version for looping over a single statement as we have already see in the initial example:

```
forall(s in SHARES) frac(s) <= 0.3
```

and a second version `forall-do` (`while-do`), that may enclose a block of statements, the end of which is marked by `end-do`:

```
forall(s in SHARES) do
 frac(s) <= 0.3
 frac(s) is_semcont 0.1       ! Turn frac(s) into a semi-continuous variable:
end-do                        ! frac(s)=0 or 0.1<=frac(s)<=0.3
```

12.2.5 Set Operations

Subsets may be added or removed from a set using the operators += and -=. (Note that a set may not decrease in size once it is used as an indexing set.) `Mosel` also defines the standard operations on sets: union, intersection and difference (operators +, *, -). The following code fragment gives some examples of operations on sets.

```
declarations
 SHARES = 1..10
 RISK = {"hardware", "theater", "telecom", "software", "electronics"}
 NA = {"treasury", "hardware", "theater", "telecom"}
 S: set of integer
end-declarations

writeln("Union: ", RISK+NA)
writeln("Intersection: ", RISK*NA)
writeln("Subtraction: ", RISK-NA)

S:= SHARES; S += {12}; S -= 3..7
writeln("S: ", S)
writeln(-2, " is in SHARES: ", -2 in SHARES)
writeln("SHARES is not empty: ", SHARES <> {})
```

This produces the following output:

```
Union: {'hardware','theater','telecom','software','electronics','treasury'}
Intersection: {'hardware','theater','telecom'}
Subtraction: {'software','electronics'}
S: {1,2,8,9,10,12}
-2 is in SHARES: false
SHARES is not empty: true
```

12.2.6 Subroutines

`Mosel` provides a set of predefined subroutines (*e.g.*, procedures like `write` / `writeln`, arithmetical functions like `cos`, `exp`, `ln`), but it is also possible to define new procedures and functions according to the needs of a specific program. User defined subroutines in `Mosel` have to be marked with `procedure` / `end-procedure` and `function` / `end-function` respectively. The return value of a function has to be assigned to `returned` as shown in the following example. It is possible to pass parameters into a subroutine. The (list of) parameter(s) is added in parentheses following the name of the subroutine.

```
function timestwo(b:integer):integer
 returned := 2*b
end-function

procedure printstart
 writeln("The program starts here.")
end-procedure

printstart
a:=3
writeln("a = ", a)
a:=timestwo(a)
writeln("a = ", a)
```

The structure of subroutines is very similar to the one of `model`: they may include `declarations` sections for declaring local parameters that are only valid in the corresponding subroutine. Subroutine calls may be nested, and they may also be called recursively.

Forward declaration: `Mosel` enables the user to declare a subroutine separately from its definition by using the keyword `forward`.

Overloading: In `Mosel`, it is possible to re-use the names of subroutines, provided that every version has a different number and/or types of parameters. This functionality is commonly referred to as *overloading*. The user may define (additional) overloaded versions of any subroutines defined by `Mosel` and also for his own functions and procedures.

12.3 Mosel Libraries

Now consider the following version of our introductory LP example. Data is now read from a file and the output printing has been removed. The data file name and certain numerical constants are defined as *model parameters* that may be reset at the execution of the model or otherwise take their default value given in the `parameters` section.

```
model "Portfolio optimization with LP"
 uses "mmxprs"                          ! Use Xpress-Optimizer

 parameters
  DATAFILE= "folio.dat"                 ! File with problem data
  MAXRISK = 1/3                         ! Max. investment into high-risk values
  MAXVAL = 0.3                          ! Max. investment per share
  MINAM = 0.5                           ! Min. investment into N.-American values
 end-parameters

 declarations
  SHARES: set of string                 ! Set of shares
  RISK: set of string                   ! Set of high-risk values among shares
  NA: set of string                     ! Set of shares issued in N.-America
  RET: array(SHARES) of real            ! Estimated return in investment
 end-declarations

 initializations from DATAFILE
  RISK RET NA
 end-initializations

 declarations
  frac: array(SHARES) of mpvar          ! Fraction of capital used per share
 end-declarations

! Objective: total return
 Return:= sum(s in SHARES) RET(s)*frac(s)

! Limit the percentage of high-risk values
 sum(s in RISK) frac(s) <= MAXRISK

! Minimum amount of North-American values
 sum(s in NA) frac(s) >= MINAM

! Spend all the capital
 sum(s in SHARES) frac(s) = 1

! Upper bounds on the investment per share
 forall(s in SHARES) frac(s) <= MAXVAL

! Solve the problem
 maximize(Return)
end-model
```

During its development a model is typically run as a stand-alone application, either directly with `Mosel` or in the IVE graphical interface. For its deployment however, a comfortable way of embedding a model into a programming language environment is required. To do this, `Mosel` provides two libraries, the *`Mosel` Compiler Library* and the `Mosel` Run-Time Libary. The former transforms models into BIM files (portable **Bi**nary **M**odel files), the latter is required for loading and running these files and subsequently accessing the model information.

The following example shows how the previous model, saved with the name `foliodata.mos`, may be executed from a C program (other interfaces available: Java, Visual Basic). The name of the data file and of the two parameters `MAXRISK` and `MAXVAL` are changed. After the optimization run we retrieve the problem status and if a solution was found, print it out. It would also be possible to obtain the full information about the current values of all model objects, including data, decision variables, and constraints.

```
#include <stdio.h>
#include "xprm_mc.h"
#include "xprm_rt.h"

int main(int argc, char *argv[])
{
  XPRMmodel model;
  int result;
  char params[128];

  sprintf(params, "DATAFILE=fdata2.dat, MAXRISK=0.4, MAXVAL=0.25");

  XPRMinit();                             /* Initialize Mosel */
                                          /* Execute = compile/load/run a model */
  XPRMexecmod(NULL, "foliodata", params, &result, &model);
                                          /* Test whether a solution is found */
  if((XPRMgetprobstat(model)&XPRM_PBRES)==XPRM_PBOPT)
    printf("Objective value: %g\n", XPRMgetobjval(model));

  XPRMunloadmod(model);                   /* Unload the model */

  return 0;
}
```

12.4 Mosel Modules

An original feature of `Mosel` is the possibility to extend the language by means of *modules* (dynamic libraries written in the C programming language that observe the conventions set out by the `Mosel` Native Interface). A module may extend the `Mosel` language with new

- constant symbols
- subroutines
- types
- control parameters

In this list, subroutines and types are certainly the most important items. *Subroutines* defined by a module may be entirely new functions or procedures or overload existing subroutines of `Mosel`. A module may, for instance, provide a subroutine that calls an algorithm or a solution heuristic that is readily available in the form of a C or C++ library function.

New *types* defined by a module are treated exactly like the own types of `Mosel` (like `integer` or `mpvar`). They can be used in complex data structures (arrays, sets), read in from file in `initializations` sections, or appear as parameters of subroutines. The operators in `Mosel` can be overloaded to work with new types. The definition of new types may be required to support solvers such as finite domain constraint solvers.

Constants and *control parameters* published by a module make little sense on their own. They will typically be used in conjunction with its types or subroutines.

Modules may be seen as an appropriate means for rapid prototyping of solution algorithms that involve a combination of solution strategies [44] or solvers originating from different areas of research. For example, using *mmxprs* and a module that provides access to a finite domain constraint solver it is possible to define a MIP search that at every node in the branching tree solves subproblem(s) using constraint propagation algorithms and depending on the results, generates cuts for the MIP problem (*e.g.*, [232]).

12.4.1 Available Modules

At present, the following modules have been implemented:

Solvers: mmxprs, mmquad, mmxslp, mmocl

Data handling: mmodbc, mmetc

System: mmsystem

Graphics: mmive

In the preceding examples the module *mmxprs* has already been used to solve Linear and Mixed-integer Programming (LP and MIP) problems with Xpress-Optimizer. *mmquad* extends the `Mosel` language for formulating and solving Quadratic Programing (QP) problems with Xpress-Optimizer. Non-linear problems can be formulated and solved with the Successive Linear Programming module *mmxslp* or with the module *mmocl* that provides access to the OptQuest Callable Library (metaheuristics).

Modules may define additional interfaces to data files: *mmetc* defines the procedure `diskdata` that emulates the data in- and output of `mp-model`; module *mmodbc* provides access to any data source for which an ODBC interface is available, using standard SQL commands.

The module *mmsystem* provides functions like `gettime` and file handling facilities. It even makes it possible to use operating system commands from within the `Mosel` language – the latter quite obviously at the expense of the portability of the model file.

The module *mmive* allows users working with `Mosel` models through the graphical user interface Xpress-IVE to draw their own graphs.

12.4.2 QP Example with Graphical Output

The investor from Section 12.2.1 may also look at his portfolio selection problem from a different angle: instead of maximizing the estimated return and limiting the portion of high-risk investments he now wishes to minimize the risk whilst obtaining a certain target yield. He adopts the Markowitz idea of getting estimates of the variance/covariance matrix VAR_{st} of estimated returns on the securities.

Starting again with the LP model in Section 12.2.1 we need to make the following changes:

- New objective function: mean variance instead of total return.
- The risk-related constraint disappears.
- Addition of a new constraint: target yield.

In addition to these modifications, we wish to solve the problem with different values of the targeted return, introducing a loop around the definition of the target yield constraint and the optimization statement. At every iteration we save the solutions we have obtained (arrays SOLRET and SOLDEV). At the end of our model we plot the resulting graph (known as the *efficient frontier*) with Xpress-IVE, adding to it also a repesentation of the input data.

```
model "Portfolio optimization with QP"
 uses "mmxprs", "mmquad", "mmive"     ! Use Xpress-Optimizer with QP solver
                                      ! and graphical output with IVE

 declarations
  SHARES: set of string               ! Set of shares
  RISK: set of string                 ! Set of high-risk values among shares
  NA: set of string                   ! Set of shares issued in N.-America
  RET: array(SHARES) of real          ! Estimated return in investment
  VAR: array(SHARES,SHARES) of real   ! Variance/covariance matrix of
                                      ! estimated returns
  SOLRET: array(range) of real        ! Solution values (total return)
  SOLDEV: array(range) of real        ! Solution values (average deviation)
 end-declarations

 initializations from "folioqp.dat"
  RISK RET NA VAR
 end-initializations

 declarations
  frac: array(SHARES) of mpvar        ! Fraction of capital used per share
 end-declarations

! Objective: mean variance
 Variance:= sum(s,t in SHARES) VAR(s,t)*frac(s)*frac(t)

! Minimum amount of North-American values
```

```
 sum(s in NA) frac(s) >= 0.5

! Spend all the capital
 sum(s in SHARES) frac(s) = 1

! Upper bounds on the investment per share
 forall(s in SHARES) frac(s) <= 0.3

! Solve the problem for a range of returns: this is the efficient frontier
 target:= min(s in SHARES) RET(s)
 RMAX:= max(s in SHARES) RET(s)

 while(target < RMAX) do
   Return:= sum(s in SHARES) RET(s)*frac(s) >= target    ! Target yield
   minimize(Variance)                      ! Solve the problem
   if (getprobstat = XPRS_OPT) then        ! Save the optimal solution value
    ct+=1
    SOLDEV(ct):= getobjval
    SOLRET(ct):= target
   else
    writeln("No solution for target return >= ", r, "%")
    break
   end-if
   target += 1
 end-do

! Drawing a graph to represent results ('plot1') and data ('plot2' & 'plot3')
 declarations
  plot1, plot2, plot3: integer
 end-declarations

 plot1 := IVEaddplot("Solution values", IVE_BLACK)
 plot2 := IVEaddplot("Low risk", IVE_YELLOW)
 plot3 := IVEaddplot("High risk", IVE_RED)

 forall(r in 1..ct) IVEdrawpoint(plot1, SOLRET(r), SOLDEV(r))
 forall(r in 2..ct)
  IVEdrawline(plot1, SOLRET(r-1), SOLDEV(r-1), SOLRET(r), SOLDEV(r))

 forall (s in SHARES-RISK) do
  IVEdrawpoint(plot2, RET(s), DEV(s))
  IVEdrawlabel(plot2, RET(s)+5, DEV(s)-1, s)
 end-do

 forall (s in RISK) do
  IVEdrawpoint(plot3, RET(s), DEV(s))
  IVEdrawlabel(plot3, RET(s)-6, DEV(s)-1, s)
 end-do
end-model
```

The resulting graph is shown in Fig. 12.4.2.

12.4.3 Example of a Solution Algorithm

In this section we show how to implement a solution algorithm for the so-called *Fiveleaper Problem* [41] where we wish to find a sequence of moves

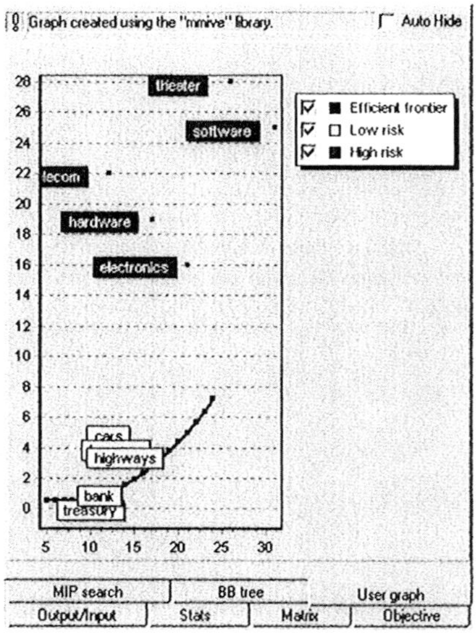

Figure 12.4.2. Graph of the efficient frontier

with Euclidean distance of five squares that visits every square of a chess board exactly once. A legal move for a fiveleaper is either traveling five squares into the same direction or traveling three squares into one direction and four squares in an orthogonal direction.

The Fiveleaper Problem can be seen as an instance of the Traveling Salesman Problem (TSP) that is notoriously difficult to solve with standard Mathematical Programming methods due to its combinatorial nature. We show how to implement with `Mosel` the following *subtour elimination algorithm* for solving this problem:

procedure `breaksubtour`:

> Construct the tour starting at square 1
> *If* a complete tour is found, stop
> *Otherwise*
>> Add a subtour elimination cut for this subtour
>> Find new subtours starting at each as yet unvisited square in turn
>>> and add subtour elimination cuts
>>
>> Resolve the problem
>> Iteratively call again the procedure `breaksubtour`

As an alternative to the recursive form of this subtour breaking algorithm we could transform the procedure `breaksubtour` into a function that returns the number of subtours found and call this function repeatedly until it returns 1.

For all permissible moves we introduce binary decision variables $x_{i,j}$ that take the value 1 if we leap from square i to square j and 0 otherwise. With these variables, the mathematical model can be expressed through three sets of constraints:

1 Each square precedes one other in the tour.

2 Each square is preceded by one other.

3 The tour does not contain any subtours.

The first two sets of constraints are easily expressed by sums over all squares of the chessboard, but excluding all possible subtours would require a large number of constraints. We do not state these intially but add subtour breaking constraints of the following form for any subtour T contained in the current solution to the problem:

$$\sum_{\substack{i,j \in SQUARES \\ \text{move } i \to j \in T}} x_{i,j} \leq size(T) - 1$$

where $size(T)$ denotes the number of leaps forming the tour T.

The complete implementation of the problem is shown below. The `Mosel` model starts by calculating the permissible moves, followed by a simple check for feasibility of the problem instance (the smallest possible square board size is 8×8). Notice that this implementation also works with rectangular boards. The size of the board can be changed via the parameters `NCOL` and `NROW` at every execution of the model.

```
model '5leaper on rectangular board'
  uses 'mmxprs', 'mmsystem'

  parameters
    NCOL = 8                           ! Number of colums
    NROW = 8                           ! Number of rows
    DISPLAY = true                     ! 'true' for printout at every iteration
  end-parameters

  forward procedure break_subtour
  forward procedure find_subtour(start: integer, TOUR: set of integer)
  forward procedure print_sol

  declarations
    NSQ = NCOL*NROW                    ! Number of squares
    SQUARES = 1..NSQ                   ! Range of squares
    COLS = 1..NCOL                     ! Range of columns
    ROWS = 1..NROW                     ! Range of rows
    nmoves, ntours, niter: integer     ! Counters
    nmovesfrom: array(SQUARES) of integer   ! Counts possible moves leaving sq
    x: dynamic array(SQUARES,SQUARES) of mpvar  ! 1 if we leap from sq1 to sq2
    NEXTX: array(SQUARES) of integer   ! Next square after sq in the solution
```

```
  end-declarations

! For each pair of squares with a valid move...
  nmoves:= 0
  forall(i1,i2 in COLS, j1,j2 in ROWS | (i1-i2)^2+(j1-j2)^2 = 25) do
    square1:= j1+NCOL*(i1-1)
    square2:= j2+NCOL*(i2-1)
    create(x(square1,square2))          ! Create the 'x' binary decision variable
    x(square1,square2) is_binary
    nmovesfrom(square1) += 1            ! Note a possible move from square
    nmoves += 1
  end-do
  writeln("There are ", nmoves, " possible moves")

! Check that each square is reachable
  forall(sq in SQUARES | nmovesfrom(sq)=0) do
    ii:= ceil(sq/NCOL)
    jj:= sq-NCOL*(ii-1)
    writeln("**** Cannot go anywhere from square (", ii, ",", jj, ")")
    exit(0)
  end-do

! Objective (arbitrary)
  AnyObj:= sum(sq1,sq2 in SQUARES | exists(x(sq1,sq2)) ) x(sq1,sq2)

! Each square precedes one other and each square is preceded by one other
  forall(sq1 in SQUARES) sum(sq2 in SQUARES | exists(x(sq1,sq2)) ) x(sq1,sq2) = 1
  forall(sq2 in SQUARES) sum(sq1 in SQUARES | exists(x(sq1,sq2)) ) x(sq1,sq2) = 1

! Optimizer settings to speed up the iterations
  setparam("XPRS_CUTSTRATEGY", 0)
  setparam("XPRS_PRESOLVE", 0)

! Solve the problem
  minimize(AnyObj)

! Check relaxation status
  if getprobstat = XPRS_OPT then
    break_subtour                       ! Start the subtour elimination alg.

    print_sol                           ! Solution printout
  else
    writeln("Problem is infeasible.")
    exit(0)
  end-if
  writeln("Total time: ", gettime-starttime, " sec")

!---------------------------Subtour breaking algorithm----------------------------

  procedure break_subtour: integer
    declarations
      TOUR, ALLTOURS: set of integer
    end-declarations

    forall(sq in SQUARES)
      NEXTX(sq):= integer(round(getsol(sum(sq2 in SQUARES) sq2*x(sq,sq2) )))
```

```
    if (DISPLAY) then print_sol; end-if

  ! Get (sub)tour containing square 1
    find_subtour(1, TOUR)

    if (getsize(TOUR) < NSQ) then
      ALLTOURS:= TOUR
    ! Add a subtour breaking constraint for the subtour
      sum(t in TOUR) x(t,NEXTX(t)) <= getsize(TOUR) - 1
      ntours+=1
      forall(startsquare in SQUARES | startsquare not in ALLTOURS) do
      ! Find (sub)tour containing 'startsquare'
        find_subtour(startsquare, TOUR)
        ALLTOURS += TOUR
        ntours += 1
      ! Add a subtour breaking constraint
        sum(sq in TOUR) x(sq,NEXTX(sq)) <= getsize(TOUR) - 1
      end-do

    ! Print a log every 10 iterations
      niter += 1
      if (niter mod 10 = 0) then
        writeln(niter, " (", gettime-starttime, " sec) tours eliminated: ", ntours)
      end-if

    ! Re-solve the problem
      minimize(any)

      break_subtour
    end-if
  end-procedure

!--------Find a subtour starting at square 'start'; result returned in TOUR--------

  procedure find_subtour(start: integer, TOUR: set of integer)
    TOUR:= {}
    square:= start
    repeat
      TOUR += {square}
      square:= NEXTX(square)
    until (square=start)
  end-procedure

!----------------------------Print the current solution----------------------------

  procedure print_sol
    declarations
      SEENSOFAR: set of integer
    end-declarations

    SEENSOFAR:={}
    forall(startsquare in SQUARES | startsquare not in SEENSOFAR) do
      write(startsquare)
      square:= startsquare
      repeat
        SEENSOFAR += {square}
        square:= NEXTX(square)
```

```
      write(" - ", square)
    until (square=startsquare)
    writeln
  end-do
end-procedure

end-model
```

With this algorithm, the problem solves to optimality within seconds for square board sizes up to 14 $\times$ 14 and within a few minutes for sizes up to 20 $\times$ 20 (on a PC Pentium4, 1.5GHz, 512Mb). Fig. 12.4.3 shows a graphical representation of a solution for the 8 $\times$ 8 instance produced with Xpress-IVE.

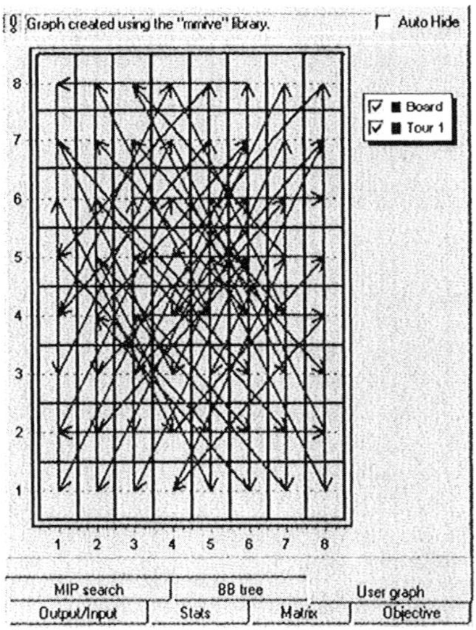

Figure 12.4.3. Solution of the Fiveleaper problem (8 $\times$ 8 board)

12.5 Writing User Modules

From the operating system point of view, a module is a dynamic library (Dynamic Shared Object, DSO) written in the C programming language. The `Mosel` Native Interface is a set of conventions that a DSO must respect to be accepted as a module by `Mosel`. Any functionality that can be implemented in the form of a C library may be made accessible from within the `Mosel` language. In this section, we show how to define (1) a new subroutine and (2) a new type.

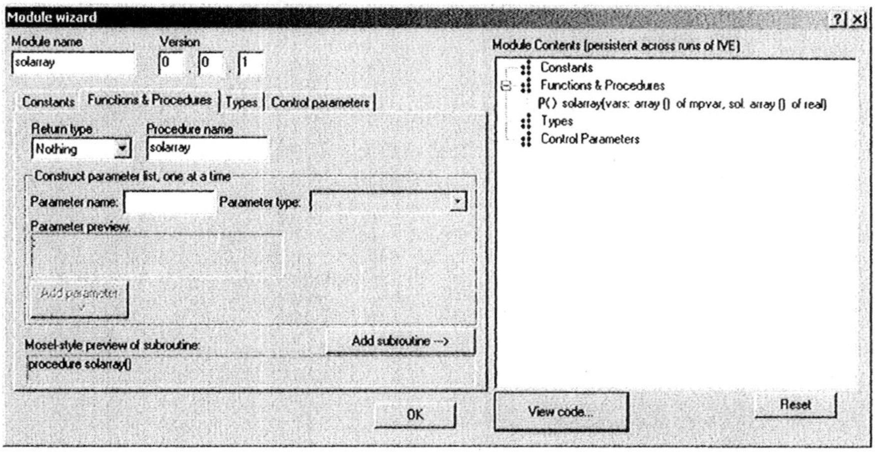

Figure 12.5.4. Module definition with IVE

12.5.1 Defining a New Subroutine

After solving a problem, users often wish to retrieve the solution and store it. `Mosel` provides the function `getsol` to access solution values one-by-one, but there is no subroutine for accessing and saving the solution to an array of decision variables all at once. However, such a subroutine can easily be implemented in the form of a module. In our initial example, we would then be able to print out the solution as follows:

```
model "Portfolio"
 uses "solarray", "mmxprs"
 ...                                  ! Data initialization
 declarations
  frac: array(SHARES) of mpvar        ! Fraction of capital used per share
  sol: array(SHARES) of real          ! Solution values
 end-declarations
 ...                                  ! Formulate and solve the problem
 solarray(frac,sol)                   ! Retrieve the solution for all variables
 writeln(sol)                         ! Print the solution
end-model
```

The code for implementing a module that provides this new function is shown in the following. The various interface structures, the initialization function, and also the prototype of the library function `ar_getsol` implementing "solarray" can be generated automatically with the module generation functionality in Xpress-IVE (Fig. 12.5.4). The only work left over to the user is to fill in the body of the function `ar_getsol`. For simplicity's sake we have left out the error handling that should be performed by this function, we merely show the required operations:

1 Get the references to the two arrays (one array of decision variables and one array of reals) from the Mosel stack.

2 Enumerate all defined entries in the array of variables (it may be sparse or dense, with up to MAXDIM dimensions).

3 Get the solution value for each entry and copy it into the array of reals.

```
#include <stdlib.h>
#include "xprm_ni.h"

#define MAXDIM 20

static int ar_getsol(XPRMcontext ctx,void *libctx);

/* List of subroutines */
static XPRMdsofct tabfct[]=
    {
     {"solarray", 1000, XPRM_TYP_NOT, 2, "A.vA.r", ar_getsol}
    };

/* Interface structure */
static XPRMdsointer dsointer=
    {
     0, NULL,
     sizeof(tabfct)/sizeof(XPRMdsofct), tabfct,
     0, NULL,
     0, NULL
    };

/* Structure for getting function list from Mosel */
static XPRMnifct mm;

/* Module initialization function */
DSO_INIT solarray_init(XPRMnifct nifct, int *interver,int *libver,
                            XPRMdsointer **interf)
{
 mm=nifct;                    /* Get the list of Mosel functions */
 *interver=XPRM_NIVERS;       /* Mosel NI version */
 *libver=XPRM_MKVER(0,0,1); /* Module version: must be <= Mosel NI version */
 *interf=&dsointer;           /* Pass info about module contents to Mosel */
 return 0;
}

/* Implementation of 'solarray' */
static int ar_getsol(XPRMcontext ctx,void *libctx)
{
 XPRMarray varr, solarr;
 XPRMmpvar var;
 int indices[MAXDIM];

/* Get variable and solution arrays from stack in the order that they are
  used as parameters for 'getsol' */
 varr=XPRM_POP_REF(ctx);
 solarr=XPRM_POP_REF(ctx);
```

```
/* Error handling:
   - compare the number of array dimensions and the index sets
   - make sure the arrays do not exceed the maximum number of dimensions MAXDIM
*/

/* Get the solution values for all variables and copy them into the solution
   array */
 if(!mm->getfirstarrtruentry(varr,indices))
  do
  {
   mm->getarrval(varr,indices,&var);
   mm->setarrvalreal(ctx,solarr,indices,mm->getvsol(ctx,var));
  } while(!mm->getnextarrtruentry(varr,indices));

 return XPRM_RT_OK;
}
```

This code needs to be compiled into a dynamic library giving it the extension `.dso`. Afterwards its location must be made known to `Mosel` (by setting the environment variable `MOSEL_DSO`) and then it may be used like any module of the `Mosel` distribution.

12.5.2 Creating a New Type

In this section we show how to implement a new type *complex* to represent complex numbers so as to make possible writing models like the following with `Mosel`:

```
model "Complex numbers"
 uses "complex"

 declarations
  c:complex                     ! Define a single complex number
  t:array(1..10) of complex     ! Define an array of complex numbers
 end-declarations

 forall(j in 1..10)
  t(j):=complex(j,10-j)         ! Initialize with 2 integers or reals
 t(5):=complex("5+5i")          ! Initialize with a string

 c:=prod(i in 1..5) t(i)        ! Aggregate PROD operator
 if c<>0 then                   ! Comparison with an integer or real
  writeln("Product: ", c)       ! Printing a complex number
 end-if

 writeln("Sum: ", sum(i in 1..10) t(i)) ! Aggregate SUM operator
                                        ! Arithmetic operators
 c:=t(1)*t(3)/t(4) + if(t(2)=0, t(10), t(8)) + t(5) - t(9)

 initializations to "complex_out.dat"   ! Output to a file
  c t
 end-initializations
end-model
```

Besides some standardized initialization and type creation functions, the module implementing this new type defines constructors, basic arithmetic operations, the equality comparison operator and a printing function.

Once a type is defined, it can automatically be used in Mosel data structures (set, array) as shown in the example. From given defitions of the basic arithmetic operations Mosel deduces the definition of aggregate operators like aggregate products or sums, and where applicable, commutations of the operands or negations. Similarly, the definition of an equality comparison suffices to derive the inequality and from basic logical operators (none defined in this case) the definition of aggregate operators is generated. With the definition of a printing function, the output to a file with `initializations to` is also available.

Since the complete code of this module[1] occupies several pages of text, we shall restrict ourselves here to highlighting some key features of this module:

- module context
- type creation and deletion
- type transformation to and from string
- overloading of arithmetic operators

As with the previous example, we assume that the module code generation facility of IVE has been used to create the required interface structures so that we only need to fill in the corresponding function bodies.

12.5.2.1 Module Context. A module needs to keep track of all objects created during the execution of a model so that all allocated space may be freed when the execution is terminated. This function is fulfilled by the module *context*. In this example, the context may be nothing but a chained list of complex numbers:

```
typedef struct
    {
     s_complex *firstcomplex;
    } s_cxctx;
```

which assumes that a complex number is represented by the following structure:

```
typedef struct Complex
    {
     double re, im;
     struct Complex *next;
    } s_complex;
```

[1] Complete source code available from the authors.

A module context can also be used to store the current values of control parameters or any other information that needs to be preserved between different calls to the module functions during the execution of a model.

A reset *service function* is called at the beginning and the termination of the execution of a `Mosel` program that uses the module. At its first call, the reset function creates and initializes a context for the model, and deletes this context (and any other resources used by the module for this model) at the second call.

12.5.2.2 Type Creation and Deletion. The objective of the type instance creation and deletion functions is to handle (create/initialize or delete/reset) the C structures that represent the external type and to update correspondingly the information stored in the module context. In this example we implement just a rudimentary memory management for the objects (complex numbers) created by the module: every time a number is created, we allocate the corresponding space and deallocate it when it is deleted. More realistically, a module may allocate chunks of memory and recycle space that has been allocated earlier by this module.
We define the *creation function* for a complex number as follows:

```
static void *cx_create(XPRMcontext ctx, void *libctx)
{
 s_cxctx *cxctx;
 s_complex *complex;

 cxctx=libctx;
 complex=(s_complex *)malloc(sizeof(s_complex));
 complex->next=cxctx->firstcomplex;
 cxctx->firstcomplex=complex;

 complex->re=complex->im=0;          /* Initialize the complex number */
 return complex;
}
```

The *deletion function* frees the space used by a complex number and removes it from the list held by the module context.

12.5.2.3 Type Transformation to and from String. To be able to use `initializations` blocks with the new type `complex` we define two functions for transforming the number into a string and initializing it from a string. The writing function is also used by the `write` and `writeln` procedures for printing this type. The reading function also gets applied when the type instance creation function is given a string.

The format of the string will obviously depend on the type. In this example the obvious format is "re+im*i*". The following function prints a complex number:

```
static int cx_tostr(XPRMcontext ctx,void *libctx,void *toprt,char *str,int len)
{
 s_complex *c;
```

```
 if(toprt==NULL)
 {
  strcpy(str,"0+0i");
  return 4;
 }
 else
 {
  c=toprt;
  return sprintf(str,"%g%+gi",c->re,c->im);
 }
}
```

The next function reads in a complex number from a string:

```
static int cx_fromstr(XPRMcontext ctx,void *libctx,void *toinit,const char *str)
{
 double re,im;
 s_complex *c;

 if(sscanf(str,"%lf%lf",&re,&im)!=2)
  return XPRM_RT_ERROR;
 else
 {
  c=toinit;
  c->re=re;
  c->im=im;
  return XPRM_RT_OK;
 }
}
```

12.5.2.4 Overloading of Arithmetic Operators. The only type conversion that is carried out automatically by Mosel is from integer to real (but not the other way round), and no conversions involving external types. It is therefore necessary to define all the operations between two numbers for two complex numbers and also for a complex and a real number. However, for commutative operations (addition, multiplication, comparison) it is only required to define one version combining the two types, the other sense is deduced by Mosel.

Taking the example of the *multiplication*, we have to define the multiplication of two complex numbers: $(a + bi) \cdot (c + di) = ac - bd + (ad + bc)i$

```
static int cx_mul(XPRMcontext ctx,void *libctx)
{
 s_complex *c1,*c2;
 double re,im;

 c1=XPRM_POP_REF(ctx);
 c2=XPRM_POP_REF(ctx);
 if(c1!=NULL)
 {
  if(c2!=NULL)
  {
   re=c1->re*c2->re-c1->im*c2->im;
```

```
    im=c1->re*c2->im+c1->im*c2->re;
    c1->re=re;
    c1->im=im;
   }
   else
    c1->re=c2->im=0;
  }
 cx_delete(ctx,libctx,c2);
 XPRM_PUSH_REF(ctx,c1);
 return XPRM_RT_OK;
}
```

and also the multiplication of a complex with a real: $(a + bi) \cdot r = ar + bri$

```
static int cx_mul_r(XPRMcontext ctx,void *libctx)
{
 s_complex *c1;
 double r;

 c1=XPRM_POP_REF(ctx);
 r=XPRM_POP_REAL(ctx);
 if(c1!=NULL)
 {
  c1->re*=r;
  c1->im*=r;
 }
 XPRM_PUSH_REF(ctx,c1);
 return XPRM_RT_OK;
}
```

It is not necessary to define the multiplication of a real with a complex since this operation is commutative and `Mosel` therefore deduces this case. The *addition* of two complex numbers and of a complex and a real number is implemented in a very similar way to multiplication. Once we have got the two types of addition, we simply need to implement the negation (–complex) in order for `Mosel` to be able to deduce *subtraction* (real – complex and complex – complex). For *division*, we need to implement all three cases since this operation is not commutative: complex/complex, complex/real and real/complex.

Furthermore we need to define the identity elements for addition and multiplication respectively:

```
static int cx_zero(XPRMcontext ctx,void *libctx)
{
 XPRM_PUSH_REF(ctx,cx_create(ctx,libctx));
 return XPRM_RT_OK;
}

static int cx_one(XPRMcontext ctx,void *libctx)
{
 s_complex *complex;

 complex=cx_create(ctx,libctx);
 complex->re=1;
 XPRM_PUSH_REF(ctx,complex);
```

```
 return XPRM_RT_OK;
}
```

Once addition and the 0-element have been defined, Mosel deduces the aggregate operator SUM. With multiplication and the 1-element, we obtain the aggregate operator PROD for our new type.

Other operators implemented by this module are constructors, the assignment and the comparison operators.

12.6 Summary

Mosel is an open, modular environment for modeling and solving problems. The examples in this chapter take the reader from a simple Linear Programming model through a series of additions to complete solution algorithms implemented directly with the Mosel language. The last part shows how the user may extend the capabilities of the language according to his needs.

The examples discussed show different possible uses of Mosel: practitioners in Operations Research typically are interested in implementing their models quickly and in a form that can easily be maintained. This is delivered by the fact that the formulation of mathematical models in the Mosel language is close to their algebraic form. Mosel also provides ample support for data handling making it possible to read and write data from various sources with just a few lines of code. When models grow larger the possibility to write solution algorithms and heuristics directly in the Mosel language is often helpful. Advanced users and researchers appreciate the possibility to add themselves whatever new feature may be required by their application, such as access to specific external data sources, new solvers, or external solution algorithms.

This last point also hints at one major direction for the future development of Mosel: according to the users' needs, the language will continue to be extended via modules for specific purposes (*e.g.*, providing access to different types of solvers or data sources).

Besides the Mosel language that forms the main topic of this paper, the Mosel environment also includes support for debugging models and libraries for embedding models into a programming language environment for their deployment. The interested reader is referred to the collection of examples in [101] and to http://www.dashoptimization.com for more documentation, case studies, and research papers about Mosel, IVE, and Xpress-Optimizer.

Chapter 13

THE MPL MODELING SYSTEM

Bjarni Kristjansson & Denise Lee
Maximal Software, Inc.
2111 Wilson Boulevard
Arlington, VA 22201
{bjarni,info}@maximalsoftware.com

Abstract Optimization is today one of the most important tools in implementing and planning efficient operations and increasing competitive advantage. Organizations need to make intelligent decisions to obtain optimal use of their available resources, such as manpower, equipment, raw materials and capital. The discipline of optimization, through the use of advanced mathematics and computer science techniques, can be used to assist organizations with solving their complex business problems in areas such as manufacturing, distribution, finance, and scheduling. Typically, these optimization problems contain hundreds, thousands, or even millions of interconnected variables and require an advanced set of software tools to solve.

Keywords: MPL, modeling, deployment, component library, modeling systems, optimization, Linear Programming, OptiMax 2002, database connection, Mathematical Programming, modeling language, Maximal Software, Internet optimization services, programming language

13.1 Maximal Software and Its History

Maximal Software as one of the leading providers of model development software is strategically positioned to assist organizations use optimization to solve their business problems.

Today, the field of optimization entails highly advanced software applications that integrate sophisticated mathematical algorithms and modeling techniques with intelligent software programming and data processing capabilities. There are a variety of business problems where optimization is applicable. For instance, organizations have been implementing, with enormous effort and expense, ERP (Enterprise Resource Planning) and SCM (Supply-Chain Management) systems in order to access corporate data to make important decisions that can improve productivity and increase their competitive edge. Optimization can now be used to perform in-depth analysis on this data in order to identify the implications of different options before making key decisions.

Optimization projects begin with the development of a mathematical model that defines the business problem. Individual business decisions are represented as *variables*, and the connections between them are represented by a series of mathematical equations termed *constraints*. The *objective* represents the goal of the business problem, for example, to maximize profitability or lower costs. Identifying the variables, the constraints and the objective is known as the *modeling* process and is an essential task for every optimization project. After the model has been formulated, it is then solved, using an optimization solver, which, at its core, has highly sophisticated algorithms adept at intelligently sorting through huge amounts of data and analyzing possible approaches to come up with an optimal solution. Maximal Software is one of the leading providers of model development software in the field of optimization. Maximal is the developer of the `MPL` Modeling System and the *OptiMax 2000* Component Library, and has many years of practical experience providing innovative solutions for our customers and assisting them in achieving the successful completion of their optimization projects.

Maximal Software was founded in 1987 by Bjarni Kristjansson and currently has locations in United States, United Kingdom, and Iceland. Since its inception, Maximal Software has consistently been on the leading edge of optimization modeling technologies.

Year	Innovation	Benefit
1987	Turbo-Simplex for DOS and Macintosh	LP for PC/Mac
1989	MPL Modeling System introduced	Optimization Modeling
1990	Maximal Software opened in the US	Closer to Market
1991	MPL Modeling System 2.0	Extended Memory
1992	MPL for Windows and Motif	GUI, Windows 3.0
1993	Database Connection	Open Data Access
1994	Maximal Web site, DLL solvers in memory	Web, Callable
1995	MPL for Windows 4.0, 32-bit	Windows 95
1996	Run-time, embed into end-user applications	Optimization Applications
1997	MPL for Windows 4.1, Full scalability	Very Large Models
1999	OptiMax 2000 Component Library	RAD/Components
2001	Distributed Applications with MPL	Web Services
2002	Stochastic Extensions for MPL	MPL/SPInE

Maximal has over the years invested significant R&D efforts in continuing to improve its software and offerings to our customers. Many technologies that are now considered standard in the field of optimization were originally developed by Maximal Software. For example here are several of the highlights: Maximal designed the first graphical user interface (GUIs) in 1992 for an optimization modeling language; in 1993 Maximal introduced for the first time a few years ahead of the market direct database connection to a modeling language. Maximal then designed in 1994 the only direct link to solvers through memory at that time. With the release of Windows 95 in 1995, 32-bit computing became possible in Windows and, again, Maximal Software was at the forefront of bringing its modeling language up to speed with current technology. Following that the run time version of `MPL`, which allowed developers for the first time to embed optimization models into end-user applications, was released. *OptiMax 2000*, introduced in 1999, took this concept even further by providing a full object-oriented component library with interface to programming languages such as VBA and C++. In 2002, Maximal Software in collaboration with OptiRisk Systems introduced `MPL/SPInE` that provided stochastic extensions for `MPL`.

13.2 Algebraic Modeling Languages

Over the years there have been a number of efforts to create modeling systems that allow you to formulate optimization models. One such type of software package is Algebraic Modeling Languages (AML) that allow the model developer to create optimization models using algebraic equations. The first AML brought to the market was the `GAMS` Modeling Language in the early 1980s that was developed by a group led by Alex Meeraus at the World Bank. Today there are seven different commercially available AMLs on the market. These seven languages can be divided into two different groups that have common characteristics. The first group, which sometimes is referred to as the independent modeling languages, consists of four languages:

- `GAMS` from GAMS Development Corporation,
- `AMPL` from AMPL Optimization LLC.,
- `MPL` from Maximal Software Inc., and
- `AIMMS` from Paragon Decision Technology.

The independent modeling languages have in common that they were developed by companies that specialize in optimization modeling, and these companies typically do not develop other types of software. Each language has one or two key people that were the main designers and developers of the language and have committed major portions of their career towards developing their language. Each language typically supports multiple different optimization solvers and can support almost any type of optimization models (for example, nonlinear, etc.). These languages tend to be very strong in the areas of indexing, sparsity, as well as other advanced modeling techniques. On the other hand, some of these languages are rather weak when they need to be integrated with other software systems, for example databases and graphical user interfaces. In addition their memory management is often not very efficient. Two of the older languages, GAMS and AMPL, also have some weaknesses when it comes to deployment and embedding optimization models inside end-user applications.

The second group, which is often referred to as the solver modeling languages consists of the remaining three languages:

- mp-model and Mosel from *Dash Optimization*,
- OPL Studio from *ILOG Inc.*, and
- LINGO from *Lindo Systems*.

The solver modeling languages are, as the name suggests, developed by companies that all have their own optimization solver on the market (XPRESS from Dash Optimization, CPLEX from ILOG, LINDO from *Lindo Systems*). Generally, the development of these modeling languages started several years after the development of their respective solver product. Their main product is still the optimization solver and the modeling language is offered as an option. The development of these modeling languages was typically over considerably fewer man-years than for that of the independent modeling languages. Each solver modeling language usually integrates efficiently with their own solver as they are produced by the same vendor, but they do not support any other solvers, which limits the choice for model developer. The solver modeling languages tend to be weaker than the independent modeling languages when it comes to indexing, sparsity and advanced modeling techniques. On the other hand, they tend to be quite strong when it comes to connection to databases and graphical user interfaces, as well as deployment and embedding into end-user applications.

13.2.1 Comparison of Modeling Languages

All modeling languages are not created equal. When choosing which modeling language is best for a particular project there are several different criteria that can be used depending on needs of the model developer and the end-user. Overall there are 10 main criteria that can be considered:

1.	Modeling Language	(completeness, easy to learn)
2.	Multiple Platforms	(Windows, Motif, Macintosh)
3.	Open Design	(multiple solvers/databases)
4.	Indexing	(sparse index and data handling)
5.	Scalability	(number of rows/columns)
6.	Memory Management	(dynamic/heap manager)
7.	Speed	(nonzeros per minute)
8.	Robustness	(stable code, years of testing)
9.	Deployment	(embedded into applications)
10.	Pricing	(development, deployment, multiple copies)

13.2.1.1 Modeling Language. Although all modeling languages have basically the same purpose – convert written equations to a matrix for the solver – each language has its own features and traits. They all handle summations, vectors, and indexes, but may use different symbols for syntax to represent different parts of the formulation. It is important that the language has a fairly easy-to-learn syntax and it allows the model developer to formulate the model logically and efficiently. Due to that fact the syntax for each language can often be quite different, it is not always easy to switch from one language to another. The independent modeling languages tend to be much stronger when it comes to the completeness of the language, in other words supporting all the different types of models that are possible.

13.2.1.2 Multiple Platforms. Optimization tools are available for almost any operating system platform on the market today. For most projects Microsoft Windows is the preferred platform. There are currently six different 32-bit versions of Microsoft Windows (95, 98, Me, NT, 2000, XP). As they all have very similar application programming interfaces (API), all versions of Windows are supported by each modeling language. There is also a new 64-bit version of Windows available for the new Intel Itanium and AMD Hammer processors. These 64-bit systems offer superior performance and scalability and already some of the solvers and modeling languages on the market are available for that platform.

The second most common platform for optimization are the many different UNIX platforms (HP, SUN, IBM etc). Typically solvers can provide superior performance on a fast UNIX machine, but since modeling languages work mainly with integer numbers they are usually faster on Windows platforms. Some of the solvers and modeling languages are also available for the Linux operating system on the PC.

A few years ago Macintosh was also a popular operating system for optimization, but they have now been discontinued and are no longer available.

13.2.1.3 Open Design. Optimization is usually part of a larger system in real-world applications and, therefore, it is important to be able to connect modeling languages to databases, spreadsheets, report generators and etc. Many of the modeling languages offer ways to import data directly from databases through ODBC. It is important to make sure that the modeling language supports connecting to the database of choice efficiently, so problems do not arise late in the development process.

Another important issue to take into consideration is whether the modeling language supports multiple different solvers and, in particular, the solver that most efficiently solves the problem at hand. The independent modeling languages are of course stronger when it comes to this point because they allow the model developer the flexibility to test multiple solvers in order to determine which is most applicable to the project.

13.2.1.4 Indexing. How a modeling language handles sparse indexes and data is one of the most important aspects of comparing different languages. All of the modeling languages can effectively define vector variables and constraints based on indexes. What makes some of the languages stronger than others, is how they handle special cases and exceptions and further how they allow the model developer to connect different indexes together.

For example, let us assume the model developer is writing a supply-chain optimization model and wants to define a variable that represents how many of each product to produce for each plant, machine, product and time period. In almost every case this variable vector would be very sparse so that only certain machines are in each factory, only certain products are done on each machine, and for each time period. How the modeling language manages vector variables, such as these, can have a significant impact on the model generation time and if it is not done efficiently this may result in the model becoming impossible to generate.

13.2.1.5 Scalability. Another issue that is important when it comes to comparing modeling languages is the scalability of the language. In particular the question is how large a model can the modeling language handle efficiently. Solvers typically use about five to ten times more memory than modeling languages to store the matrix, but when reading in the model the solver knows beforehand how large the matrix is for the model. Modeling languages, on the other hand, do not know in advance the size of the model and, therefore, need to dynamically allocate the memory necessary to store the matrix while reading the model formulation. Today, the largest models that are solved in real-world situations are usually supply-chain and manpower related models. These models often have millions of variables and constraints and, therefore, require huge amounts of memory. Because of the dynamic nature of model generation, most modeling languages (but not all) have problems dealing with models that are larger than a few hundred thousand variables.

13.2.1.6 Memory Management. As mentioned in the previous paragraph, when discussing scalability, memory management is a very important aspect of the design for modeling languages. Large-scale models often need to store huge matrixes with millions of nonzeros dynamically. If the standard run-time library for the programming language (typically the C language) is used when storing these large matrixes, the model generation will often run into problems. The solution for this is to write a specialized heap memory manager that is optimized for the particular aspects of the model generation. The memory manager also needs to be able to release the memory efficiently and completely after each solving run so the software does not waste time between each run.

13.2.1.7 Speed. In recent years solvers have made great strides towards decreasing the solution times solving very large and difficult models. Model generation time, by the modeling language, can be just as important as how long it takes for the optimization solver to solve the model. As a rule of thumb the modeling language should never take more time generating the matrix for the model than for the optimizer to solve it. Typically model generation times are measured in the number of nonzeros generated per time period (seconds or minutes).

13.2.1.8 Robustness. Modeling languages are extremely complex types of software and can take many man-years (or decades) to develop. Typically, the longer the modeling language has been on the market the better it is when it comes to robustness. It can take many years for a modeling language to reach enough stability and robustness for it to be used in mission critical applications.

13.2.1.9 Deployment. Modeling languages are very effective when used for prototyping models, as in formulating and developing models. Another important aspect of optimization projects is how to embed the optimization model into real- world business applications. One of the difficulties that arise is there is a fundamental gap of understanding between the programmers that create the application program and the model developers that formulate the models. Good modeling languages offer tools or libraries that make this connection or process easier and help bridging the gap between the model developers and the computer programmers.

13.2.1.10 Pricing. Optimization software is often very expensive and can run into tens of thousands of Dollars/Euros per developer. This is often offset by the fact that successful optimization modeling projects can easily result in savings of hundreds of thousands or even millions of Dollars/Euros. Since optimization software is typically sold per seat, how pricing for deployment and multiple copies is calculated, can have a major impact on the total cost of the software. Some modeling languages or solvers also offer size limited versions at a considerable discount.

13.3 MPL Modeling System

The main component of the `MPL` modeling system is the algebraic modeling language that allows the model developer to create optimization models using algebraic equations. The model is used as a basis to generate a mathematical matrix that can be relayed directly into the optimization solver. This is all done in the background so that the model developer only needs to focus on formulating the model. Algebraic modeling languages, such as MPL, have proven themselves over the years to be the most efficient method of developing and maintaining optimization models because they are easier to learn, quicker to formulate and require less programming.

MPL offers a feature-rich model development environment that takes full advantage of the graphical user interface in Microsoft Windows, making MPL a valuable tool for developing optimization models. MPL can import data directly from databases or spreadsheets. Once the model has been solved, MPL also has the ability to export the solution back into the database. MPL models can be embedded into other Windows applications, such as databases or spreadsheets, which makes MPL ideal for creating end-user applications.

The main purpose of a modeling language is to retrieve data from a structured data source, such as a database, and generate a matrix that the optimization solver can handle. For large optimization models, this matrix generation requires a modeling language with highly advanced capabilities, such as sparse indexing and database management, as well as high scalability and speed. Many details need to be taken into account when choosing a modeling language for optimization projects:

- Model Development Environment
- Robustness and Flexibility of the Modeling Language
- Indexing and Data Management
- Scalability and Speed
- Database Connections
- Connection to Solvers
- Deployment into Applications

In the rest of this chapter we will explore each of these areas and demonstrate how the MPL modeling system implements them.

13.3.1 MPL Integrated Model Development Environment

The MPL Modeling System is a state-of-the-art optimization modeling software. MPL, through the use of advanced graphical user interface features, creates a flexible working environment that enables the model developer to be more efficient and productive. MPL provides in a single system all the essential components needed to formulate the model, gather and maintain the data, optimize the model, and then analyze the results.

MPL was designed to be portable and to be run on multiple platforms. MPL for Windows is the most popular platform but an OSF Motif version is also available for various UNIX flavors including: HP 9000, IBM RS-6000, Sun

SPARC, Silicon Graphics and Linux systems. MPL models are portable so a model created for one platform can always be read on any other supported platform.

The model developer uses the built-in model editor to formulate the model in MPL and then selects the optimizer directly from the menus to solve the model. The solution results are automatically retrieved from the solver and displayed, providing the user with instant feedback. Each item defined in the model is also displayed in a tree window allowing the model developer to browse through them easily.

13.3.1.1 Solve the Model. After the model is entered into the model editor the model developer would choose *Solve <solvername>* from the Run menu to optimize the model. While optimizing MPL displays a Status Window that provides the model developer with information on the solution process as shown here below:

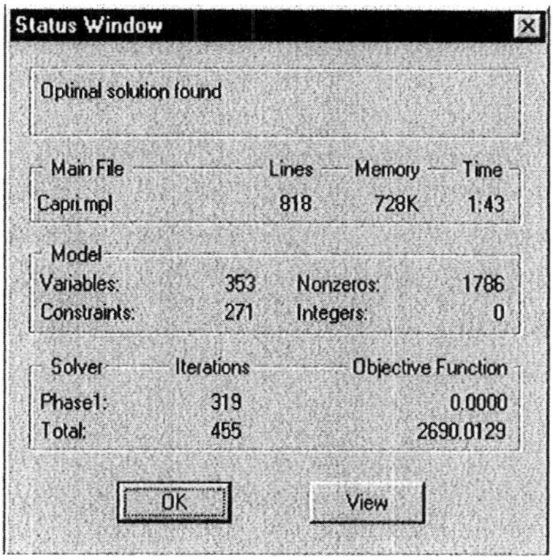

Figure 13.3.1. The Status Window

While MPL reads in the model file, the status window displays various statistics such as the name of the file being read, the number of lines read so far, the number of variables and constraints in the model, and how much memory has been used. While the optimizer is solving the model, the number of iterations and the current value of the objective function are also shown.

If MPL finds a mistake or an error in the model, an error message will be displayed. It shows the erroneous line in the model along with short explanation in plain language of what the problem is. The cursor will automatically be positioned at the error in the model file, with the mistake highlighted.

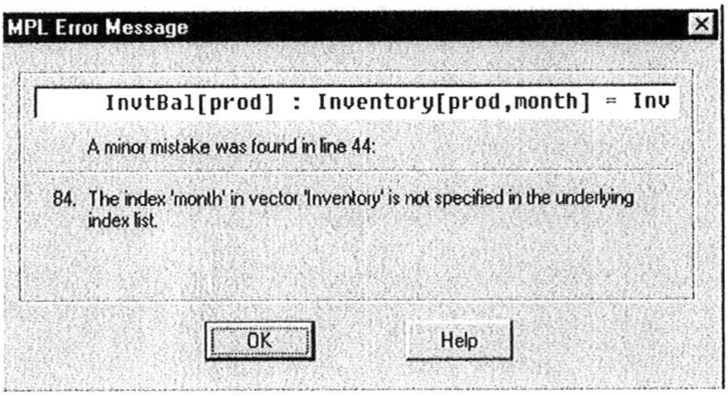

Figure 13.3.2. The MPL Error Message Window

13.3.1.2 View the Solution Results. After the model has been solved MPL will generate a solution file. The model developer would select the *Files <solution filename>* command in the *View* menu to display the solution file in a *View* window.

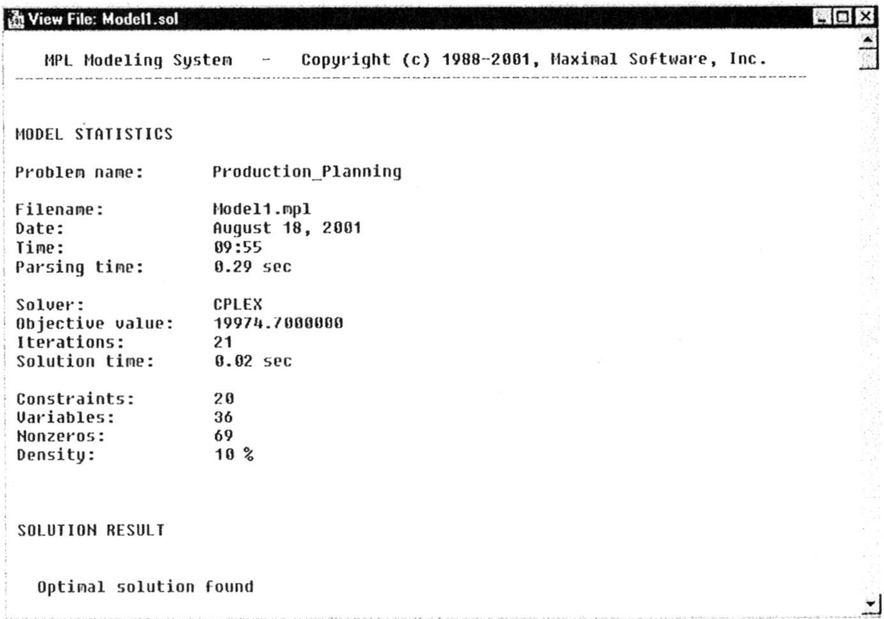

Figure 13.3.3. Viewing the Solution File in a Window

The *View window* stores the file in memory allowing the model developer to quickly browse through the solution on the screen using the scroll bars. The model developer can also use various commands in the *View* menu to view other generated files, the model statistics, and the solution values. The *Model Definitions* window also allows the model developer to view all the defined items in the model in a easy-to-use hierarchical tree structure.

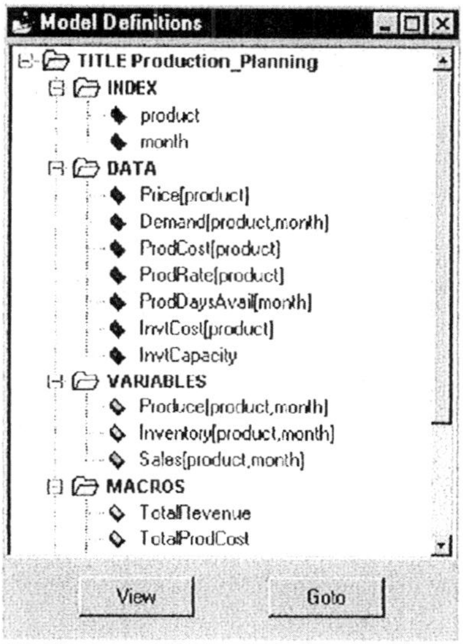

Figure 13.3.4. Model Definition Window

Each branch in the tree corresponds to a section in the model. The model developer can expand and collapse each branch to show only the elements they are interested in or double-lick on any item to get the solution values for that item. The model developer can leave the *Model Definitions* window open for the duration of the `MPL` session since it will be automatically updated every time the model is run.

13.3.1.3 Display Graph of the Matrix. `MPL` also allows the model developer to display a graph of the matrix nonzero elements, by choosing Matrix from the Graph menu.

The positive numbers are shown using blue dots and negative numbers using red dots. If the nonzero is either 1 or -1 the dot will displayed using a little bit brighter color.

The model developer can use the mouse to zoom into the matrix. For example, place the cursor in the upper-left corner of the part of the matrix to zoom in. Hold the left mouse button down and drag the mouse down and to the right. This will draw a box that shows the area that will be zoomed. When the box is the

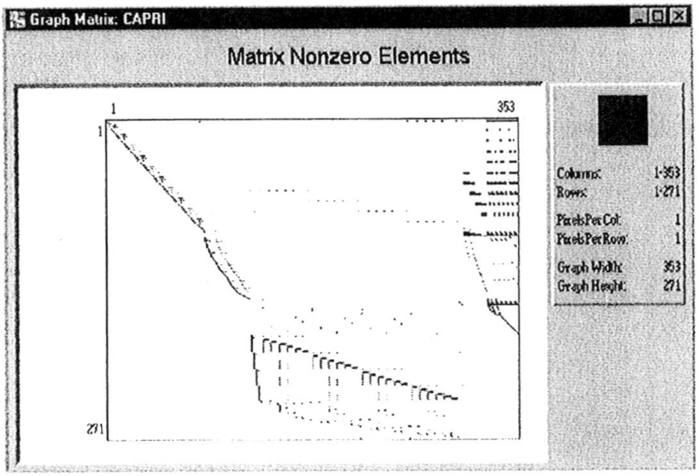

Figure 13.3.5. The Graph of Matrix Window

correct size release the left mouse button to zoom. The zoom can be repeated several times until it reaches the spreadsheet view.

The model developer can also display a graph of the objective function values for each iteration by choosing *Objective Function* from the *Graph* menu. The graph will show the objective function values with a dark red line and the amount of infeasibilities for Phase 1 iterations with a green line.

13.3.1.4 Change Option Settings. MPL offers a range of options dialog boxes in the Options menu that allow the model developer to change various default settings and preferences for MPL as well as option parameters for supported solvers. For example, to display the simplex algorithmic options for CPLEX, *choose* CPLEX *Parameters | Simplex* from the *Options* menu.

13.4 MPL Modeling Language

The MPL modeling language offers a natural algebraic notation that enables the model developer to formulate complex optimization models in a concise, easy-to-read manner. Among modeling languages, MPL is unrivaled in its expressive power, readability, and user-friendliness. The MPL modeling language was designed to be very easy to use with a clear syntax making the process of formulating models in MPL efficient and productive. MPL is a very flexible language and can be used to formulate models in many different areas of optimization ranging from production planning, scheduling, finance, and distribution, to full-scale supply-chain optimization.

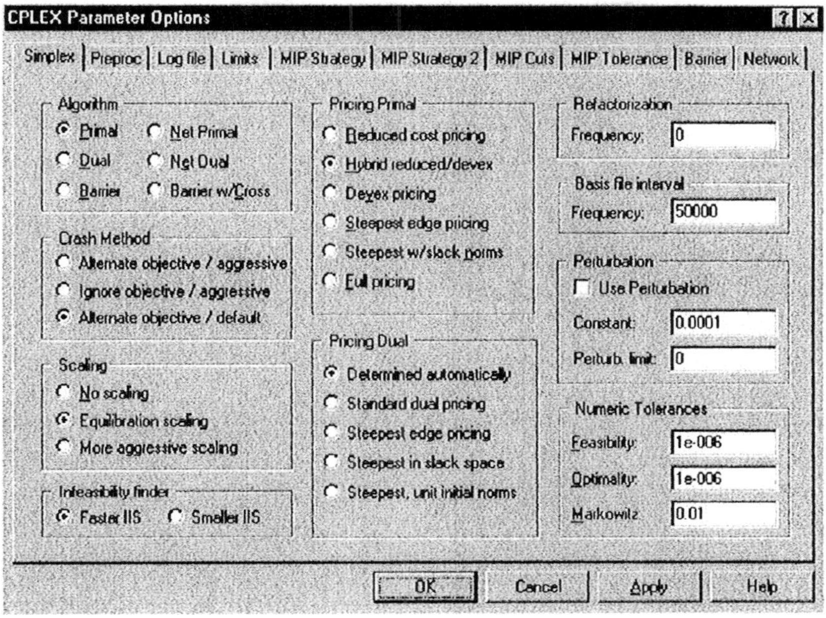

Figure 13.3.6. CPLEX Parameter Options Dialog

MPL is a very robust and stable software whose core modules have been through extensive use and testing over more than a decade. This assures that the MPL software is both reliable and dependable and can be used in mission-critical projects. Some of the more notable features of the MPL language include:

- Separation of the data from the model formulation.
- Import data from different data sources.
- Independence from specific solvers.
- Use of macros for repetitive parts of the model.
- Include files to make handling of large problems easier.
- Exclusion of parts of the model using conditional directives.
- Special Ordered Sets and Semi-continuous variables
- WHERE/IF conditions to handle special cases.
- Readable and helpful error messages.

13.4.1 Sparse Index and Data Handling

One of the most important features of any modeling language is how it handles large amounts of data. What makes MPL so powerful is its ability to effectively handle very large sparse index and data sets. In addition, MPL has extensive flexibility when working with subsets of indexes, functions of indexes, and compound or multi-dimensional index sets. This allows the model formulator, for example, to index only over products that are made by each machine in a specific plant instead of having to go through all the products for all the machines and all the plants, which would be considerable slower.

13.4.2 Scalability and Speed

MPL can easily handle very large matrices with millions of variables and constraints. This is especially important when dealing with large supply-chain optimization models over multiple time periods that can grow very quickly. MPL has its own memory manager that can dynamically store models of any size, giving it a full scalability. The only limitation the model developer faces is how much memory is available on his or her machine. Typically, MPL uses only 1-2MB of memory per 10,000 variables, which puts a minimal additional burden on the machine capacity needed to generate and solve the matrix.

The matrix generation in MPL is extremely fast and efficient which is important since it contributes to the overall time needed to obtain the solution of the model. Maximal has over the years invested significant R&D efforts on continuing to improve the speed of the matrix generation. As a result, we can now run models with millions of variables and generate a matrix for them in less than one minute. This is very important, because if the matrix generation takes too long it can seriously add to the time needed to reach the solution even if the fastest optimization solvers are used. MPL provides the fastest and most scalable matrix generation capabilities available in a modeling system on the market today.

13.4.3 Structure of the MPL Model File

The MPL model file is divided into two main parts; the definition and the model. In the definition part you define various items that are then used throughout the model. The model part, on the other hand, contains the actual model formulation. Each part is further divided into sections, which are as follows:

The Definition Part

```
TITLE           - Names the problem
    INDEX       - The dimension of the problem
    DATA        - Scalars, datavectors and datafiles
    DECISIONR   - Define vector variables
    MACRO       - Reusable macros for expressions
```

The Model Part

```
MODEL           - Description of the problem
    MAX or MIN  - The objective function
    SUBJECT TO  - The constraints
    BOUNDS      - Simple upper and lower bounds
    FREE        - Free variables
```

```
        INTEGER        - Integer variables
        BINARY         - Binary (0/1) variables
    END
```

The objective function and the constraints are required; all other sections are optional. The keywords TITLE, INDEX, DATA, DECISION VARIABLES, MACRO, BOUNDS, FREE, INTEGER, and BINARY are used only when the model contains their respective sections. The keywords MODEL, SUBJECT TO, and END are used to aid readability. As an example of how simple the model in MPL can be, here is one tiny but still perfectly legal model:

```
MAX x + y;
x + 2y < 10;
```

13.4.3.1 Sample Model in MPL: A Production Planning Model.

This section will attempt to convey the flavor of using MPL, by going through the model *planning.mpl*, step by step and explaining the input.

The aggregate production planning problem is how to distribute production capacity between products, and to determine production and inventory levels, given the demand for the products at each month. There are three products and the planning is done for a whole year, or twelve months. Prices, demand, production capacity and cost are fixed, while production, sales and inventory are the decision variables. The problem will be optimized with respect to profit. Here is the actual model file *planning.mpl*:

```
    {     Planning.mpl     }

    {     Aggregate production planning for 12 months      }

TITLE
    Production_Planning ;

INDEX
    product  =  1..3;
    month    =  (January,February,March,April,May,June,July,
                 August,September,October,November,December);
DATA
    Demand[month,product]          :=  1000 DATAFILE(demand.dat);
    ProductionCapacity[product]    :=  1000 (105, 42, 14);
    ProductionCost[product]        :=  (64.30, 188.10, 653.20);
    InventoryCost                  :=  ?

VARIABLES
    Inventory[product,month]       ->  Invt;
    Production[product,month]      ->  Prod;
    Sales[product,month]           ->  Sale;

MACRO
    Revenues  :=  SUM(product,month: price * Sales);
    TotalCost :=  SUM(product,month: InventoryCost * Inventory
                                   + ProductionCost * Production);
MODEL
```

```
    MAX   Profit  =  Revenues - TotalCost;

SUBJECT TO
    InventoryBalance[product,month]  ->  InvB :
    Inventory  =  Inventory[month-1] + Production - Sales;

BOUNDS
    Sales  <  Demand;
    Production  <  ProductionCapacity;
    Inventory[product,month=January..November]  <  90000;
    Inventory[product,December]  =  20000;
END
```

13.4.3.2 Going Through the Model File.

```
    {    Planning.mpl    }

    {    Aggregate production planning for 12 months     }
```

Any text that appears within curly braces is a comment. Comments may be anywhere, even inside a statement. The exclamation mark '!' is an alternative style for comments that extend only from that point to the end of the line.

```
TITLE
    Production_Planning;
```

The title is optional, but a convenient place to name the problem. It is put in the generated input file, when possible.

```
INDEX
    product   =  1..3;
    month     =  (January,February,March,April,May,June,July,
                  August,September,October,November,December);
```

In the index section the dimensions of the problem are declared. Here, we have products number 1, 2 and 3, and the period from January to December. The vector variables and the constraints can be conveniently specified when these dimensions have been given, encapsulating the sizes of the dimensions to this single place of declaration. Notice there are two ways of defining indexes:

1 Subrange of integers as in the case of product, meaning all the integers in the given range – inclusive. Both numbers can be any non-negative integer.

2 Enumeration of items or simply a list of names such as 'month'. This is a convenient way of assigning meaning to indexes and the vectors that use them. Because of name size limitations of most LP solvers, only the first few characters are used in the generated input file (default three). The actual number can be set in the *Options* menu.

The data section defines data vectors and single value scalars.

```
DATA
   price[product]  :=  (105.09, 234.00, 800.00);
```

Here *price* is a data vector over one dimension, product, and as there are three products, list of three numbers are given. The numbers need to be enough to satisfy the sizes of the dimensions of the vector, but additional numbers are allowed and just ignored.

```
  Demand[month,product]  :=  1000 DATAFILE(Demand.dat);
```

Here the *Demand* vector has 36 values, 12 months times 3 products, which are stored in a separate file, *Demand.dat*. A datafile is a free-format textfile, where alphabetic text is ignored and numbers are read in the order they appear. The numbers are separated with commas or spaces and comments are allowed too.

```
   ProductionCapacity[product]  :=  1000 (105, 42, 14);
   ProductionCost[product]      :=  (64.30,188.10,653.20);
   InventoryCost                :=  ?
```

The number 1000 in the first line, means a constant with which all the items in the list are multiplied with. Thus the actual *values* for the vector are (105000, 42000, 14000).

A question mark denotes an interactive value. The user is prompted at runtime to type in a value for the *InventoryCost*.

```
VARIABLES
   Inventory[product,month]     ->  Invt;
   Production[product,month]    ->  Prod;
   Sales[product,month]         ->  Sale;
```

The decision section contains declaration of the subscripted variables of the problem. Here, Inventory, Production and Sales variables all range over products and months, total 36 normal decision variables for each vector variable.

The name that appears after the '->' (read *becomes*) sign is an optional abbreviation of the vector name to combat the variable name size limitations of most LP solvers. They allow you to use long and descriptive names in your MPL model, while maintaining sufficient conciseness of terms needed by your LP package. Same abbreviations can also be given for constraint names and named indexes.

Macros are a convenient and a powerful way of naming and reusing complex expressions and parts of expressions.

```
MACRO
   Revenues  :=  SUM(product,month: price * Sales);
   TotalCost :=  SUM(product,month: InventoryCost * Inventory
                                  + ProductionCost * Production);
```

Two macros have been set up: *Revenues* is the expression for the sum of the revenues of all three products over the whole period. *TotalCost* is the cost of the inventory plus cost of production over all months and all products. Now the model developer can refer to these sums using the macro names in the model. This makes the model more readable and sometimes even smaller, by eliminating unnecessary variables.

Now comes the actual model. The keyword MODEL is optional, denoting the end of declarations and start of the model part.

```
MODEL
    MAX   Profit  =  Revenues - TotalCost;
```

The first section in the model part is the definition of the objective function. It is either maximized MAX or minimized MIN. The function may have a name, in this case *Profit*, and finally the formula for the objective function. In this case, the objective is simple because of the macros used: maximize profit equal to the difference between revenues and the total cost.

```
SUBJECT TO
    InventoryBalance[product,month]  ->  InvB :
    Inventory  =  Inventory[month-1] + Production - Sales;
```

Here are the constraints for this problem. This single equation is expanded to 36 normal constraints, one for each product and month. The intuitive meaning of the constraint is simple: The inventory of a given month (for any product) has to equal the previous month's inventory plus what we have produced minus what we've sold. For the special case of January the constraint is: inventory is equal to production minus sales.

A constraint is specified by:

```
constraint_declaration :  formula  comparison  formula ;
```

A constraint declaration consists of a name, a list of the indexes that it ranges over (unless it is a simple single constraint), and an optional abbreviation for output generation purposes. The constraint *InventoryBalance* which shortens to *InvB*, for each product and each month.

The formulas of the constraint can be combinations of constants, single variables, data vectors, variable vectors and sums, with the restriction that all the indexes of a vector must be accounted for; either as a part of the constraint declaration or as part of sum indexes, or by fixing it to a given value.

Note that dimensions of vectors need not be specified, but are implicit. And even if they are listed, they can be placed in any order, unless it has dimensions with a fixed value, in which case order is important to correctly recognize.

In the term Inventory[*month-1*], the month index has a so-called slide factor, which alters the index of month used. For example, the last constraint would look like:

```
InvB1Dec : Invt1Dec = Invt1Nov + Prod1Dec - Sale1Dec;
```

because of the slide of minus one. In the constraint for product number one, month of December, the November element is therefore used. This method works for any integer slide factor, and entries that fall out of bounds are ignored, thus for certain constraints the variable may fall out.

Simple upper and lower bounds on variables are specified separately, in a manner similar to constraints. They are preceded with the keyword BOUNDS.

```
BOUNDS
    Sales  <  Demand;
    Production  <  ProductionCapacity;
    Inventory[product,month=January..November]  <  90000;
    Inventory[product,December]  =  20000;
```

One side of a bound must be only a normal variable or variable vector, and the other side a scalar constant or data vector. It is possible to specify both lower and upper bound for variable in one single statement in the usual manner $(1 < x < 3)$.

The meaning of the above bounds is very intuitive: they cannot sell more than there is demand for, they cannot produce more than they have capacity for, and they cannot keep more in inventory than our storage space allows. With the last bound the model developer effectively wants to constrain the solution to give an inventory of 20,000 at the end of the year, thus giving the variable a fixed value.

These bounds illustrate a special feature: the use of subindexes. Sometimes a constraint/bound should only be given for a subrange of the originally specified index. Here, the given subrange *month=January... November* constrains the month index to be within this range specified, so the *MaxInventory* bound is not generated for December.

The *CloseInventory* bound is similar and complementary, even as it looks different. Here we have an example of a fixed dimension: *Inventory* has the month dimension fixed to December. Alternatively, this bound could also have been given by:

```
    Inventory[product,month=December] = 20000;
```

The last sections of the model specification involve labeling some variables as special. Decision variables can be marked FREE, INTEGER or BINARY, as a status classification for the LP solver. Here the model developer wants the *Production* variables to be integer variables.

```
    INTEGER
        Production
    END
```

The END keyword is optional and marks the end of the model.

13.4.4 Connecting to Databases

Importing data from a variety of corporate database systems into optimization models is frequently an essential requirement for optimization projects. One of the advanced features of MPL is the database connection option that directly links MPL with relational databases and other data sources. This option enables the model developer to gather both indexes and data values from various data sources and import them directly into the model. After the model has been optimized, the solution output can be exported back into the database. This, along with the run-time features of MPL, allows the model developer to easily create customized end-user applications for optimization using the built-in data entry and reporting capabilities of the database.

The database connection in MPL has the ability to access data from many different sources, such as databases, *Excel* spreadsheets, external data files, and the Internet. This gives the model developer the flexibility to choose the most efficient and convenient way to incorporate the data into the model. Among the data formats that are supported by MPL are: *Microsoft Access* and *Excel*, *ODBC*, *Paradox*, *FoxPro*, *Dbase*, *SQL Server*, and *Oracle*.

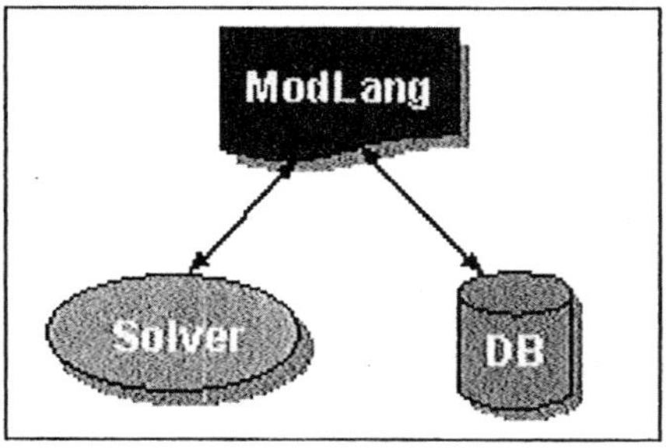

Figure 13.4.7. MPL Database Connection

Relational databases are designed to store and retrieve structured data enabling the user to access corporate data such as plants, products, machines, and etc., on the symbolic level. The purpose of the modeling language is to take this structured data and generate a mathematical matrix from the model that the optimization solver can process.

One alternative to using a modeling language is to write a customized matrix generator in a programming language. This kind of programming is very difficult and causes the resulting application to be highly dependent on specific methods and libraries, of both the database and the solver. MPL offers clear benefits to this approach, because with the database connection it can automatically map the columns of the database tables to the indexes and data vectors of the model without involving any programming. This gives the model developer unprecedented flexibility and expressive power and the ability to focus mainly on formulating the model and connecting the data instead of programming.

To show how simple it is to import data from a database here is an example that reads in a two-dimensional datavector for the shipping cost from factories to the depots.

```
DATA
     FactoryDepotCost[factory,depot] := DATABASE("FactoryDepot","ShipCost");
```

MPL will map the three database columns, *Factory*, *Depot*, and *ShipCost* of the table *FactoryDepot* to the indexes *factory* and *depot* and the datavector *FactoryDepotCost* in the model, and then automatically read in the data.

13.4.5 Reading Data from Text Files

As an alternative to importing data from databases, MPL can also read in data from external text files. Text files are mainly used when the data for the model is stored locally, generated by other programs or by running SQL queries from a database. The data can be read either in a dense format, where all numbers are specified, or in a sparse column format, where only the numbers than have non-zero values are included. MPL can also read single scalar numbers for data constants. The data file can be any free-format text file, which gives the developer the ability to read the data from different sources without first requiring a conversion to a standard format. Reading data from text files is one of the fastest ways available to import data into MPL.

13.4.6 Connecting to Excel Spreadsheets

Optimization in *Microsoft Excel* has over the years become popular because of the built-in solver in *Excel*. Spreadsheet optimization allows users to create models that are easy to use, enabling the user to quickly update the data and solve the model. Spreadsheets are efficient at handling and managing two-dimensional dense data (rows and columns) and single scalar values.

Excel has the *Visual Basic* programming language built in, which enables easy pre- and post-processing of the data. The solution from the optimization can then be used to update the spreadsheet, and to create graphs that visually represent the solution.

But there are some problems with the standard use of the built-in solver in *Excel*. First, there are limitations on the number of rows and columns *Excel* can handle, and it lacks the structured data handling capabilities other applications such as databases have. Secondly, importing data from the spreadsheet into the solver is very slow compared to other faster matrix generation methods. Thirdly, and probably most importantly, the optimization model itself is hidden in the formulas for each cell and has none of the advanced indexing techniques available in modeling languages, such as MPL. Finally, the solver in *Excel* does not scale very well and is much slower than the advanced industrial-strength solvers available on the market today.

The key issue to realize is that while *Excel* is may be not the best modeling tool, it does provide some excellent user-interface capabilities for optimization models. MPL builds on this by offering the ability to import and export data directly from *Excel* ranges. This allows the developer to use an *Excel* spreadsheet for the user interface and data manipulation while using MPL to specify the model formulation. The model can then be solved with any solver supported by MPL with no limits on the size, speed or robustness of the solution.

To give an example of how data is imported from an *Excel* range, here is an MPL statement that is used to read in the pattern cuts generated for a cutting stock model:

```
DATA
    CutsInPattern[patterns, cuts]
        := EXCELRANGE("Cutstock.xls","Patterns");
```

In the above example, MPL will open the *Excel* spreadsheet *Cutstock.xls*, locate the range *Patterns*, and then retrieve the entries for the MPL data vector *CutsInPattern*.

Furthermore, MPL is available as a part of the *OptiMax 2000 Component Library*, which is specially designed to embed optimization models into application programs, such as *Excel*. *OptiMax 2000* allows MPL models to be linked directly with the *Visual Basic for Applications* language in *Excel* enabling the developer to create large-scale optimization models that can be solved directly from an *Excel* spreadsheet.

13.4.7 Optimization Solvers Supported by MPL

MPL works with the world's fastest and most advanced solver optimization engines, such as CPLEX and XPRESS and many other industrial strength solvers. MPL is designed to have an open architecture and is not restricted to only one solver. This enables the model developer to choose the solver that best suits his specific project needs.

A unique feature of MPL is that it links to solvers directly through memory. As no files are involved, this seamless connection is considerably faster and more robust than the traditional use of files in other modeling systems. In the event it is necessary to change any of the algorithmic options for the solver, MPL provides easy-to-use option dialog boxes.

MPL has extensive solver support and offer advanced features such as:

- Direct link to solvers through memory
- Status window with progress information
- Fast and efficient correction of errors
- Automatic infeasibility finder
- Log information and warnings displayed
- Setting of solver options through dialog boxes

MPL can work[1] with most commercial solvers available on the market today, including the following

Solver	Reference	Supported Algorithms
CPLEX, XPRESS	[101]	LP, MILP, MIQP, BARRIER
FortMP/SPInE, OSL		LP, MILP, MIQP, BARRIER, SP
XA, OML		LP, MILP
LINDO, FrontLine		LP, MILP
COIN, GLPK, LPSolve	[160]	LP, MILP
PCx		BARRIER
CONOPT	[60]	LP, NLP
LSGRG2	[151], [152]	LP, NLP
LGO	[188]	NLP (GLOBAL)

CPLEX is one of the most advanced and popular optimization solvers on the market today. It offers a complete solution that contains almost every feature that the model developer would need in an optimization solver. XPRESS from Dash Optimization, is a world-renowned solver, which strength lies in its ability to solve very large optimization problems especially mixed integer. OSL from IBM is also a very strong optimizer that has the ability to solve many different types of models, including, for example, quadratic mixed integer problems.

There are numerous other solvers on the market today that offer different features that are sometimes not supported by the market leaders; FortMP for example from *OptiRisk Systems* offers quadratic (MIQP) and the *Stochastic Programming* (SP) solver, SPInE for continuous linear problems. Sunset Software Technologies offers a reasonably priced, relatively fast solver, XA. Another middle range solver is OML from Ketron Management.

Lindo System's product, LINDO, is popular with academic users. FrontLine is from FrontLine Systems, the same company that provides solvers for Microsoft *Excel*.

COIN is a new solver from the COIN open source initiative (www.coin-or.org) [160]. GLPK is another free open source solver that is part of the gnu software library (www.gnu.org). LPSolve is a free MILP solver that is downloadable from the web. Argonne National Laboratories' solver PCx is also downloadable from the web and has a very good implementation of Newton Barrier method.

MPL currently supports three nonlinear solvers: CONOPT [60], a large-scale solver from *ARKI Consulting* in Denmark that is highly specialized in solv-

[1] *Supported algorithms* in the table means: available in that solver and seen and supported by MPL.

ing difficult nonlinear models; and the LSGRG2 solver made by Leon Lasdon (*cf.*, [151] and [152]) at *Optimal Methods* that is used, for example, by *Excel*. LGO [188] by János Pintér at *Pinter Consultancy Service* computes global optima of nonconvex NLP problems.

13.5 Deployment into Applications

The main purpose of a modeling language is to retrieve data from a structured data source, such as a corporate database, and generate a matrix to the optimization solver. The process of building the model and testing it is often referred to as the development phase of the optimization project. Following the development phase there is typically a second phase that consists of deploying the optimization model into end-user applications. This *deployment phase* is just as important to the success of the project as the development phase.

Phase I - Model Development

- Model Development Environment
- Formulating the Model
- Connecting to External Data
- Solver Testing and Tuning

Phase II - Application Deployment

- End-User Applications
- Run-times / Component Libraries
- Embedding Optimization into Application
- Packaging/Distribution.

To address the need to support application deployment, *Maximal Software* designed the *OptiMax 2000 Component Library* that is specifically created to embed MPL models into end-user applications. This bridges, for the first time, the gap between model developers, whose expertise lies in developing models, and IT professionals, whose focus is working on databases and building graphical user-interfaces and end-user applications.

This section will cover various factors involved in deploying optimization applications and how *OptiMax 2000* can assist the developer in building business applications.

13.5.1 Deployment Phase: Creating End-User Applications

The Deployment Phase involves deploying the model for use by the end-user or the customer. After the model has been built and tested, an end-user application typically needs to be built, using a web-based or graphical user interface, allowing the end-user to edit the model data and view the results in an easy-to-use manner. This is where a component library becomes very useful, allowing highly customized end-user applications to be built, to solve business optimization problems.

Application Programming	Optimization Modeling
Programming Languages (Procedural)	Modeling Languages (declarative)
Application/User-Interface	Optimization Packages
Database Programming	

End-user applications are typically written in a high level programming language such as Visual Basic, Java, and C++. Unlike modeling languages, which are declarative programming languages are inherently procedural. Application programs are written as a sequence of statements, which tells the program exactly in what order to perform the tasks. Most programmers are familiar with various aspects of software development and database programming. On the other hand, most programmers tend to know rather little, and do not have the background, to be effective at formulating optimization models. A programmer that has the skill to write programs using procedural languages can not necessarily translate that skill into being able to formulate models.

MPL was not only designed to help the model developer to create optimization models, but also to make it easy to deploy the model by embedding it into business applications. With the introduction of the *OptiMax 2000* Component Library, MPL models can now be easily embedded into end-user applications including *Excel* and Access from Microsoft Office, using programming languages such as Visual Basic and C++. These end-user applications are invaluable to the model developer when presenting their model data to the end-user. The ability to create graphical front-ends directly in the application enables the model to be presented in a user-friendly environment.

13.5.2 OptiMax 2000 Component Library Application Building Features

The *OptiMax 2000* object library is built as a hierarchy of objects that details all the knowledge that MPL contains about the model and also contains an extensive set of collections, objects, methods and properties, covering all aspects of implementing optimization models in end-user applications. The *OptiMax 2000* Component Library includes the following capabilities:

- General Model Handling
- Model Parsing
- Model Querying
- Modifying Models / Solving Models
- Write Standard Output Files
- Database Import/Export
- Multiple Models/Data Sets/Runs
- Callbacks/User Exits
- Access to MPL Internal Indexing Data Structures

This technology accomplishes the separation of the optimization modeling and the application software development. Experienced programmers are very used to working with Component Libraries and do not need to have any specific knowledge of the model or how it was formulated in order to be effective using *OptiMax 2000*. This enables the application developer to concentrate on building the end-user application without having to be concerned with the details of the model.

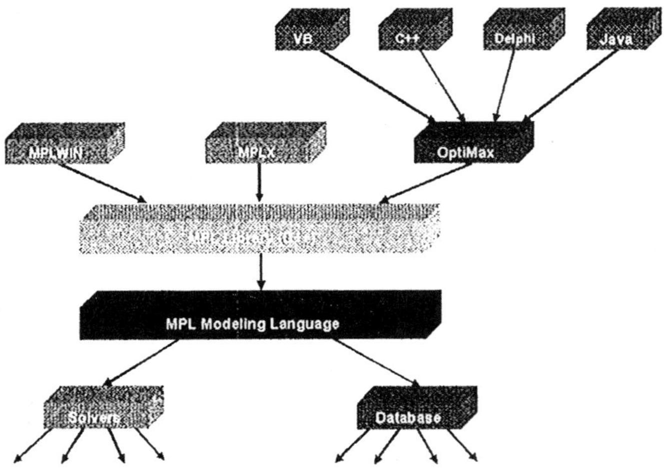

Figure 13.5.8. MPL OptiMax Design Overview

To give an idea how applications using *OptiMax 2000* are laid out here is a short sample in Visual Basic:

```
'Initialize OptiMax Library
Dim MPL As MPLLib.OptiMax
Dim csModel As Model
Set MPL = New OptiMax
```

```
' Change some properties
MPL.WorkingDirectory = "c:\mplwin4"
MPL.Options("UseAdvBasis").Value = False

' Initialize CPLEX solver
MPL.Solvers.Add "CPLEX"
Set csModel = MPL.Models.Add("CutStock")

' Read and solve model
MPL.StatusWindow.Visible = True
csModel.ReadModel("cutstock.mpl")
csModel.Solve

' Display optimal solution result
MsgBox csModel.Solution.ResultString & ":TotalCost = " & _
             csModel.Solution.ObjectValue
```

The following is a list of some of the languages that can be used with the OptiMax library.

- *Visual Basic* from Microsoft is the programming language that is the easiest to work with when creating OptiMax applications. This is due to its high integration with Automation interfaces and easy-to-learn approach.

- *Visual Basic for Applications (VBA)* also comes as a standard with Microsoft Office applications, such as *Excel* and Access, as well as multiple other popular applications from other companies. This version of Visual Basic is called VBA or Visual Basic for Applications and due to its ready availability, is often the most convenient way to create programs using OptiMax.

- *Visual C++* When creating applications with OptiMax where speed is important or full flexibility is needed, Visual C++ is often the language of choice. It is not as quick and easy to program as Visual Basic, but still fairly straightforward to use. In most cases the standard COM interfaces for OptiMax are used when programming with Visual C++, as compared with the Automation interface which is most often used in Visual Basic.

- *Visual J++* The OptiMax library can be used from Visual J++ which is the Microsoft version of the Java language. The Microsoft SDK for Java can be downloaded from the Microsoft web site: www.microsoft.com/java.

- *Delphi* Another popular language that can be used with OptiMax is Delphi from Inprise/Borland. This language is based on PASCAL and can be very effective to create end-user applications.

The OptiMax Object Model

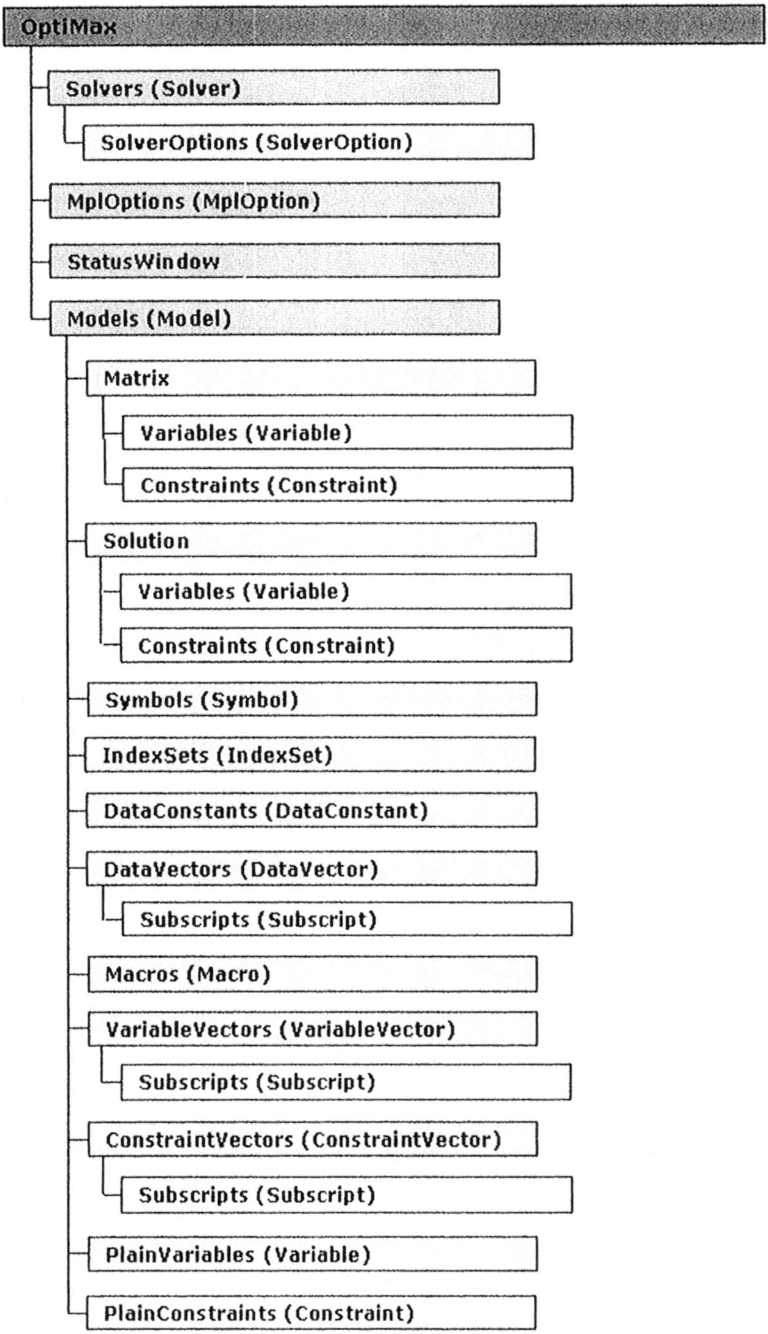

Figure 13.5.9. OptiMax Object Model

Chapter 14

THE OPTIMIZATION SYSTEMS MPSX AND OSL

Kurt Spielberg
Independent Consultant
100 N, 22nd St., Unit 134
Philadelphia, PA 19103
kurtspie@yahoo.com

Abstract We present important information, past and present, about IBM Program Products MPSX and OSL, which have set many standards for mathematical programming and modeling over the years and were crucial for the expansion of the marketplace. We also review some old and new work which may be important in treating difficult (0,1) MIP models.

Keywords: MPSX, Mixed Integer Programming, logical processing, double contraction, preconditioning of models

14.1 Introduction

MPSX and OSL would not normally be viewed as modeling languages. Thus the reader might ask the question why a chapter is devoted to them. While mainly focussing on algorithmic functions, MPSX and OSL have also often provided devices for modeling, such as MGRW (various matrix generation tools) and RPG (report generation tools); and have provided important user interfaces (such as ECL; an *extended control language* built on the structures of PL1) and the successor product OSL has an extensive user interface capability which permits many algorithmic and modeling modifications. Hence they played an important role before the advent of algebraic modeling systems and many companies are now benefiting from standards set by algebraic modeling tools introduced in MPSX and OSL 30 or 15 years ago. Therefore, it may be worthwhile and beneficial for the reader to learn about the following three topics and to put things in proper context:

First, we discuss the history of IBM's MPSX (mostly MPSX/370) from about the mid 1960's to the present. It is probably useful to give a brief history, with some details (on techniques, marketing, customer relations, projects, and some applications and packages) which may still be good to know by the current modeling and user communities.

A short second topic is IBM's OSL which indeed had many important innovations, that can be briefly sketched and may have, via the open literature, found their way in some (possibly modified and extended) form into currently leading solvers such as CPLEX and Xpress-Optimizer.

A third topic arises from the author's decades-long work on MIP techniques (and of course from the work of many others as well). It is becoming increasingly clear, especially from current results and reflections about past work, that the effective solution of difficult problems, especially of optimization problems involving binary variables, still requires both innovative systems and modeling work.

It seems necessary to improve current solvers (as excellent as, indeed, they already are), especially in their user interfaces and in *logical processing* techniques. One of the best ways of achieving progress may be to integrate some *problem enhancement* techniques into the modeling languages rather than into the solvers.

This conclusion has in part been confirmed in two *small business innovation research* projects completed in phase 1, and another proposed two years ago jointly with GAMS Development Corporation. Unfortunately, it was not accepted on the basis of an insufficient business case, but had excellent technical reviews and may well deserve another chance with new additional technical results.

14.2 MPSX from its Origins to the Present

14.2.1 Initial Stages Leading to MPSX/370

During 1963-1965, IBM developed the 7040/7044 LP System II-III to run on IBM's 7040/44 series computers that followed the 7094. This code, also based on LP90/94, contained systems updates and structured optimization extensions such as generalized upper bounding, column generation, Dantzig-Wolfe decomposition, and Alan Manne's discrete production resources allocation model. Bill Orchard-Hays' involvement as a consultant was crucial. He was a superb systems programmer, with intimate knowledge of the simplex method, and an extremely congenial colleague – a true gentleman, adds Grigoriadis, who was a project leader for IBM (and later manager in a Scientific Center).

When IBM developed the revolutionary 360 series of computers in the mid 1960's, various oil companies in particular asked for a powerful LP system and sent specialists to IBM to help insure the code was practical. The 7040/44 LP

System formed the basis for the outstanding effort mounted for the 360 including the latest algorithms, ultimately resulting in the world's best generally available LP solver known as MPS/360 (later MPSX), where MPS stands for *Mathematical Programming System*; the X for *extended* [see, for instance, Tomlin & Welch (1990,[233])]. Haverly implemented `MaGen` on the 360 (later 370) computers and it became used worldwide in conjunction with MPS/360 and MPSX/370.

According to Milt Gutterman, who was an intensive user of LP at Standard Oil of Indiana (now part of BP), the input formats for the MPS/360 (see example on page 19) were identical to those of its predecessors. Eight character names were used instead of six, and bounds and ranges were added. MPS/360 was not such an easy product to maintain. Orchard-Hays, Grigoriadis, and increasingly the IBM-France/Paris group (Gauthier, Hentges, and others) served as consultants to the follow-on IBM software development group for the 360. This group rewrote the code for disks, but again in Assembler Language.

From the late 1960's to about 1978 the IBM Development group was in Paris (France) headed first by Gerard Hentges, who had worked in White Plains, and then Jean-Michel Gauthier; with excellent staff Gerard Ribiere (NLP) and Michel Benichou (MIP). Scientific Marketing at the IBM DP HQ in White Plains had world-wide marketing and managing responsibility for MPS (then MPSX/370); a separately marketed and priced Module MIP was added with many new features, probably influenced somewhat by similar work in England (Martin Beale, John Forrest, John Tomlin). Integer specifications were added to the MPS format. An *extended control language* (ECL) in PL1 was introduced to permit efficient modeling and optimization for difficult problems. A "Separable Programming" facility helped with some nonlinear problems.

It is probably true that for a long period (primarily 1972 to 1985) MPSX/370 was a, if not *the*, leading solver in the field, both in function and penetration, even though of course there were always worthy competitors. Perhaps surprisingly to some, there is still considerable use of it, mostly in big application packages which were presumably developed at great cost to the end user and are still reliably running in various customer locations.

Over these years, many customers have expressed the reaction to us that MPSX is "bullet proof", which was good to hear. Of course, we know that now there are many excellent new products which not only have improved performance, but are also available on a wide variety of platforms (MPSX remained essentially a mainframe product), and occasionally we are approached to help in porting the applications.

At any rate, it seems useful to briefly relate the stages in the MPSX/370 development story technically and in the marketplace. The transition from LPS to MPS/360 was briefly outlined in the historical section; some of the earlier parts in the 1960's have become forgotten over time, even for those of us who were around then, but as to the 1970's we can be more certain.

14.2.2 The Role of the IBM Scientific Centers

MPS/360 was developed (strongly patterned after LPS and LP95) by IBM in White Plains, starting in 1965. Members of the *IBM Scientific Centers*, (such as the author) attended some talks about it in White Plains. The *Philadelphia Scientific Center* had strong expertise in LP and IP, and related subjects, which is often underrated. Most credit is usually given to staff at IBM Research, who were indeed outstanding, but there was practical work at the scientific centers which was of similar importance.

One can name Mike Grigoriadis, John Greenstadt, and Yan Bard as important in NLP, Mike Held as one of the leaders in Lagrangean Optimization (seminal work jointly with Dick Karp of Research), Bill White and Abel Bomberault in LP, Susan Hahn, Jerry Rubin, Lee Papayanopoulos, and Kurt Spielberg in IP and MIP.

Especially in IP, a number of prototype products were made available in the so-called SHARE library mostly in FORTRAN: *e.g.*, DZIP1, DZIPL (enumeration, with and without LP), SPLT1 and SPRN1 (*simple plant location* programs; with and without search restarts; see two papers, *e.g.*, [224]; BBMIP a B&B program, coded by Shareshian (no publication), probably also SCA, a *set covering* code; and some NLP codes.

There were also distinguished visitors: Carlton Lemke, one of the top people in the field, Egon Balas of Carnegie-Mellon, Monique Guignard of the University of Pennsylvania, and members of other IBM centers such as those in Paris and Heidelberg. Interchanges of staff-members between the *New York Scientific Center*, which around 1971 became the *Philadelphia Scientific Center*, and Paris were frequent and influenced the eventual good work done by the French development center with seamless cooperation from White Plains. There were several week-long courses given in London, Paris, and Tokyo. In IP, a basic paper (with some generalizations) on Balas and Benders related enumeration, and introducing *logical processing*, was Lemke & Spielberg (1969,[155]).

An amusing side story in retrospect: when the author was on a three month assignment in Paris, it so happened that the period coincided entirely with the French Revolution of 1968: no public transportation or garbage pickup; overturned cars and a burning bourse (on the way, by foot, from an apartment in Passy to Rue Reaumur, not far from the Hotel de Ville). It limited the work but was in a sense fascinating: a huge parade on the Champs Elysees *for* the government (the opposition paraded around the Bastille) with Georges Pompidou, etc., calling for a return of General de Gaulle was memorable and ended the strike in effect.

At the *Philadelphia Scientific Center*, the author headed a strong Management Science group (which had been led at the *New York Scientific Center* by Mike Grigoriadis). The center was oriented mostly towards the interpretive language

APL (K. Iverson, A. Falkoff). It was natural that there was developed an APL-(M)IP (complete IP, partial MIP) by Spielberg and M. Guignard, with some help from the APL experts, which may have been the first system with *logical processing (probing)*; it gave rise to (much aided by the interactive interpretive nature of APL) ideas for many algorithmic explorations (there were dozens of papers and internal reports), a good survey can be found in [94]; a recent extensive overview of *logical processing* is Johnson & Spielberg (2004,[132]). APL-MIP also used cutting planes, but not as well as others did later.

While the above may sound too algorithmic, we believe that there is strong relevance to modeling, at least in the logical processing work. That indeed it may be best brought to bear on difficult problems in a modeling language; always also assumed that the used solvers are improved to make the subsequent processing easily accessible to the users and efficient. The now prevalent tool of *priorities for branching* is a good example of this. It brings great benefits at little cost to the users who model properly.

14.2.3 An Important Product: Airline Crew Scheduling

One could make a case that some of the work at the centers was potentially as important, professionally and financially, as the MPSX story. We describe the clearly most important instance (developed in cooperation with IBM's Airlines Marketing Group): TPACS.

TPACS may perhaps be the most successful stand-alone practical application ever, fully solved and marketed to industry. It was the first successful treatment, on a large scale, of the *airline crew scheduling* problem, developed from about 1967 to 1972, based on a FORTRAN *set covering* and *packing* code [36] and incorporated in the model, TPACS, developed by Jerrold Rubin ([36],[201]).

In two versions (at $100,000 for the first; at $200,000 the second) it was sold to the nine major US airlines. The code handled problems with several thousand constraints and more than two million binary variables representing crew assignments.

Essentially the same modeling is still used (of course improved in many ways), with vastly superior machines and algorithms replacing the early machines and the set covering code, and handling much larger problems.

The industry is now saving costs on the order of $500 million a year for one carrier (as per a talk held at the University of Pennsylvania a few years ago when the invention of the computer was celebrated, by a manager of American Airlines). Ellis Johnson, formerly Fellow at IBM Research, has been an important consultant to American Airlines over the years; and the work has also expanded, in natural fashion, to other optimization tasks, *fleet scheduling* and *yield management* for example. A current review is [131].

14.2.4 MPSX Management in White Plains and Transition to Paris

There was a major effort, starting in White Plains (*e.g.*, advocated by Joel Cord, with Skip White, a manager) and then ably continued in Paris (Development had been transferred from White Plains to Paris around 1970), towards improvements in LP and especially the development of a first-rate MIP module: compact LU methodology: F-T, SOS type 1, or the use of pseudo costs.

John Tomlin (at the time in England) recalls the above and its excellence, as well as its relations to work at SCICON/BP. Instead of giving references here we suggest that the reader consult [70], a "Note on Important Topics" in a projected volume of AOR on *Progress in IP*, which contains material on the seminal work of Martin Beale and colleagues. There were a number of important papers by the French group, summarized in [13], with references to other papers published as early as 1971.

MPSX Management was in White Plains: the *Data Processing Headquarter* (DPHQ) within a Scientific Marketing Group handling also other products. The management (Requirements, Interface to Development in Paris, Marketing Customer Interfaces at SHARE, through classes and by phone) was first handled capably by H. V. Smith, systems questions by Laurel Slate who joined in 1973 (after the closing of some scientific centers; probably a big mistake) and took over the mission of H. V. Smith when he retired. Laurel Slate who died tragically was replaced by Dick Shareshian (formerly *New York Scientific Center*).

14.2.5 A Major Growth Period in LP and MIP: MPSX/370; 1972-1985

With the new functions in MPS/370 (sketched above), and especially an excellent MIP module (separately priced) this period led to a sharp expansion in usage and revenue.

The marketing group organized regular week-long classes on % MPSX and MIP in the *New York Scientific Center* (normally 3-4 times per year) or various places in the New York area; held mainly by Shareshian, including one day on MIP by the author. There were regular SHARE meetings (twice a year) including an active Mathematical Programming user's group (IBM was represented by the White Plains group; there were also competitors, such as Ketron Inc., regularly, and other companies occasionally. Modeling companies (Haverly Systems and Chesapeake, for example) also attended.

The active give-and-take was mutually beneficial in leading to some project enhancements and joint studies. There were many resulting brochures on joint studies or uses of MPSX at companies; only a few still exist in old files, *e.g.*, "Logistic Modeling System Using IBM's MPSX/370 at International Paper", "MPSX in use at IBM Montpellier Plant".

During 1978 to 1980 the development mission was transferred to Tito Ciriani and colleagues (Rome, Italy), not without a long period of dispute at high management levels. There were improvements in MILP based on new *logical processing* and also *cutting plane* ideas from two influential papers by the authors Johnson, Padberg, Kostreva, and Suhl, enhancing earlier work by Lemke, Spielberg (in 1967-9) and Spielberg, Suhl (in 1978), and an experimental prototype pure integer code PIPEX (Ellis Johnson cooperating with the Rome group).

A highly important matter was that MIP use increased dramatically, from only about 5% to 90% of LP (measured in licenses), over roughly the period 1974-85, aided by much emphasis on it by the author and by applied work of Johnson and others at IBM Research.

Another important contribution was ECL, a new *extended control language* (beyond and instead of the standard MPSCL control language). It was suggested as a requirement and programmed in Paris with PL1. It gave users unprecedented ease in setting up data; input and output, also intermittent results, in standard PL1 data structures which would be convenient for a particular application. There was a special ECL manual with examples, and references: Slate & Spielberg (1978,[221]), and Spielberg (1980,[225]). We discuss some of its uses below.

14.2.6 MPSX as an Engine in Research and Applications

There was much applied modeling work. We were directly involved in three cases. In case A, a customer contacted us, usually because of poor results with an initial model and asked for advice on how to do better. A particularly interesting case A came up during a class in the *New York Scientific Center*.

14.2.6.1 CASE A: Algorithmic Tools for Solving Difficult Models.

A FRITO-LAY (subsidiary of Pepsi Cola) model of the "Production Allocation and Distribution" type (having also a truck scheduling flavor) could not find any solution with MIP after 20 hours of CPU time. Hearing the description and looking briefly at the MPSDATA file it appeared there was a possibility of *disaggregation* (known already from the sixties and seventies for plant location type problems). It was clear there, that an *aggregated* constraint on shipments from a source i, say to destinations j, of the form $\sum_{j=1}^{J} x_{ij} \leq My_i$, M sufficiently large, y_i a binary variable which would include or exclude plant i with the values 1 or 0 could be *very much weaker* than its replacement by as many as J constraints of the form $x_{ij} \leq X_{ij}^{+} y_i$, where X_{ij}^{+} represents an upper bound on x_{ij}. The second model with what we called *disaggregated constraints* was much more likely to produce integral or near-integral values for the y_i (*cf.*, Spielberg (1969,[224]).

After suggesting this as a possible improvement, I also contacted Ellis Johnson, who was flying on business to Dallas, to visit the customers. He managed to find one more item, the replacement of some constraints by a simple expression.

The end result was highly gratifying: not only could one now find a good solution in a few minutes (the LP solution already came close to an integer solution), but one was also able to solve much larger problems in reasonable time; see, for instance, Mairs *et al.* (1978, [167]). We believe this may have been the first application of the concept *disaggregation* to a large-scale practical problem. This also raises one of the issues we shall discuss in the short section on OSL: should we not design modeling systems which help the user to apply simple known concepts semi-automatically? One can not expect the user (often a novice) to be familiar with certain concepts which the specialist usually knows, they should be aided with the modeling systems!

14.2.6.2 CASE B: New Solver Programs with ECL. In case B we look to some new solver feature and consider whether it might, in our judgment, be useful to tackle some problems better than before. We illustrate such an approach by two examples related to the new feature ECL (discussed above).

MIP Improvement (Employing Logical Processing: Probing) implemented efficiently via the use of ECL and MPSX LP. In the mid 1970's the author arranged with IBM Germany a cooperation (such as had been undertaken several times before in the Scientific Centers). IBM Germany would finance the employment (for a year or two) of a promising young employee or student at an IBM Scientific Center. A successful earlier instance had been the stay of Franz Schober of DPHQ in Stuttgart at the *Philadelphia Scientific Center* (1972-73). His work in my Management Science group was immensely successful. He implemented a full econometric system in APL (EPLAN) which turned out to be highly successful. It was eventually employed by Nobel Laureate Lawrence Klein of the University of Pennsylvania in his project of tying together some of the top official banks: Deutsche Bank, World Bank, etc., into a network operation for which EPLAN turned out to be a perfect tool. Franz was also eventually one of the few users, in Germany, of APL-MIP, both at IBM and later as a Professor at the University of Freiburg.

The DPHQ White Plains did not provide an office for such an assignment so that it was necessary to ask the Director of *Mathematical Sciences at IBM Research* for one. That was to have many drawbacks, but the invitation of Uwe Suhl from the Freie Universität Berlin was still a major success. He had the right background and skills to implement two large-scale versions of (0,1) IP solvers (ENUM and MIPEX), patterned in part on ideas in APL-IP, but using MPSX-370-LP via ECL highly efficiently.

It led to an important IBM research report [226] which was essentially accepted by MP (very good reports and just minor changes suggested); but as IBM Research had by that time engaged Uwe Suhl on work with research staff, the revised report was not finished; a big mistake.

However, some of the ideas then became part of two outstanding later papers by Johnson, Padberg, Kostreva, Suhl, see [226] for the relevant references (Crowder, Johnson, Padberg) and (Johnson, Kostreva, Suhl), which were to be highly influential for MIP.

An Application of ECL to Distribution. DPSS, a *multi-plant multi-layer multi-commodity distribution system*, was designed and implemented with a Ph.D. Student of Prof. M. Guignard during a summer stay at White Plains (with similar office problems). It used FORTRAN and ECL, as well as MPSX-MIP/370, with a high degree of success. The final system could also be counted under CASE C below, *i.e.*, as an "application package", but was not released as such and only used for demonstrations, *e.g.*, at SHARE meetings and presented at a yearly German symposium in Passau, [93].

A few years later, we recommended Siwhan Kim to a small company (whose owner had earlier participated with us in a joint study involving Uwe Suhl, and hired him as head of Research). One of his many successes at "Bender Management" was a similar package which was to play a part in making the company a big success (it was later taken over by a larger company).

14.2.6.3 CASE C: Application Packages – Precursors to Modeling.

In case C we focus on the subject of Application Packages presented in Business Cases to top management. From varied customer contacts and participation in some advisory functions, we were involved in many projects which were usually unique, but suggested to us the possibility of expanding the market and generating revenue via application packages for certain problem classes. Indeed, IBM had for a while two types of such offerings: IUP's and FDP's (Installed User and Field Developed Programs).

We describe just a few of such attempts at a technical level (for various reasons these were only moderately successful; probably management did not provide enough resources for marketing). Yet all of these efforts were in some sense precursors of what was later done more widely and generally after the advent of good general modeling languages (as described in this book). We cite a few examples (there are many more):

1 VSP, a *vendor selection program* based on MPSX, was built by a group located at Poughkeepsie, with the aid of the Bender Management Group, a small company headed by Paul Bender. They cited multi-million dollar savings obtained from minimizing the costs to IBM over a set of vendor proposals for purchase of computer components. We made an effort to turn this into a generic product, but did not obtain support for it.

2 DPSS described above (and the later "Phydias" of Bender Management).

3 FAS "*Funds Allocation System*" (principal author Abraham Serfaty)[1].

14.2.7 Business Cases for MPSX

Partly of their own volition, partly in response to management requests, the author and then also Ellis Johnson at Research constructed analyses of the MPSX market, and projections of future growth, given certain actions.

A main item was the analysis of given data and contacts with the IBM Branch Offices as to what computer utilization was caused by LP and, especially MIP usage. This was fairly intricate and involved a number of assumptions which could influence results. Our end results tended to show substantial hardware usage with each application; we estimated that overall machine usage expenditures would amount to 30 to 60 times the software costs. With such factors estimated (attained by what we thought were reasonable assumptions) we could project excellent business potentials over the years, even with only modest investments into the optimization products. Management told us that hardware drag estimates could not be used in business cases (not allowable under a consent degree signed by IBM), but we felt that they were giving us at any rate valid insights, purely professionally, as to the usefulness of our work in IP and Optimization Software.

A moderate modeling effort was a system named `MPDEMO` (a "Demonstration Guide Marketing Documentation": "Mathematical Programming Sample Problems and Solutions", with a marketing brochure), which contained 10 problems and described the relations among the needed FILES and EXECS and the uses of MGRW and MPSRG to go from raw data to MPSDATA and solutions with desired outputs. There was also a similar *network* system, and later a number of larger proto-products (such as on *lotsizing*).

[1]Financial applications became quite fashionable in the seventies and eighties. The author gave invited talks on uses of MIP in Financial Applications. This led directly to the installation of a large IBM computer devoted to such work at a big Wall Street firm.

Later there was a large symposium held in Westchester with broad participation of Financial Experts; they were being called *quants* in the financial literature and had substantial influence on trading around the period which included the 1987 market crash. (whether beneficial or not is not so clear). In consequence of the above, IBM formed a little optimization group situated in Wall Street (manager: Julie Landstein).

FAS used MPSX for asset allocation, debt reduction and so on. The principal architect of FAS was Mr. Abraham Serfaty; Dick Shareshian helped with the FORTRAN interfaces to MPSX. Mr. Serfaty later had several important contracts with foreign governments (in South America and elsewhere) to help them re-schedule their debts.

We had high hopes for FAS which were not fully realized, in spite of many seminars held at financial centers and at IBM throughout the US. A main obstacle seemed to be that the Finance Industry Marketing representatives were not, with some exceptions, able to explain the (financial and mathematical) technical aspects involved. This work continued with and without the Wall Street Optimization group. Mr. Serfaty (who has vast experience, and was associated with the Federal Reserve System in the past), revised FAS to be suitable for *debt restructuring* of countries such as Venezuela and Mexico, via a *debt management* and *restructuring system*. A recent reference is [219], a paper on the subject in *The Handbook of Fixed Income Technology*. LP (and possibly MIP), is still a major tool.

14.2.8 Changes in Computing, Development and Marketing Groups

As always, life changes. In 1978-80, due entirely to IBM Europe's desires to place more emphasis on activities in Italy, Development was transferred from Paris to Rome. In the mid 1980's, Technical Marketing decided to move the MPSX function to Dallas, Texas.

The first transition was contested, on the basis that it might be detrimental to abandon a highly experienced group in Paris. Eventually, top management agreed to let the Paris group have a five year period for developing new features of their choosing. It was a big disappointment to many of us that the group members had in the meantime taken on other jobs and declined the offer. It is hard to guess what good could have come out of such a period. In retrospect it might have made the inevitable transition to an entire new hardware market (PC's and workstations) much smoother and more productive for IBM.

There was eventually good work done in Italy (on a technical level, the cooperation with Stefano Gliozzi and Giorgio Carnevale was always productive; both for MPSX and OSL) , but we did not manage to achieve an easy transition to the new hardware market, in spite of good things, for example, the modeling system `GIANO` developed in Rome, which promised some of the features that `GAMS` and other modeling languages were providing (it was available in the US, probably only briefly in the 1990's for experimental use, as: `EASY MODELER`).

14.2.9 Transition to OSL

Around 1987, a general scientific marketing directorate had good success with new *vector features* and *parallel processing* for mainframe machines. In particular the engineering scientific computing group had obtained fine results from a system called *Engineering Scientific Subroutine Library*.

The head of the *Los Angeles Scientific Center* asked me to convene jointly with him a meeting of IBM Research staff and Marketing to discuss Vector and Parallel computation in Optimization. This meeting was held and, with John Forrest, a new member of IBM Research, important progress was made in taking advantage of these computational modes.

In consequence the scope of inquiry was extended to discuss a replacement of MPSX by OSL (an *Optimization Subroutine Library*). This was most likely a good decision, but unfortunately management did not decide to retain the valuable Trademark name MPSX in some form (even though, OSL finally developed more into a new product than into an Optimization Subroutine Library). Fortunately the OSL User Interfaces did eventually constitute a highly valuable consequence of the original subroutine library concept. Marketing also decided to more or less market OSL as a replacement of the ("old") MPSX, which may have led to a quicker erosion of MPSX than would have been necessary, without quite attaining for OSL the dominant position that MPSX had held for so long. Nevertheless we should not forget the major achievements of OSL:

- OSL overcame the 32,000 constraints size limitations of MPSX/370. Even today, OSL is being used on huge problems arising from internal IBM needs, when often hundreds of millions of dollars can be saved by optimization on problems with many millions of constraints. The LP presolve can often reduce such huge models to problems with only tens of thousands of constraints. Similarly, for MIP presolve; *e.g.*, a big classroom scheduling problem was solved entirely in the preprocessing phase! (by Dwight Calkins, Dallas Marketing group).
- New features of OSL permitted the imbedding of an efficient network code (Mike Grigoriadis), a stochastic module (Alan King), and an efficient quadratic and NLP code (David Jensen).
- A *New Open Source Initiative* (COIN)has become successful for the implementation of complicated algorithms involving cutting planes and so on. Moreover, the codes are not limited to OSL, but can be general (*cf.* [150] and [95]).
- OSL User Interface work provides a modeling environment for MIP involving binary variables[2].

[2]The author has in recent years made substantial use of the OSL User Interfaces for innovative work on improved algorithms for MIP involving binary variables: Short starts are generated for easy known solutions and the search is then pursued further (more effectively) starting from the last short start. For certain very difficult models this has been shown to lead to improvement factors of roughly 10- to 20- fold. OSL is particularly suited for the generation of the short starts; for the continued search CPLEX (and probably other solvers) can also be used.

There are some especially difficult models for which further work is needed, such as the generation of B&B Implications (special cuts) which are best obtained by careful runs from specially constructed short starts (depth-first B&B with a special partitioning of the variables). This work needs to be pursued further on a larger variety of difficult problems (and via semi-automatic implementation). This is probably best done in the framework of modeling systems such as GAMS (we have had substantial cooperation with the staff of GAMS: Alex Meeraus and Michael Bussieck), and hope to carry this further. A good reference is Guignard & Spielberg (2003,[95]). An additional point of interest discussed in that paper is that industrial solvers should probably be enhanced further to make the above procedures particularly effective (essentially by improving their Logical Processing capabilities).

Chapter 15

THE NOP-2 MODELING LANGUAGE

Hermann Schichl*
Institut für Mathematik der Universität Wien
Strudlhofgasse 4, A-1090 Wien, Austria
Hermann.Schichl@esi.ac.at

Arnold Neumaier
Institut für Mathematik der Universität Wien
Strudlhofgasse 4, A-1090 Wien, Austria
Arnold.Neumaier@univie.ac.at
http://www.mat.univie.ac.at/~neum

Abstract We present a short overview over the modeling language `NOP-2` for specifying general optimization problems, including constrained local or global nonlinear programs and constrained single and multistage stochastic programs. The proposed language is specifically designed to represent the internal (separable and repetitive) structure of the problem.

Keywords: NOP-2,modeling language, global optimization, NOP-2

15.1 Introduction

Solving global optimization problems effectively in a rigorous way requires a lot of analytical knowledge about the functions involved. The speed of a B&B approach is, *e.g.*, directly correlated to the amount of overestimation produced by interval enclosures of linear, quadratic, or convex relaxations.

*supported by EU project COCONUT IST-2000-26063

The origin of the NOP-2 modeling language does not lie in the development of modeling languages but in global optimization. The NOP input format was constructed in 1997 by Arnold Neumaier mainly as an interface to the global optimization package GLOPT [182, 50], a global solver for nonlinear partial separable global NLPs. NOP provided a good means for entering specifications in a compact line oriented syntax; it had the possibility to mix element functions and mathematical notation but it did not have the strength of a true modeling language, it did not provide named variables, extensibility of element functions, or matrices. The first parser was written in FORTRAN 77, as was the GLOPT solver.

In the following year 1998, GLOPT turned out to be too inflexible, so the system was re-designed, and for the implementation it was decided to switch from FORTRAN to C. Since a new input interface was required anyway, NOP-2 was designed as input language for GLOPT-2, and it would be a convenient modeling language, at least for mathematicians. The new parser was written in Bos (an enhanced version of Ox), a kind of preprocessor language for Lex and Yacc.

At the time NOP-2 was developed, its element concept (see Section 15.3.1) seemed to be one of the most promising approaches to global optimization, but nowadays more flexible problem representations have been developed (*e.g.* directed acyclic graphs [206]).

The development of GLOPT-2 has been stopped in the meantime, and the authors work in the COCONUT [46] project towards a new solver platform for global optimization. This new platform uses AMPL [76] (see Chapter 7) as its primary input format, but a GAMS [32] (see Chapter 8) interface will be available as well.

These facts have greatly reduced the need for the NOP-2 language, and so its *development is stopped*, and it is *not maintained* any longer.

However, the language still contains a few unique concepts, and we hope that the providers of the commercial modeling languages can be persuaded to include some, if not all, of them into their systems.

15.2 Concepts

NOP-2 was primarily designed for modeling *global optimization* problems, and to be a general purpose modeling language which contains the possibility to split a problem specification into elementary functions, *explicitly display* its *block separable structure* and which has similar flexibility and extensibility as the other modeling languages presented in this volume (see Chapter 4).

Most solvers which are connected to modeling systems like AMPL or GAMS are local optimization codes who do not need analytical knowledge for a fast solving process. Extracting analytical information needed for global optimization (*e.g.*, range enclosures with minimal overestimation, almost optimal linear enclosures of functions, convexity areas, etc.) from flat models which are typical for algebraic modeling languages cannot easily be done without human intervention or an enormous amount of computer algebra. Most rigorous global optimization packages, however, make use of this kind of information.

In NOP-2 a model has to be specified in a highly structured way; the emphasis is on the block-separability of the model structure. The element function approach, the extensibility of NOP-2, and a library of element functions make it possible to specify optimization problems in such a way that global optimization packages can be provided with as much analytical information as possible, ee Subsection 15.3.1. However, the modeler needs a decent knowledge in mathematics and a will to twist the model, performing mathematical transformations, until it nicely fits with the provided elements.

Another important design parameter was the *minimality of modeling overhead.* The modeling language SIF [48] (the input language of Lancelot, and a nonlinear extension of MPS) was a negative example.

Furthermore, *stochastic programs and multistage models* should be representable, and the language should be easily expandable. Finally, as outlined in Chapter 4, Section 4.2, for global optimization and verified computing it is extremely important that all levels of rigor for data are representable.

Still, the main interest in developing NOP-2 was to provide a convenient input format for our solver. The design was an extension of the original NOP idea and was not developed as a new modeling language, and the wuthors were not specialists in modeling language design. Therefore, some features generally available in algebraic modeling languages are missing from NOP-2.

We did *not* care about *set indexing*, so variables can only be *indexed by integers* not by sets. This does not reduce the number of models which can be specified, but it is a major difference on how models are developed.

We did *not* think about *connecting* many *different solvers* to NOP-2, since it was primarily intended as an input language to GLOPT-2. There was no API written for solver connection. In addition, we did not care about passing data back from the solver to the modeling system; we could control the output of the solver anyway.

There was *no clear separation between models and data.* Data had to be specified in NOP-2 syntax, and the input statement was the only chance to separate data and models somewhat. No elaborate data reading was implemented, no database connections, no spreadsheet interface, etc.

Finally, there is (almost) no automatic generation of derived data, especially *no automatic differentiation.* Since the amount of analytical information in global optimization is huge, we kept the generation of this information, hence the automatic differentiation, in the solver.

By the way, there is absolutely no procedural part (not even a `solve` statement) in `NOP-2`. In some sense, it is a very pure declarative language.

15.3 Specialties of NOP-2

This section is devoted to presenting special features unique to `NOP-2`. Instead of giving details on how models in general are designed in our language, modeling the Rosenbrock [130] function

$$\begin{aligned} \min\ & 100(x_1^2 - x_2)^2 + (x_1 - 1)^2 \\ \text{s.t.}\ & x_1, x_2 \in [-2, 8]. \end{aligned} \tag{15.3.1}$$

shall serve as a illustrative example.

In (15.3.1), the objective function is a polynomial in two variables of degree 4. For multivariate higher degree polynomials, direct calculation of good interval range enclosures and quadratic relaxations are in general a non-trivial task. We can improve on that in this case, making the least squares structure apparent by introducing the variable

$$x_3 = 10x_1^2 - 10x_2, \tag{15.3.2}$$

thereby reducing the objective function to

$$x_4 = x_3^2 + (x_1 - 1)^2. \tag{15.3.3}$$

This shows that we can rewrite the optimization problem equivalently as

$$\begin{aligned} \min\ & x_4 \\ \text{s.t.}\ & x_4 = x_3^2 + (x_1 - 1)^2 \\ & x_3 = 10x_1^2 - 10x_2 \\ & x_1, x_2 \in [-2, 8]. \end{aligned} \tag{15.3.4}$$

In this form, we have increased the dimension by one (adding x_3) but we have reduced the objective function and constraints to quadratic functions. Quadratic relaxation and good interval enclosures have become easy; we can even use inverse functions for improving on the solution process.

If we now count the total number of variables and remember the bounds, we end up with the following `NOP-2` file. (The lines starting with // are comments added only to make the file more readable.)

```
// Rosenbrock function
min x[4];
bnd x[1 2] in [-2,8];
```

```
// element list
qu4 x[1 2]; 0 10 -10 0 = x[3];
qu2 x[3 1]; 0 1 = x[4];
```

Here the line `min x[4]` tells the solver that the variable x_4 should be minimized, the line starting with `bnd` fixes the bounds on the variables x_1 and x_2, and the remaining lines give the constraints.

`qu2` and `qu4` are so called element functions (see 15.3.1). Many of them are predefined and specially selected such that most of the analytic information and estimates can be computed efficiently for them. Although it is possible to define new element functions, a modeler is advised to perform mathematical transformations until the model consists of predefined element functions only. This new structure will enable the global optimization code to use most of the built-in analytical knowledge.

15.3.1 Specifying Structure — The Element Concept

The most special element in `NOP-2` is the very explicit way of specifying the structure of the problem. Every model must be decomposed into elementary functions. These *element functions* are then entered line by line.

In general, `NOP-2` allows to describe an optimization problem of one of three basic forms

1 A classical global optimization problem

$$\begin{aligned} \min\ & \omega x_s \\ \text{s.t.}\ & E_\nu(x), \quad \nu = 1, \ldots, N \\ & x \in [x_0], \end{aligned} \tag{15.3.5}$$

possibly with additional integrality constraints. The bound constraints $x \in [x_0]$ define componentwise restrictions $\underline{x}_0 \leq x \leq \overline{x}_0$, and may contain $\pm\infty$ as bounds to allow for one-sided bounds and free variables.

2 A single stage stochastic optimization problem

$$\begin{aligned} \min\ & f(x, \xi) \\ \text{s.t.}\ & E_\nu(x, \xi), \quad \nu = 1, \ldots, N, \\ & x \in [x_0], \\ & \xi \sim g(b, x), \end{aligned} \tag{15.3.6}$$

as in 1. possibly with additional integrality constraints. The variables ξ are stochastic variables with (probably unknown or partially unknown) distribution functions $g(b, x)$.

3 Most general, a multistage optimization problems of the form

$$
\begin{aligned}
\min \quad & f(x^{(k)}, \xi)^{(k)} \\
\text{s.t.} \quad & E_\nu^{(k)}(\xi^{(k)}, \xi^{(<k)}, x^{(<k)}), \quad \nu = 1, \dots, N^{(k)}, \\
& x^{(k)} \in [x_0^{(k)}], \\
& \xi^{(k)} \sim g^{(k)}(b^{(k)}, x),
\end{aligned}
\tag{15.3.7}
$$

as in 1. and 2. possibly with additional integrality constraints. The variables $\xi^{(k)}$ are stochastic variables with (probably unknown or partially unknown) distribution functions $g^{(k)}(b^{(k)}, x)$, and are valid in stage k of this multistage problem.

The so-called *elements* $E_\nu(x)$ are expressions of one of the forms

$$\bigodot_k f(a_k, x_{J_k}) \in b\mathbf{q} + c, \tag{15.3.8}$$

$$\bigodot_k f(a_k, x_{J_k}) = b x_{K_j} + c, \tag{15.3.9}$$

$$\bigodot_k f(a_k, x_{J_k}) \in S, \tag{15.3.10}$$

and a few irregular variants, that consist of only one operand, allowing simple coding of Boolean expressions, polynomials, trigonometric polynomials, and a limited form of branching.

Here f is a so-called *element function*, a_k, b, c are parameters, parameter vectors, parameter matrices, higher order tensors, or lists of such. x_{J_k}, and x_{K_j} are subvectors of x indexed by the index lists J_k and K_j, $\mathbf{q}$ is a possibly unbounded box, possibly restricted to integers, and S is a union of finite sets and possibly unbounded boxes. $\odot$ specifies one of the following operators: $\sum$, $\prod$, max, min, $\sum_k (-1)^k$. The contributions $f(a, b_k, x_{J_k})$ are referred to as the *pieces* of the element. (Elements containing a single piece only are, of course, permitted.)

The element functions that we found most useful in coding a large number of problems are listed in Tables 15.1 and Table 15.2, and are among others collected in a standard library, that is by default included into every `NOP-2` specification. For all these functions it is possible to get a complete analytic overview over ranges and inverse ranges, which makes these elements suitable for applications in a branch and bound framework. Other element functions can easily be defined using algebraic statements in a syntax similar to most algebraic modeling languages.

name	element function	element shape
`sum`	x	$\sum x_i$
`lin`	px	$\sum p_i x_i$
`abs`	$\|x\|$	$\sum \|x_i\|$
`abs2`	$p\|x\|$	$\sum p_i \|x_i\|$
`sqr`	x^2	$\sum x_i^2$
`qu1`	px^2	$\sum p_i x_i^2$
`qu2`	$(x-p)^2$	$\sum (x_i - p_i)^2$
`qu3`	$p_2(x-p_1)^2$	$\sum p_{i+m}(x_i - p_i)^2$
`qu4`	$p_1 x + p_2 x^2$	$\sum (p_i x_i + p_{i+m} x_i^2)$
`pow`	x^p	$\sum x_i^p$
`log`	$\log \|x-p\|$	$\sum \log \|x_i - p_i\|$
`xlog`	$x \log x$	$\sum x_i \log x_i$
`exp`	$p_1 e^{-p_2 x}$	$\sum p_i e^{-p_{i+m} x_i}$
`gss`	$p_3 e^{-\frac{p_2}{2}(x-p_1)^2}$	$p_{i+2m} e^{-\frac{p_{i+m}}{2}(x-p_i)^2}$
`atan`	$\arctan(x)$	$\sum \arctan(x_i)$
`pr0`	$x_1 x_2$	$\sum x_{2i-1} x_{2i}$
`pr1`	$x_1 x_2$	$\sum x_i x_{i+m}$
`pr2`	$(x_1 - p_1)(x_2 - p_2)$	$\sum (x_i - p_i)(x_{i+m} - p_{i+m})$
`div`	x_1 / x_2	$\sum x_{2i-1} / x_{2i}$
`bil`	$x_1 x_2 + p_1 x_1 + p_2 x_2$	$\sum (x_i x_{i+m} + p_i x_i + p_{i+m} x_{i+m})$
`qf1`	$p x_1 x_2$	$\sum p_i x_i x_{i+m}$
`qf2`	$p_1 x_1^2 + p_2 x_1 x_2 + p_3 x_2^2$	$\sum (p_i x_i^2 + p_{i+m} x_i x_{i+m} + p_{i+2m} x_{i+m}^2)$
`dsq`	$(x_1 - x_2)^2$	$\sum (x_i - x_{i+m})^2$
`atan2`	$\pm\arctan(x_1/x_2)$ (sign as in C, FORTRAN)	$\sum \pm\arctan(x_{2i-1}/x_{2i})$
`pol`		$\sum_{i=1}^n p_i {x_j}^{n+1-i}$
`sin`		$\sum_{i=1}^n a_i \sin(i\omega x_j - p_i)$
`cos`		$\sum_{i=1}^n a_i \cos(i\omega x_j - p_i)$
`if1`		if $x_i \le p$ then x_j else x_k
`if2`		if $x_i \ge p$ then x_j else x_k
`if3`		$\begin{cases} x_j & \text{if } x_i \le p, \\ x_k & \text{if } p < x_i \le q, \\ x_l & \text{if } x_i > q \end{cases}$
`prod`	x	$\prod_{i=1}^{n} x_{j_i}$
`min0`	x	$\min_{i=1}^n x_{j_i}$
`min1`	$\|x\|$	$\max_{i=1}^n \|x_{j_i}\|$
`min2`	$p\|x\|$	$\max_{i=1}^n p_i \|x_{j_i}\|$
`max0`	x	$\max_{i=1}^n x_{j_i}$
`max1`	$\|x\|$	$\max_{i=1}^n \|x_{j_i}\|$
`max2`	$p\|x\|$	$\max_{i=1}^n p_i \|x_{j_i}\|$

Table 15.1. Predefined elements with one dimensional result.

name	element shape	short description
mv	Av	left multiplication of vector with matrix
vm	$v^T A$	right multiplication of vector with matrix
vmv	$v^T Av$	evaluation of the quadratic for specified by A
lookup		interpolate in a specified way a function given by a table
trace	$\mathrm{tr}(A)$	the trace of the matrix A
det	$\det(A)$	the determinant of the matrix A
cond	$\mathrm{cond}(A)$	the condition number of the matrix A

Table 15.2. Elements having higher dimensional results or involving matrices and tensors.

15.3.2 Data and Numbers

Data for global optimization is a tricky thing. We have seen in Chapter 1, Section 1.2.8 and in Chapter 4, Section 4.2 that various levels of rigor are interesting when data has to be specified. There are several possibilities, and the modeling language has to act according to the type of solver the model is sent to. Basically, for GLOPT-2 one could distinguish between a rigorous mode and an approximating mode. In NOP-2 data can be entered in various ways.

Data can contain an *approximate number*, *e.g.*,

```
const A = sin(4/3*pi);
```

in rigorous mode constructs a small interval enclosing the number, and in non-rigorous mode takes an approximate floating point value (rounded towards nearest).

Sometimes, data can be an approximate number, even in rigorous mode, but if it must be rounded, the *rounding direction* is important:

```
const A = sin(4/3*pi)<+>; round towards +∞,
const A = sin(4/3*pi)<->; round towards −∞,
const A = sin(4/3*pi)<0>; round towards 0 .
```

Especially important for mathematical proofs is that some numbers are not tampered with. *Exact data* is entered as follows:

```
const A = sin(4/3*pi)!;
```

If the data is not exactly representable as a floating point number, then do one of the following depending on the mode of operation: issue a warning, construct a small interval containing the true number, or pass the value as a literal string to the solver.

In some cases, a number is really approximate

const A = 3.56647?

It is represented it by a rounded floating point number in approximate mode, and converted it to an interval. The radius of that interval should be $\frac{1}{2}$ in the first unspecified digit or $0.5ulp$, whichever is bigger.

For many practically relevant optimization problems, data cannot be specified exactly. So in `NOP-2` it is possible to specify uncertain data in three widely used ways:

```
const B = [ 3.23245, 3.23268 ];
const C = 1.3345 +- 0.0013;
const D = 1.334562(34);
```

Here B is in interval notation, C is defined using center and radius, and the declaration of D, in this case $D = [1.334528, 1.334596]$, is using the engineering notation,

Of course, data items can be infinite as well:

`const I = inf;` or `const MI = -inf;`

and numbers can be unknown

`const U = ?;`

Finally, `NOP-2` allows several data types: integer, real, complex integer, complex, interval, and complex interval (disk representation).

15.3.3 Sets and Lists

Sets in `NOP-2` are completely different from sets in most other modeling languages; some of the set-theoretic constructions can be performed in `OPL`, see Chapter 17, as well.

In `NOP-2` sets represent possible domains of variables, constraints, and indices (integers). After specification, a range set is transformed to a normal form, a union of closed boxes, and an integer set is converted into a list of integer tuples.

In addition to sets, *lists* can be produced by analogous constructors. The only difference is that sets never contain duplicate entries, but lists do.

There are several ways for constructing sets in `NOP-2`. The easiest way is by giving explicit intervals

`[-3, 4.5],`

which is possible for reals and for integers.

Ranges follow their intuitive meaning. The value in the middle specifies the increment; if it is missing the increment is 1.

`1:2:12`, `0.5:0.2:3.4`, and `1:3`

will produce the sets $\{1, 3, 5, 7, 9, 11\}$, $\{0.5, 0.7, 0.9, \ldots, 3.3\}$, and $\{1, 2, 3\}$, respectively.

Everybody with experience in using numerical software (*e.g.* MATLAB) will know that roundoff problems sometimes make increments unreliable, especially if a range has to be dissected into a given number of pieces of equal sizes. So in NOP-2 there is a constructor for that:

```
0.1~10~0.9
```

constructs the set $\{0.1, 0.18, 0.26, 0.34, \ldots, 0.82, 0.9\}$ of 11 points representing 10 subintervals of (approximately) equal size.

Repeaters just produce lists of specified length.

```
1**27
```

produces a list containing 27 entries of 1.

Of course, sets can be *enumerated explicitly*

```
{ 1, 2, 3, 7, 8.2, 9, 3*e^4 },
```

but it is also possible to use *simple descriptions* like

```
{ 3+4*i | i = 1:3:22 },
```

which are very close to mathematical notation. For complicated sets more *complex description* syntax is available:

```
{{ int k,int l,real r | 4*l*m^k | k=1:n, l=1:m,
                                   r=0.5:0.03:0.8 }}.
```

All the lists can be constructed sorted or unsorted, and there are the usual set operations *union* (|), *intersection* (&), *set difference* (\), *set power* (^), and *cartesian product* (*). Finally, there is the membership relation in:

```
A in ([3,4] | [1,2])*(0~4~1)^2.
```

15.3.4 Matrices and Tensors

Since many mathematical models contain matrices, we thought it natural to include matrix and tensor operations in NOP-2.

Vectors are constructed automatically from lists or from sparse definitions, *e.g.*,

```
v = ( 97 $ 1.4 $ 3.7@1,3,9 4.8@44 );
```

which defines a vector of dimension 97. Almost all components are 1.4, except there is a 3.7 in v_1, v_3, and v_9, and a 4.8 in v_{44}.

There are many ways to specify *matrices* and *tensors* (arrangements of numbers with more than two indices). You have the choice between several shapes:

dense, sparse, banded, symmetric (or hermitian), antisymmetric, unsymmetric, and triangular. The matrices can be entered using MATLAB notation, row index notation, or as a stream of numbers.

Many operators for vectors, matrices, and tensors are defined, which can be used in the models:

- Sums for vectors (x, y), matrices (A, B), and tensors (S, T) of the same shape:

  `A+B,   T-S,   x+y.`

- Products of vectors (x, y), and matrices (A, B), if the shapes are compatible

  `A.x,   A.B,   x.y.`

- Tensor products:

  `A(*)B`

  if A and B are matrices, this definition gives the tensor $A \otimes B$ with 4 indices.

- Indexing is very similar to MATLAB, the : stands for the whole range of the given index. If there are more than two indices, the notation : (list) can be used to take the whole range for all indices specified in the list, *e.g.*,

  `A[1,3],   A[:,4],   T[:,3,:,5],   S[:(1:3),6].`

- As in MATLAB, a whole submatrix can be assigned:

  `A[1:3,:] = B[6:8,:].`

- Finally, additional useful matrix operators such as

  `A', det(A), tr(A), cond(A,2),`

  namely the transpose, determinant, trace, and condition numbers may be used; and there are matrix constraints like

  `A is psd,`

  which codes a positive semidefiniteness requirement on A.

15.3.5 Stochastic and Multistage Programming

In NOP-2 stochastic programs can be specified just like ordinary optimization problems. They just involve some stochastic variables and special constraints.

Stochastic variables and their distribution functions can be specified using the following syntax:

```
stoch real xi[5];    stoch int yi[3];
```

defines vectors of 5 continuous and 3 discrete stochastic variables. For these variables distribution information can be added:

```
distr xi[1] ~ N(0,1);
```

makes ξ_1 normally distributed with mean 0 and variance 1, while

```
distr xi[2] ~ N(3.44,?);
```

specifies that ξ_2 is normally distributed with mean 3.44 and unknown variance. It is even possible to specify covariance matrices, such as

```
distr xi[3:5] ~ covar(A);
```

where the matrix A would have to be defined somewhere else in the model.

Discrete variables work just like continuous ones:

```
distr yi[1] ~ binom(12,0.3);
```

defines a binomial distribution, while

```
distr yi[2] ~ discrete((1,0.1),(2,0.3),(3,0.6));
```

declares an arbitrary discrete distribution with values and associated probabilities, and some or all of the probabilities might even be unknown:

```
distr yi[2] ~ discrete((4,?),(5,?),(9,?));
```

In order to provide the capability to model stochastic programming NOP-2 allows the use of stochastic operators in constraints such as

`Expect(xi[3])`, `Prob(xi[2])`, or `Variance(yi[1]*xi[4])`.

Coding multistage problems is done by assigning a stage to each variable.

```
stage x = 3;
```

If no stage is assigned, the variable is implicitly supposed to be in stage 0. Stages are propagated through all expressions, and an expression belongs to the maximal stage of all variables occurring in it. Note that allowed stagenumbers are all integers, so variables could be declared to be in *e.g.* stage -4.

15.3.6 Recursive Modeling and Other Components

Every NOP-2 model containing free variables can be used as a function within another model. This is done via the `call` operator, as in

```
call "subprob.nop"(x, 3.0, 1.2);
```

supposing that the subproblem has 3 free variables. Here, the first variable is kept and x is passed there, and the two other variables are fixed at 3.0 and 1.2, respectively. The return values to the `call` operator are in the subproblem specified by the `solution` record:

```
solution x[1:30];
```

In principle, even recursive modeling is possible, if `call` is combined with *conditionals* and *loops* in the model.

In addition to the presented features NOP-2 provides an interface to *black-box functions*, and *integrals*.

Finally, bounds and constraints can be defined using arbitrary sets, *e.g.*,

```
bnd x[1 3 19] in {0} | [3.1,19.223];
bnd x[2 4] in {i^2 | i=0:6 } | [36,112];
```

could be used to specify semi-continuous and partially integer variables.

15.4 Conclusion

NOP-2 is a modeling language specifically designed for global optimization. It is between an advanced input language and a modern modeling language. There are many features not readily available in other modeling languages. On the other hand, many features contained in most algebraic modeling languages are missing. The element concept for capturing the structure is an interesting idea. However, nowadays newer representations for global optimization problems are available, making the element approach less appealing.

The language is unsupported now, and its development and use have ceased. In spite of that, it contains many useful features, and we hope that some, if not all, of them will in the near future be included in some commercial modeling systems. Especially important would be rigorous data handling and matrix support.

Chapter 16

THE OMNI MODELING SYSTEM

Larry Haverly
Haverly Systems, Inc.
12 Hinchman Avenue
P.O. Box 1110
Denville, NJ 07834
larryh@haverly.com, http://www.haverly.com

Abstract The OMNI system is presented from four different approaches. The first is a discussion of the OMNI features as they developed historically. This gives insight into the practical reasons behind key features.

The second is a discussion of the development cycle of large applications and how OMNI evolved to meet the needs. The third shows an example while the fourth gives more language details.

Keywords: OMNI, modeling, behavior of LP models, refinery planning & scheduling

16.1 OMNI Features as they Developed Historically

16.1.1 Early History

The origins of `OMNI` go back to a practical need. It was early days for Linear Programming (LP) around 1960. We were doing an LP model for the Esso Everett refinery (which no longer exists). The experts could read the computer output from the IBM7094 but not the refinery people. It was proving too slow to produce a written report every time we did a computer run. So Bob Koehl decided we had to have a computer program to turn LP output into a readable report automatically. Your author worked on the design and it was implemented in FORTRAN and met an early practical need.

By 1963, the author had left the Operations Research group at Esso and formed Haverly Systems Inc. He developed at first an LP solver for the IBM 1401 computer. The first major enhancement was a new design for an automated report writer. A user set up the commands once and then it executed for each run. It was an interpretive program but served the early needs very well.

Esso purchased the LP system on the 1401 and then provided encouragement that I try to create a model generator - or as we called it in those days – a matrix generator. Practical consideration said that it had to generate both the input to an LP and then be a report writer to turn the output into a user-friendly report. By then, the author had a lot of experience building LP models by hand and setting up the reports so it was an exciting challenge to find a creative practical approach that would really help. It was clear that models had a lot of logically definable structure but also had exceptions to the general structure that had to be recognized. Models grew and shrunk as crudes, products and specifications were added or removed. Also it was clear that both matrix generation and reporting used a lot of common information and logic.

For a few months nothing seemed to come together and there were many false starts. But one day while sailing and not thinking about the matter there came an Ah-Ha inspiration. I guess it was like Kekulé in 1865, who tells how he suddenly saw that a C_6H_6 molecule could be organized into the benzene ring structure we know today.

It was clear that data and commands needed to be separate and that there was a need for classes to drive the commands. Thus the entities of TABLEs and CLASSes were born and became key design elements, which have endured. It was recognized that the FORM word felt right for the commands. The earliest implementation had FORM ROW, FORM VECTOR, FORM LINE, FORM TABLE and COPY. These are still used today. The first two generate the matrix in a file to be input into the optimizer. The third does the reports, the fourth some important data manipulation and the last is a way to insert headers. The members of classes then drove the execution of the FORMs. The FORMs utilize data directly from tables or use equations to create the coefficients. It was clear that there had to be modifiers to handle exceptions. Thus, WHEN, EXCEPT and differentiating between a blank and a numeric entry (blank and 0 being different) fell into place. Also, the need for new variables to be created and later used was recognized. These were all in the earliest implementation of what was called `MaGen` and have proved sound so they are in today's `OMNI` along with further extensions.

I had done a lot of programming in FORTRAN and had struggled with its limitations. To make `MaGen` a really easy tool, I decided to deal with tables and their columns and rows by name. This was unlike the FORTRAN of those days where you had arrays of specified size and you referenced values by row and column numbers. Also in FORTRAN you had to declare sizes of arrays and, if you ran short of memory (which you often did in those days) you had to program to move arrays to tapes and back in when you needed to reference them. `MaGen` was to treat data by name and as virtual. If you added a column or row to a table `MaGen` would automatically expand the table. Since rows and columns were referenced by name, you could have them in any convenient order - or change the order later. If you ran out of memory, MaGen would write the table out and then bring it back at later references, *i.e.*, it managed data storage by name and without any user action. MaGen anticipated Relational Data Base concepts, virtual memory and memory management features that now are common.

The approach was implemented. It proved to be practical, met a need and soon was in use at numerous sites. MaGen was the world's first general purpose LP model builder and included integrated matrix generation and report writing, data separation and many other useful features.

Then came the feedback from our own use and other users doing real applications. Mike Chipalone's complains about entering all those zeros in the tables to avoid the blank exception led to the K control to indicate that in a particular command a blank reference was to be treated as though it was zero. Today the K-type control has grown into a whole collection to have zeros treated as blanks, to control spacing when indexing across lines, etc.

A user at a steel company explained how he used the card duplicator to create multiple executions of groups of FORMS – particularly to treat table names as variable. We provided the FORM SECTION to accomplish this directly. Some people said they wanted more options for classes than just defining them in one place. Thus the FOR class was created for locally defining the list of members. Also the value of defining classes as the rows or the columns of a table was noted and implemented. The history of OMNI is one where users have suggested many enhancements and we have generalized and implemented them while adding our own ideas.

MaGen kept growing in capabilities and was implemented on various computers including the Honeywell 200 series, the IBM 1401, the IBM 7094 and then the IBM 360. We were always able to keep the system upward compatible as we added new options.

A brilliant summer student and later full time employee with Computer Science background, Steve Halliburton came to work for us. He asked if he could do a compiler version of MaGen and apply some new ideas he had. We had a contract from Univac for an LP system with matrix and report generation. Halliburton led the implementation of the generators. It was MaGen compatible to the user but internally was an optimizing compiler. Using a compiler means that code was generated in a way to execute very fast. The product was called PDS-MaGen. It was only slightly used but proved to be a major step forward in technology. Many new features were added to the system. The design was then implemented on the CDC 6000 line where it was used for President Ford's Project Independence and later became widely used around the world.

Our major numbers of users were on the IBM 360 line. We had linked MaGen to the widely used IBM LP solver, MPS (later MPSX). FORM BOUNDS was added to the language to meet new LP options. We then made the commitment to do a new implementation in assembly language based on the compiler design experience, the new features pioneered in PDS-MaGen and more extensions, but maintaining upward compatibility for existing user models. Several years of effort was rewarded by a strong new product with enhanced capabilities and extremely fast in execution. The product was named OMNI. LP applications were growing rapidly in size and complexity so the new OMNI was timely. The upward compatibility meant users could easily switch over and get the advantages of the much improved internals.

A new generation of computers was on the horizon. These were the PC's. At first they didn't seem capable of handling the complexities of the LP calculations and the real, large, complex applications. But as PC's rapidly evolved and particularly when they went to the 32 bit word size, it was clear they would become important in the LP world.

We undertook a new implementation to provide an assembly version OMNI on the PC's using our optimizing compiler approach. Martha Venard, Frank Esposito, Steve Ilg, with some key help from Bill Golush, worked for 2 years and delivered the new system exactly on time. It was delivered to users who could simply move their IBM 360 OMNI models to a PC without having to modify the models. Today, almost all of our users are on PC's. The largest, most complex applications run surprisingly well.

16.1.2 Activities Versus Equations

OMNI, and later GAMMA, were both oriented toward specifying a model by activities and the constraints on these activities. Most other modeling languages specify model by equations. Either approach can be used so it is worth considering which is best.

It may be worth noting that Dr. George Dantzig has said that he achieved insight to invent the LP simplex algorithm by looking at the problem as activities to be optimized using an objective function. I personally found the activity view very insightful. I had learned LP as equations and followed the details OK but I struggled to really understand modeling. One day someone pointed out that a model was just activities with constraints and objective values. It was a light turning on for me. Since then, the activity view has always been my preferred, and most productive, approach to modeling. I have since seen the light happen to others. The activity view sometimes appears in print as in Saul Gass's book, "An Illustrated Guide to Linear Programming" . Therefore, I used the activity approach in MaGen and OMNI and continue to be enthused about it. We did add some options to OMNI for the equation approach (FORM EQUATION and FORM NLE – nonlinear equations) but they have rarely been selected by our users.

I realize there is a large body of people who believe in the equation approach. LP is normally taught as equations and these are compact for teaching and documentation. Algorithm developers and optimization code implementors need to work with matrix algebra. But for modeling, I find the activity approach much better for linear, recursive (successive LP), and MIP models. I rarely do other types of nonlinear modeling so perhaps it would not work as well there.

It may be a matter of background. Our user base has always included many people with engineering background and practical problems to solve. Our models have largely been process type models such as refineries, mines, forest products, and so forth. We find that our users relate to the activity approach.

When I have had constructive arguments over the years with those favoring the equation approach it seems that most came from mathematical or academic background. Perhaps for them the equation approach works best and they understand modeling best that way. An example of the latter may be that Johannes Bisschop and Alexander Meeraus used `MaGen` for awhile and then decided they wanted to have similar capabilities based on the equation approach and developed `GAMS`.

I think that using the activity view has led me to be more interested in modeling, as distinct from algorithms. I think "Behavior of Models" is a worthy area of study. This focuses on the building of models which are stable and give sound practical results in contrast to just being mathematically correct. An example is when a model has variable ending inventories but, if not handled well, can change in different runs from maximum to minimum inventories with tiny (noise level) changes in some objective or other values. The changes in values need not even be on the inventory vectors themselves. This instability is not helpful to users relying on results to make decisions.

16.1.3 Recent and Current Trends

`OMNI` in assembly language on the PC is now our major offering. Our users are focused on Microsoft Windows and so operating system support is directed to the many versions of Windows. We have tied `OMNI` to Xpress, CPLEX and our HS/LP as the primary optimizers. We have made enhancements as warranted. The major recent enhancement was to extend all names to 16 characters and all arithmetic to double precision to handle the future demands. The system is very robust and extremely fast.

The major change we have witnessed is the change in usage. At one time many companies did their own LP model application development. Today, many look to packaged, generalized modeling systems. That is now a large part of our market. We provide the GRTMPS planning system for oil refineries. GRTMPS is written in `OMNI` but the `OMNI` code does not change from refinery to refinery. GRTMPS takes advantage of the data-driven capabilities of `OMNI`. Only the data tables change from refinery to refinery. The generalized model will set up a model for a single refinery or for a group of refineries. It will set up a single time period or multi-period model. The model can have any number of crude oils, products, qualities, process units, and so forth depending solely on how data is entered. Corresponding reports are generated automatically. The model can include MIP or recursion handling of non-linearities and the logic of recursion is handled in the `OMNI` language. Recursion is also known as successive LP. A useful reference on modeling and model behavior is [99].

Another trend we have observed is the expansion of `OMNI` to applications other than LP. The chief of these has been in scheduling and in simulation. It has been found that `OMNI` already had the capabilities to be a strong language for developing refinery scheduling and refinery simulation applications. The use of the FORM VALUE and FORM TABLE and functions such as cardinality, ordinality, MIN and MAX, SORT, full arithmetic and WHEN/EXCEPT are heavily used in these applications.

A major extension was required for scheduling. This was graphics for input and display. The graphics included dialog boxes for data input, GANTT charts for activities and tanks, time graph lines for changes in volumes, qualities and composition over time and process flow diagrams. We decided to implement the drivers for the graphics in `OMNI` tables and the program to do the graphics in Visual C++. An extensive set of features and variations are included in the graphics. The graphics are largely set up in `OMNI` tables so this becomes a generalized package as our markets demand. It integrates with `OMNI` to pass all the information back and forth electronically.

Our scheduling applications often utilize MIP for blending. Thus a scheduling run may call on MIP from time to time but it was easy to integrate using `OMNI`. The scheduling logic is in `OMNI`. If the logic calls for an MIP run then it sets up the matrix, transfers control to the optimizer, accepts back the MIP results, uses them and continues the scheduling process. In some applications, there may be many automatic runs of MIP optimization for one scheduling run.

Another interesting development is the creation of an object-oriented front end to GRTMPS. In keeping with market trends, this front end reduces the knowledge required to make GRTMPS runs. The front end essentially is a behind-the-scene, alternate way to make entries into the `OMNI` table. `OMNI` is a key component but it is less visible to the user. Our package development capabilities are greatly enhanced because we have the powerful `OMNI` as a tool.

Another trend is to imbed `OMNI` in a larger application. Therefore we have enhanced the interfacing of `OMNI` to readily connect to other programs, systems and databases. This is done both on the input and output sides.

We see the future of `OMNI` as being used for some individual LP model development by skilled developers. We see it being used by developers of application packages. We see `OMNI` as being used as the core of application packages which are then used by people with less LP background. We foresee more use of `OMNI` in simulation and scheduling applications.

16.2 Omni Features to Meet Applications Needs

OMNI has always been focused on meeting as many of the needs of providing an optimization application as could be achieved. We observe a cycle in practical model applications which include the following:

1. Recognition and definition of a modeling opportunity and obtaining authorization,
2. Formulating a suitable model structure,
3. Collection of data for the model,
4. Construction of the application system,
5. Installation of the application,
6. Use of the application for decisions, and
7. Maintenance of the application.

A skilled Operations Research person must be available to recognize and define a modeling opportunity. He might be invited by a manager who has a problem or sees the possibility for improvement. But optimization has strengths and weaknesses that must be weighted to determine if it is worthwhile to do a project. OMNI may be of some help to the expert if he wants to quickly try a prototype on the possible model. Then the OMNI prototype can easily have some reports added and used in explaining the proposal to get authorization to proceed from management.

Once the project has begun, it is important to devise the best structure for the model. This requires developing a deep understanding of the problem to be the solved and the answers that will be expected. The data that can be obtained when the application is running is a part of the consideration. My experience is that there are often many possible model structures. Since OMNI is a higher level language it is generally fast to implement a basic model structure. Approximate data can be used to run the model. The results may suggest that one or more alternate model structures should be considered. This is easy to do in OMNI.

Now one needs to concentrate on actual data. Some data will be critical while other make only a small difference in the results and perhaps no difference in the decisions that will be made. At this stage, data can be entered by hand into the OMNI data tables. The model can be run and the behavior of the model examined. This may lead to changes in the model structure. The runs should give a good idea as to which of the data will have the most impact on the decisions. Some reports may be created at this stage as a help to the expert to more quickly analyze the results of a run. OMNI is suitable for this.

The next stage is to construct the application system. You have a model structure. You have knowledge of the data that will be available. You know the objectives of the application and what questions will be asked and what decisions will be considered. You now need to think of how the final application system will be set up and used. With OMNI you may decide that one or more spreadsheets is the best input for the data. Some data may be entered manually into the spreadsheet. Some might be loaded into the spreadsheet from other sources. You need to consider what the reports will look like. They must be in a form useful to the users and in terms he understands. `OMNI` has a lot of power to produce reports. Sample reports can be created and discussed with the user and then included in the final application without wasting effort. Most aspects of creating an application can be done with `OMNI` and a suitable optimizer.

Installation of the application is not only the physical side but also training of the user. If you have involved him in the various steps and your implementation is sound, it should go well. But it is not unusual to have some items come up that impact the model structure or reports. Fortunately `OMNI` generally will make it possible to adjust the structure or reports, without a major crisis.

The whole point of the application is to provide someone with a tool to improve his decision making. Thus it is important that the expert insure that all the prior steps have resulted in a proper tool and that the user knows how to use it and is aware of its limitations. Only when better decisions are being made can the expert feel he has done his job well.

All applications will require maintenance. `OMNI` is well suited for this aspect. It leads to data driven models so changes in the `OMNI` code should be rare. Changes might be required if there is a significant change in the plant or way of operating or in the way the application is being used. Since the original expert may not always be available or his memory may have lost some details, it is strongly urged that use be made of the many documenting capabilities in `OMNI`. It already shows data in an organized way, The FORM statements are generally understandable. But liberal use of comments in the data tables and in the code are highly recommended and made easy to do in `OMNI`.

In the early days of LP applications, we thought of having a data book to document what was in the LP model. In practice, the data book rarely got set up and those that did were never kept up to date. Therefore there was an incentive in the original `MaGen` to eliminate the need for a data book. The language was to be self-documenting, the data organized in readable tables and liberal use of comments was encouraged. This remains true today. In addition, we now recommend that reports include key data items directly. Some data is passed through from input to reports, without ever being directly involved in the model, where this will help the user.

16.3 OMNI Example

The general approach to use of `OMNI` will be illustrated by a simple two-period, two-product, blending example, in which two grades of gasoline are to be blended from components subject to starting inventories of products and components, demands of products to be met and receipts of component in the two time periods, storage capacities to be observed and blending limits.

First we set up tables for the data. They can be organized in many possible convenient formats. The following show the structures of the tables that were used in this example:

```
TABLE PRODUCTS Demand and Price by time periods
          >     DEM1 DEM2 PRICE1 PRICE2
   REGULAR REG
   PREMIUM PRE
*
TABLE COMP Component data. Rundown rates per period. Costs. Qualities
          > RD1 RD2 COST1 COST2 OCT ...
   BUTANE C4S
   ...
*
TABLE INVCAP starting inventories and storage capacities for products and
*                         components
             INV CAP
   P.REG
   P.PRE
   C.C4S
   ...
*
TABLE ALLOW Logic control. 1 if component allowed in product blend, else blank..
            REG PRE
   C4S       1   1
   ...
*
TABLE BLENDER capacity of blender by time periods
          1  2
   BCAP
*
TABLE SPECS Specifications for product by time periods
         >       TYPE REG1 REG2 PRE1 PRE2
   OCTANE OCT 'N'
   ...
*
```

To make the model entirely data driven in this example, we generate the classes and their members from columns or rows of specific tables. The first FORM CLASS below creates a class T with members 1 and 2. This class is saved and form statements can then reference class T by use of (T) to get members 1 and 2.

If later you want a third time period you will get it by adding a column 3 to table BLENDER and data to other tables.

```
FORM CLASS T Time Periods
    (Y)      FOR Y=TABLE BLENDER (#,)
FORM CLASS PRD List of Products
    (YYY)    FOR YYY=TABLE PRODUCTS (,#)
FORM CLASS CMP List of Components
    (YYY)    FOR YYY=TABLE COMP (,#)
FORM CLASS QUA List of Qualities
    (YYY)    FOR YYY=TABLE SPECS (,#)
*
FORM ROWS
    OBJ = OBJ !defines the objective row as named OBJ
```

```
    (T)BL(PRD) = FIX !generates a blend balance row for each product-time period
    (T)Q(QUA) = MIN WHEN(TABLE SPECS(TYPE*1-1,(QUA))='N')
    (T)Q(QUA) = MAX WHEN(TABLE SPECS(TYPE*1-1,(QUA))='X')
    (T)D(PRD) = FIX !generates fixed rows 1DREG,1DPRE,2DREG,2DPRE
    (T)R(CMP) = FIX
    (T)BP(PRD) = FIX
    (T)BC(CMP) = FIX
    (T)CAPB = MAX
    (T)CAPP(PRD) = MAX
    (T)CAPC(CMP) = MAX
*
* for each time period create vector for each component to
each product if allowed
FORM VECTOR (T)(CMP)(PRD) WHEN TABLE ALLOW ((PRD),(CMP))
    OBJ = TABLE COMP (COST(T),(CMP)
    (T)BL(PRD) = -1
    (T)BC(CMP) = 1
    (T)Q(QUA) = TALE COMP ((QUA),(CMP))
*
* for each time period create vector for each product being blended
FORM VECTOR (T)BL(PRD)
    (T)BL(PRD) = 1
    (T)BP(PRD) = -1
    (T)Q(PRD)(QUA) = - TABLE SPECS ((PRD)(T),(QUA))
    (T)CAPB = 1
*
* for each time period create a vector for the run down of each component to
* storage activity
FORM VECTOR (T)REC(CMP)
    OBJ = TABLE COMP (COST(T),(CMP))
    (T)BC(CMP) = -1
    (T)R(CMP) = 1
*
* for each time period create a vector for each product delivery from
* storage activity
FORM VECTOR (T)DEL(PRD)
    OBJ = -TABLE PRODUCTS (PRICE(T),(PRD))
    (T)D(PRD) = 1
    (T)BP(PRD) = 1
*
* for each time period show final inventory of components and transfer to
* the next period except last
FORM VECTOR (T)FIC(CMP)
    (T)BC(CMP) = 1
*\                              !$=ordinality,#=cardinality
    (Y/T)BC(CMP) = -1 EXCEPT ($T=#T) FOR $Y=$T+1
    (T)CAPC(PRD) = 1
*
* for each time period show final inventory of products and transfer to
* the next period except last
FORM VECTOR (T)FIP(PRD)
    (T)BP(PRD) = 1
    (Y/T)BC(CMP) = -1 EXCEPT ($T=#T) FOR $Y=$T+1
    (T)CAPP(PRD) = 1
*
* do the right hand side values
FORM VECTOR RHS
```

```
1BC(CMP) = TABLE INVCAP (INV,C.(CMP))
1BP(PRD) = TABLE INVCAP (INV,P.(PRD))
(T)CAPP(PRD) = TABLE INVCAP (CAP,P.(PRD))
(T)CAPC(PRD) = TABLE INVCAP (CAP,P.(CMP))
(T)CAPB = TABLE BLENDER ((T),BCAP)
```

It can be seen that there are six types of activities in the model structure and these are represented by six FORM VECTORS. The seventh FORM VECTOR creates the right hand side values which include initial inventories of components and products, receipt amounts of components, delivery amounts of products and the capacities of the component and product storage and the capacity of the blender.

The first FORM VECTOR generates the component activities into blends. Each unit of this activity adds one unit to the product blend, consumes one unit of the component and adds quality-volumes to the blend. The second FORM VECTOR generates the activity for the product being blended. The amount will be the total of the components. It adds that amount to the product balance. It keeps the blend within specification constraints and within the blender capacity. The third FORM VECTOR generates the activities for receipt of components at the amount specified on the right hand side, pays the cost of the component received and adds to the inventory of the component. The fourth FORM VECTOR generates the activity of delivery of product at the amount specified on the right hand side, receives the revenue and subtracts from the inventory of the product. The fifth (sixth) FORM VECTOR generates the ending inventory of each component (product), keeps it within storage capacity and makes it the starting inventory for the next time period (except the last).

Once the data are entered, one runs `OMNI` and the optimizer to get a model which is "optimal" if not "infeasible". An optimal solution shows in which periods to make the blends for which products, how much and the recipes for the blends. Changes in data will result in different "optimal" solutions. As well as changing specific data items, one can add additional products, components, time periods and quality constraints. The model is data driven and no change in the `OMNI` coding will be required to create the expanded model.

But let us consider "Model Behavior" on this model. You assume that you are doing economic optimization as you have entered costs of components and prices of products. In this case doesn't LP always do economic optimization? The answer is; "not always." In this model the cost of the component is on the receipt of the component but the amount of component received is fixed. The price of the product is on delivery of the product but the amount of the product delivery is fixed. Thus the LP can not consider the costs or prices. It will give the same "optimal" decisions regardless of what you enter for costs and prices.

So what did the LP optimize? It did a material balance and when it first found a solution which balanced and did not violate any constraints it quit and reported "optimal" . It had no guidance to do better. Thus there can be many possible alternate solutions depending on the solution path the algorithm chose and the starting basis. This, although mathematically correct, is not satisfactory for most real users.

Maybe you add a small charge for the cost of the blending activity operation. Now you find that the LP tries to avoid this cost and will minimize the use of the blender. It will blend just enough to avoid component overflow or failure to meet product delivery. Beyond that the LP will minimize final product inventory.

So may be we put the prices on final product inventories and costs on final component inventories. Lets assume, as is normal, that product prices are greater than component costs. Now the LP will maximize final product inventories and minimize final component inventories within the constraint limits. If this is what the user considers reasonable, then you have a model with good behavior. If he likes a more balanced answer, then you have more model development work to do. If you put costs and prices on initial inventories, they will have no impact as the initial amounts are fixed. If you put them only on the inventories at the end of time period 1, you will cause maximum blending in period 1.

But look at another aspect. If there is lots of blender capacity, then the LP is indifferent as to when it does the blending, as long as it does not violate any constraints and maximizes final product inventory. Many possibilities may exist for when blending is done and when the LP finds one, it says "optimal". As an example, the data may allow the product demand in time period 1 to be met from inventory and the incoming components to find space in storage. No blending is done in time period 1 but the blender is heavily used in time period 2. Another set of data may lead to doing all the blending in period 1 and none in period 2. Some users prefer a more steady use of facilities. In these cases, the modeler must find ways to make the LP behave as desired. He might add a constraint that blender use in both periods is the same. But this alone is dangerous as it sometimes may not be feasible. Penalty vectors can be introduced to allow deviation from the equal-use constraint. These carry a small "cost" or penalty to direct the LP to avoid them when possible but still allow them into the solution to prevent an infeasible result.

One needs to also recognize that the model is only doing two time periods of what is a continuous operation. There will be blends to be made, rundowns to be received and products to be delivered in the time after the final time period. We need to consider whether the inventories at the end of our model will be satisfactory for the continuing operation. We must not let the model use up unnecessarily all "good" components to leave an unreasonable balance or to leave product inventories in a difficult position for the operations people. Depending on the situation, one may need to add penalty vectors to discourage exceeding specification on selected qualities (giveaway), or to include some quality balancing on the pool of final component inventories or to add restrictions on the balance of the ending product inventories.

The modeler is not done until he has studied the model and ran tests to understand how it will behave. He needs to "Think like an LP" . Then he needs to encourage the LP to do what he wants and discourage it from doing what he does not want it to do and put constraints where necessary. Optimization is more than just putting in costs and prices. Yes, LP will give you a solution labeled "Optimal" (if not infeasible) but whether it is the true one you need for decision making requires some consideration for practical applications. `OMNI` is a good tool for adjusting a model to have useful behavior.

16.4 Omni Features

The DICTIONARY section is a place where global CLASSes can be defined. These classes have a name, a list of members and may have up to two descriptive texts with each member. The member name is likely to be used in referencing information in a table and in building a name to be used in the LP optimizer input. The text may be a short name and a long name for use in the reports or might be a reference to a company product code list of mixed digits and letters.

The DATA section is a place where data tables are entered. Tables have a name, column names and row names. References are made to specific data items as TABLE tname (colname,rowname). Any names may include literal characters and class references up to sixteen characters. An example might be TABLE COMP (COST(T),(CMP)). The class references, such as (T) or (CMP) are replaced by the current active member of that class before the data is looked up.

Intersection values in a table may be a blank, a decimal number with or without the decimal point or a literal string of characters. In addition, each row has an option to include leading text fields which can be referenced in the reports. Examples of these were shown in the earlier example.

`OMNI`'s feature, OLS, allows tables to be in several spreadsheets and imports the tables from the spreadsheets into `OMNI` format. FORM TABLE allows a complex set of operations to create a table or new rows or columns in a table or to modify data in a table. SORT will reorder the rows of a table. FORM CLASS will create a global class for the run.

The generation of the extended MPSX input is done automatically when you use the FORM ROW, FORM VECTOR and FORM BOUND commands. Headers are included by use of COPY. FORM EQUATION can be used to generate a model by equations. The output is automatically reordered to MPSX input order. FORM NLE (nonlinear equations) was a special option which can only be used with MINOS to include nonlinear equations (written as equations) in a model. Full reference to data tables, to class looping and other `OMNI` features is provided. The derivatives of the equations are automatically generated in MINOS format.

Almost any non-MPSX text input file can be created using the FORM LINE and COPY commands. FORM LINE allows formatting a line as you wish. It has many options to place numbers and text on the line where you want them. The information can come from tables or calculations and may include full logic. FORM SECTION allows class indexing over groups of FORMS including nested FORM SECTIONS. FORM VALUE is used to calculate values or what we call new variables. This command is widely used in simulation and scheduling applications.

The same arithmetic is available in all places where it applies. This includes operators such as add, subtract, multiply, divide and functions such as SUM, PROD, MAX, MAXA (maximum absolute), MIN, MINA, INT, MODulus, RANDom, FIBONichi and VALUE (to convert a text string of digits to a floating value). Operands include constants, table values, ordinality or cardinality of a looping class, LP solution or matrix values and new variables created by FORM VALUE.

All parts of the system are under control of WHEN and EXCEPT logic. This logic may include tests for a blank value (vs a number), presence of a member in another class, or complex boolean expressions involving GT, GE, EQ, NE, LT, LE and OR, XOR and AND with possible parenthesis partitioning. The comparisons can be numeric or text.

There are CONTROL options to direct input and output when non-standard and to control factoring and other compiler choices. SWITCH TO allows input to be assembled from multiple sources while REASSIGN allows output to send to multiple destinations.

16.5 Summary

OMNI is a model management system with integrated data management, model generation, report writing function. OMNI allows a structured, data-driven approach to the design and building of modeling applications or generalized modeling packages. It interfaces with optimizers such as, CPLEX, HSLP, Xpress-Optimizer and MINOS. OMNI includes modules for processing, storing and updating input data, a fast optimizing compiler for the language and utilities for model management. The report writer features allow production of easily read reports of model results and assumptions used.

Before OMNI, the development, use, maintenance and interpretation of optimization models was a complex and time-consuming task. With OMNI, and its highly structured language and consistent grammatical rules the analyst needs only to describe the model and report structures and provide the data. OMNI does the rest presenting the results in readable reports. OMNI is excellent for setting up easy-to-use optimization applications. There are hundreds of OMNI users worldwide, representing industries as diverse as petroleum, chemicals, forest products, mining, banking, agriculture and transportation using OMNI for optimization, simulation and scheduling. A special graphics package for input and output is provided with the scheduling and simulation applications.

Chapter 17

THE OPL STUDIO MODELING SYSTEM

Pascal Van Hentenryck
Brown University, Box 1910, Providence, RI 02912
pvh@cs.brown.edu

Laurent Michel
University of Connecticut, Storrs, CT 06269-3155
ldm@engr.uconn.edu

Frederic Paulin and Jean-Francois Puget
Ilog SA, 9 rue de Verdun, BP 85, F-94253 Gentilly, France
{paulin,puget}@ilog.fr

Abstract This chapter describes `OPL Studio`, a modeling system for combinatorial optimization. `OPL Studio` features a number of innovative features, including its modeling language `OPL`, its visualization and debugging facilities, and its integration in larger applications through code generation. In particular, `OPL` modeling language combines high-level algebraic and set notations from modeling languages with a rich constraint language and the ability to specify search procedures and strategies that is the essence of constraint programming. Its development environment also supports automatic animations of the search space reduction and exploration which are often critical in understanding model behaviors in complex application. The chapter illustrates some of these functionalities using frequency allocation, sport scheduling, scene allocation, and project scheduling applications.

Keywords: OPL Studio,modeling, combinatorial optimization, Constraint Programming, Mathematical Programming

17.1 Introduction

Combinatorial optimization problems naturally arise in many practical applications, including scheduling, resource allocation, planning, and configuration problems. These problems are computationally difficult (*i.e.*, they are NP-hard) and require considerable expertise in optimization, software engineering, and the application domain. In the last 50 years, substantial research has been devoted to finding effective solution techniques and developing high-level modeling tools to ease the solving of such applications.

The last two decades have witnessed substantial developments in combinatorial optimization and saw the emergence of constraint programming, a new entry in the repertoire of tools for combinatorial optimization. On the one hand, constraint programming languages feature rich languages for expressing constraints and search procedures. Their constraint languages go beyond traditional linear and nonlinear constraints and support logical, high-order, and global constraints. Their search languages make it possible to specify search procedures at a high level of abstraction. On the other hand, constraint programming features a computational model where constraints are used to prune the search space by removing values which cannot appear in any solution. This computational model, and the avenues it opens to exploit combinatorial substructures, is particularly effective for finding solutions in very combinatorial applications as those found in timetabling, sport and resource-constrained scheduling, car sequencing, and various resource allocation problems to name only a few. Constraint programming strenghts are orthogonal to mathematical programming and there are numerous opportunities for synergies. At the computational level, their respective foci on feasibility and optimization should make it possible to approach more effectively those problems that combine hard feasibility constraints and difficult objective functions. At the language level, the respective orgins, from computer science and operations research, led once again to orthogonal strenghts and limitations.

Historically, the `OPL Studio` modeling system originated as an attempt to unify mathematical modeling and constraint programming languages, while supporting constraint, linear, and integer programming solvers. It evolved over time to meet industrial requests, to provide natural and simple integrations of `OPL` models inside larger applications, and to simplify the development, debugging, validation, and visualization of `OPL` models. The `OPL Studio` modeling system now consists the `OPL` optimization programming language [235], its associated script language OPLSCRIPT [236], and its development environment which includes code generation and advanced visualization and debugging components.

`OPL Studio` contributed numerous innovations to modeling systems and constraint programming, several of which will be reviewed in this chapter. Perhaps its most significant innovation is the `OPL` language which unifies algebraic mathematical programming languages and constraint programming languages. By introducing rich constraint and search languages, OPL made constraint programming accessible to a much wider audience, provided novel ways of modeling combinatorial problems inside modeling languages by making the problem structure more explicit, and made integer and constraint programming cooperation possible at a high abstraction level. OPL also contributed new, very high-level, search abstractions for constraint programming. A second fundamental innovation of OPL Studio is its advanced visualization and debugging components, which provide automatic animations of the domain reduction process and the search tree exploration. A third innovation addresses the problem of integrating a model into a broader application written in a traditional language. OPL Studio solves this seemingly difficult issue through code generation generating a C++ class which can easily interact with the rest of the application.

The purpose of this chapter is to review some of these novel features. We do not describe the use of OPL Studio for mathematical programming into detail; this has been described elsewhere and is not the most innovative contribution of the system. Rather the chapter focuses on illustrating some of the main contributions of OPL Studio through a number of applications and illustrations. As a consequence, readers should expect a modeling system whose expressive power, modeling style, and environmental support go well beyond traditional modeling languages, although it supports mathematical programming as earlier modeling systems.

The rest of the paper is organized as follows. Section 17.2 and 17.3 give brief overviews of the main contributions of OPL and OPL Studio. The next sections illustrate several of these novelties through specific applications. Section 17.4 briefly illustrates the use of OPL for a mathematical programming problem, which should help readers familiarize themselves with the OPL syntax on a well-known applications.

Section 17.5 describes a model for a frequency allocation application that illustrates how to use high-level algebraic and set manipulation, how to exploit sparsity, and how to implement search procedures in OPL. Section 17.6 describes a model for a sport-scheduling applications that illustrates the use of global constraints in OPL. Section 17.7 describes a job-shop scheduling application which illustrates vertical extensions of constraint programming and search strategies. Section 17.8 describes a scene allocation application which illustrates symmetry-breaking and the cooperation between constraint and integer programming. Section 17.9 describes a more substantial scheduling application. It illustrates some advanced features of OPL, some uses of OPLSCRIPT, and the code generation facilities. Section 17.10 describes the visualization facilities of OPL Studio, including search tree visualizations, tracing and spy facilities, and synthetic views aggregating them. Some of the material is this chapter is based on ([30, 240, 239, 173]). More detail about `OPL Studio` can be found in the user manual [127].

17.2 Overview of OPL

The core of `OPL Studio` is the modeling language `OPL` which is a significant departure from earlier modeling languages. Perhaps the most fundamental novelty of `OPL` is its elegant combination of the set and algebraic notations of mathematical modeling languages with the rich constraint and search languages of constraint programming. We briefly review these various novel aspects and their consequences.

Constraints. `OPL` features a constraint language which goes much beyond the traditional linear or nonlinear arithmetic constraints typically supported in modeling languages. `OPL` supports logical and higher-order constraints, the ability to index arrays and matrices with arbitrary expressions, and global constraints which capture combinatorial substructures arising in numerous applications. As a result, models in `OPL` often differ from integer programming models drastically and rarely use 0/1 variables. They make problem structure more apparent, enabling solvers to exploit tailored algorithms whenever appropriate. Observe that global constraints are not only useful for constraint programming engines; They could also be exploited by integer programming solvers to generate polyhedral cuts.

Vertical Extensions. One of the advantages of constraint programming is its ability to offer vertical extensions for important classes of applications, such as scheduling and routing. The key idea, often called model-based computing, is to encapsulate decision variables, global constraints, and nondeterministic decomposition inside objects and methods closer to the application. For instance, scheduling applications feature high-level objects such as activities and resources. `OPL` follows the same methodology as is illustrated later in the paper.

Search Procedures. One of the fundamental features of constraint programming languages is their ability to specify search procedures.[1] This functionality is often critical in obtaining satisfactory performance in hard combinatorial optimization problems, since the nature of the application often dictates the effectiveness of various search procedures. For instance, effective search procedures for jobshop scheduling are very different from those used in sport scheduling.

To account for this critical aspect of constraint programming, an `OPL` statement generally consists of two parts: a constraint component which specifies

[1] Of course, this functionality is also beneficial for integer programming models.

the problem constraints and a search component which describes the search procedure. Traditional modeling languages specify the constraint component only.

OPL specifies search procedures at a very high level of abstraction, often in declarative terms. Conceptually, a search procedure specifies a decomposition scheme, *i.e.*, it describes how to divide a problem into sub problems that are easier to solve. This decomposition is, in general, declarative and can be expressed naturally using very high-level search constructs pioneered by OPL. In addition, OPL makes it possible to specify how the search tree must be explored, which is another fundamental aspect of search procedures. Note that search procedures may be omitted or only partially specified. OPL, in this case, analyzes the problem statement and generates an appropriate default search procedure.

Cooperation. OPL makes it possible to express both constraint and integer programming models which can then cooperate in solving a combinatorial optimization problem. This cooperation may bring significant benefits since constraint and integer programming have orthogonal strengths. On the one hand, integer programming, through the use of linear programming and possibly syntactic and polyhedral cuts, may provide tight lower bounds on the value of the objective function and prune away sub optimal solutions. On the other hand, constraint programming, through domain filtering, reduces the set of possible values for each variable (which, in turn, may strenghten the relaxation) and prune away infeasible configurations. This cooperation, often called bound cooperation, is supported in OPL and its implementation. When requested to implement bound cooperation, OPL sends all linear constraints to a linear programming solver which interacts with the constraint programming solver by exchanging bounds on the objective functions and the decision variables. Stronger forms of cooperation, *e.g.*, cut cooperation, are possible and an important topic for future research.

17.3 Overview of OPL Studio

Although the OPL language is a key aspect of the OPL Studio system, there are many other components which play a critical role in the solving of practical applications. Some of them are briefly reviewed in this section and will be illustrated in the rest of the chapter.

OPL Script. Many applications require solving several instances of the same problem (*e.g.*, sensibility analysis), sequences of models, or a combination of both as in column-generation or Benders decomposition applications. These applications may involve several models which may need to communicate data and results. OPLSCRIPT is a domain-specific scripting language for composing and controlling OPL models in order to address the needs of these applications.

OPLScript shares the same data abstractions as `OPL`, making it easy for models to use the same data and to communicate results conveniently. In addition, OPLScript provides a number of novel abstractions dedicated to optimization applications. In particular, it supports `OPL` models and linear programming bases as first-class objects as well as extensible data structures to name only a few.

Visualization and Debugging. The development environment of `OPL` and OPLScript provides, beyond traditional support for the traditional "edit, execute, and debug" cycle, automatic visualizations of the results (*e.g.*, Gantt charts for scheduling applications), as well as visual tools for debugging and monitoring `OPL` models.

The visualization support of `OPL Studio` is probably one of its most innovative aspects. Upon request, `OPL Studio` automatically and dynamically visualizes the domain reduction process for selected decision variables, enabling modelers to identify bottlenecks and deficiencies in their models. `OPL Studio` also makes it possible to visualize the search tree exploration, which is often helpful in evaluating the effectiveness of search procedures. Together with the visual trace of the model text, these visual facilities should help modelers evaluate their models conveniently. Similarly, the visual display of the results, which is also tailored to the vertical extensions, let modelers visualize their results in appropriate terms and may reveal interesting properties of the models. `OPL Studio` also makes it possible to trace the propagation itself over time, understanding the domain reduction history. Finally, many of these visualizations can be aggregated together, in a synthetic fashion, through an annotated version of the search tree called the Christmas tree.

Integration within a Larger Application. Models are rarely solved in isolation. They are often a small block in the context of a larger application (*e.g.*, a supply-chain application) in which they must be integrated as smoothly as possible. The problem of interfacing a model to an application written in a traditional programming language can be error-prone if not dealt with appropriately.

One of the main innovations of `OPL Studio` in that respect is the use of code generation whose key idea can be described as follows. Once a model has been developed, `OPL Studio` may compile it into a C++ class which gives access to its inputs and results but hides the detail of the solving process. The larger application, written in C++, can then supply inputs and collect results from an instance of the class in terms of the data structures used in the model. Code generation greatly simplifies the integration process and reduces the risk of errors in this, potentially difficult, step. Observe that it is possible to add/remove constraints dynamically and to overwrite the search procedure, even after code generation.

In addition to code generation, `OPL Studio` also provides generic and/or model-specific COM components for integration with other programming languages.

17.4 Mathematical Programming

`OPL` can be used as a modeling language for mathematical programming. This section illustrates a typical assignment problem and its modeling as a mixed integer program within `OPL`. This application should help readers familiarize themselves with the syntax of `OPL`. A more detailed treatment of the use of `OPL` for mathematical programming can be found in ([191, 161]).

Consider a company which plans to create a network of warehouses to supply its existing stores. Suppose that the company already has a number of suitable sites for building warehouses and would like to determine whether to create a warehouse on each such site. For each chosen site, the company also would like to determine the optimal capacity for the warehouse. The company assumes an identical average merchandise turnover for all stores. However, the distance among locations and the transportation infrastructure lead to varying transportation costs $cost_{s,w}$ for each pair of store s and warehouse w. The construction of a warehouse has a given cost $fixed$ and all stores must be supplied regularly by a single warehouse. Moreover, each warehouse w has a finite capacity $capacity_w$. The objective is to minimize total cost by determining for each warehouse which stores it should supply. The total is then the sum of supply costs plus the costs of building each warehouse.

A MIP model for this particular problem can be derived as follows. Denote by W the set of possible warehouse location and S the set of stores. There is one decision variable $open_w$ per warehouse location w; its value is 1 if a warehouse is built and 0 otherwise. There is also one decision variable $supply_{s,w}$ for each pair consisting of a store s and a warehouse w; its value is 1 if s is supplied by w, 0 otherwise. The objective function can be expressed as

$$\sum_{w} fixed \cdot open_w + \sum_{s,w} supplyCost_{s,w} \cdot supply_{s,w} \tag{17.4.1}$$

The constraints in this problem can be derived as follows. A store is supplied by exactly one warehouse, *i.e.*, only one variable $supply_{s,w}$ can have a value of 1 for a given store s:

$$\sum_{w} supply_{s,w} = 1, \; \forall s \in S \tag{17.4.2}$$

If a store s is supplied by a warehouse w, then there must be a warehouse at location w. In other words, if $supply_{s,w}$ is equal to 1, then $open_w$ must be equal to 1:

```
range Boolean 0..1;
range W = ...;
range S = ...;
int supplyCost[S,W] = ...;
int capacity[W] = ...;
var Boolean open[W];
var Boolean supply[S,W];
minimize
   sum(w in W) fixed * open[w] +
   sum(w in W, s in S) supplyCost[s,w] * supply[s,w]
subject to {
   forall(s in S)
      sum(w in W) supply[s,w] = 1;
   forall(w in W, s in S)
      supply[s,w] <= open[w];
   forall(w in W)
      sum(s in S) supply[s,w] <= capacity[w];
};
```

Figure 17.4.1. A MIP Model for Capacitated Warehouse Location

$$supply_{s,w} \leq open_w, \forall w \in W, \forall s \in S \tag{17.4.3}$$

The number of stores supplied by a warehouse w must be smaller than the capacity $capacity_w$ of the warehouse:

$$\sum_s supply_{s,w} \leq capacity_w, \forall w \in W \tag{17.4.4}$$

An OPL statement encoding the MIP model is depicted in Fig. 17.4.1. It features the separation between the instance data and the model illustrated in more detail in the next section. The rest of the statement is a natural encoding of the MIP model using algebraic notations and linear constraints.

17.5 Frequency Allocation

The frequency-allocation problem is a typical constraint programming model. It illustrates various constraints, as well as a typical labeling search procedure based on dynamic variable/value orderings. The model also features an interesting data representation of OPL, sparse arrays, which is useful in large-scale linear models. This section describes the problem statement and each part of the statement in detail.

The Problem. The frequency-allocation problem consists of allocating frequencies to a number of transmitters so that there is no interference between

```
int nbCells = ...;
int nbFreqs = ...;
range Cells 1..nbCells;
range Freqs 1..nbFreqs;
int nbTrans[Cells] = ...;
int distance[Cells,Cells] = ...;

struct TransmitterType { Cells c; int t; };
{TransmitterType} Trans = { <c,t> | c in Cells & t in 1..nbTrans[c] };
var Freqs freq[Trans];

solve {
   forall(c in Cells & ordered t1, t2 in 1..nbTrans[c])
      abs(freq[<c,t1>] - freq[<c,t2>]) >= 16;

   forall(ordered c1, c2 in Cells :  distance[c1,c2] > 0)
      forall(t1 in 1..nbTrans[c1] & t2 in 1..nbTrans[c2])
         abs(freq[<c1,t1>] - freq[<c2,t2>]) >= distance[c1,c2];
};

search {
   forall(t in Trans ordered by increasing <dsize(freq[t]),nbTrans[t.c]>)
      tryall(f in Freqs ordered by decreasing nbOccur(f,freq))
         freq[t] = f;
};
```

Figure 17.5.2. The Frequency-Allocation Problem (`alloc.mod`)

transmitters and the number of allocated frequencies is minimized. The problem described here is an actual cellular phone problem where the network is divided into cells, each cell containing a number of transmitters whose locations are specified. The interference constraints are specified as follows: (1) The distance between two transmitter frequencies within a cell must not be smaller than 16. (2) The distances between two transmitter frequencies from different cells vary according to their geographical situation and are described in a matrix. The problem of course consists of assigning frequencies to transmitters to avoid interference and, if possible, to minimize the number of frequencies. The rest of this section focuses on finding a solution using a heuristic to reduce the number of allocated frequencies.

The Model. Fig. 17.5.2 shows an OPL statement for the frequency-allocation problem and Fig. 17.5.3 describes the instance data. Note the separation between models and data which is an interesting feature of OPL.

Data and Variable Declarations. The model data first specifies the number of cells (25 in the instance), the number of available frequencies (256 in the instance), and their associated ranges. The next declarations specify the number of transmitters needed for each cell and the distance between cells. For example, in the instance, cell 1 requires eight transmitters while cell 3 requires six transmitters. The distance between cell 1 and cell 2 is 1. The first interesting feature of the model is how variables are declared:

```
struct TransmitterType { Cells c; int t; };
{TransmitterType} Trans = { <c,t> | c in Cells & t in 1..nbTrans[c] };
var Freqs freq[Trans];
```

In the model, transmitters are identified by their cells and a unique identifier with the cell. The above declarations make this structure explicit, which will simplify the constraint statement. In other words, a transmitter is simply described as a record containing a cell number and a transmitter number inside the cell. The set of transmitters is computed automatically from the data using

```
{TransmitterType} Trans = { <c,t> | c in Cells & t in 1..nbTrans[c] };
```

which considers each cell and each transmitter in the cell. The model then declares an array of variables

```
var Freqs freq[Trans];
```

indexed by the set of transmitters; the values of these variables are of course the frequencies associated with the transmitters. This declaration illustrates a fundamental aspect of `OPL`: arrays can be indexed by arbitrary data. In this application, the arrays of variables `freq` is indexed by the elements of `transmitters` that are records. For instance,

```
freq[<c,t>];
```

is the decision variable representing the frequency of the transmitter `t` in cell `c`. This functionality is of primary importance to exploit sparsity in large-scale models and to simplify the statement of many combinatorial optimization problems.

The Constraints. There are two main groups of constraints in this model. The first set of constraints handles the distance constraints between transmitters inside a cell. The instruction

```
forall(c in Cells & ordered t1, t2 in 1..nbTrans[c])
    abs(freq[<c,t1>] - freq[<c,t2>]) >= 16;
```

enforces the constraint that the distance between two transmitters inside a cell is at least 16. The constraint is not linear, since it uses absolute values of the expression `freq[<c,t1>] - freq[<c,t2>]` through the functions. In integer programming, absolute values can be represented by disjunctions of

```
nbCells = 25;
nbFreqs = 256;
nbTrans = [8 6 6 1 4 4 8 8 8 8 4 9 8 4 4 10 8 9 8 4 5 4 8 1 1];
distance = [
   [16 1 1 0 0 0 0 0 1 1 1 1 1 2 2 1 1 0 0 0 2 2 1 1 1]
   [1 16 2 0 0 0 0 0 2 2 1 1 1 2 2 1 1 0 0 0 0 0 0 0 0]
   [1 2 16 0 0 0 0 0 2 2 1 1 1 2 2 1 1 0 0 0 0 0 0 0 0]
   [0 0 0 16 2 2 0 0 0 0 0 0 0 0 0 0 0 1 1 1 0 0 0 1 1]
   [0 0 0 2 16 2 0 0 0 0 0 0 0 0 0 0 0 1 1 1 0 0 0 1 1]
   [0 0 0 2 2 16 0 0 0 0 0 0 0 0 0 0 0 1 1 1 0 0 0 1 1]
   [0 0 0 0 0 0 16 2 0 0 1 1 1 0 0 1 1 1 1 2 0 0 0 1 1]
   [0 0 0 0 0 0 2 16 0 0 1 1 1 0 0 1 1 1 1 2 0 0 0 1 1]
   [1 2 2 0 0 0 0 0 16 2 2 2 2 2 2 1 1 1 1 1 1 1 0 1 1]
   [1 2 2 0 0 0 0 0 2 16 2 2 2 2 2 1 1 1 1 1 1 1 0 1 1]
   [1 1 1 0 0 0 1 1 2 2 16 2 2 2 2 2 2 1 1 2 1 1 0 1 1]
   [1 1 1 0 0 0 1 1 2 2 2 16 2 2 2 2 2 1 1 2 1 1 0 1 1]
   [1 1 1 0 0 0 1 1 2 2 2 2 16 2 2 2 2 1 1 2 1 1 0 1 1]
   [2 2 2 0 0 0 0 0 2 2 2 2 2 16 2 1 1 1 1 1 1 1 1 1 1]
   [2 2 2 0 0 0 0 0 2 2 2 2 2 2 16 1 1 1 1 1 1 1 1 1 1]
   [1 1 1 0 0 0 1 1 1 1 2 2 2 1 1 16 2 2 2 1 2 2 1 2 2]
   [1 1 1 0 0 0 1 1 1 1 2 2 2 1 1 2 16 2 2 1 2 2 1 2 2]
   [0 0 0 1 1 1 1 1 1 1 1 1 1 1 1 2 2 16 2 2 1 1 0 2 2]
   [0 0 0 1 1 1 1 1 1 1 1 1 1 1 1 2 2 2 16 2 1 1 0 2 2]
   [0 0 0 1 1 1 2 2 1 1 2 2 2 1 1 1 1 2 2 16 1 1 0 1 1]
   [2 0 0 0 0 0 0 0 1 1 1 1 1 1 1 2 2 1 1 1 16 2 1 2 2]
   [2 0 0 0 0 0 0 0 1 1 1 1 1 1 1 2 2 1 1 1 2 16 1 2 2]
   [1 0 0 0 0 0 0 0 0 0 0 0 0 1 1 1 1 0 0 0 1 1 16 1 1]
   [1 0 0 1 1 1 1 1 1 1 1 1 1 1 1 2 2 2 2 1 2 2 1 16 2]
   [1 0 0 1 1 1 1 1 1 1 1 1 1 1 1 2 2 2 2 1 2 2 1 2 16]];
};
```

Figure 17.5.3. Instance Data for the Frequency-Allocation Problem (`alloc.dat`)

linear constraints, which can then be eliminated using various techniques. The constraints in the `OPL` model preserves the original structure of the problem. Note also the expressions `freq[<c,t1>]` and `freq[<c,t2>]` which illustrate array indexing again.

The second set of constraints handles the distance constraints between transmitters from different cells. The instruction

```
forall(ordered c1, c2 in Cells :  distance[c1,c2] > 0)
    forall(t1 in 1..nbTrans[c1] & t2 in 1..nbTrans[c2])
       abs(freq[<c1,t1>] - freq[<c2,t2>]) >= distance[c1,c2];
```

considers each pair of distinct cells whose distance must be greater than zero and each two transmitters in these cells, and states that the distance between the frequencies of these transmitters must be at least the distance specified in the matrix `distance`.

The Search Procedure. Another interesting part of this model is the search strategy. The basic structure is typical of many constraint programming search procedures: the OPL model considers each transmitter and chooses a frequency non deterministically. In other words, the search procedure decomposes a sub problem by choosing a variable and associating a sub problem with each possible value for that variable. The interesting feature of the model is the heuristic which dynamically chooses which variables to instantiate next and which values to try first. More precisely, the OPL model chooses to generate a frequency for the transmitter whose variable has the fewest frequencies left in its domain (*i.e.*, its set of feasible values at this stage of the computation) and, in case of ties, for the transmitter whose cell size is as small as possible. This multi-criterion heuristic is expressed using a tuple `<dsize(freq[t]),nbTrans[t.c]>` to obtain

```
forall(t in Trans ordered by increasing <dsize(freq[t]),nbTrans[t.c]>)
```

The heuristic associates with each transmitter a tuple $< s, c >$, where s is the number of possible frequencies and c is the number of transmitters in the cell to which the transmitter belongs. A transmitter with tuple $< s_1, c_1 >$ is preferred over a transmitter with tuple $< s_2, c_2 >$ if $s_1 < s_2$ or if $s_1 = s_2$ and $c_1 < c_2$.

Non deterministically. Once again, the model specifies a heuristic for the ordering in which the frequencies must be tried. To reduce the number of frequencies, the model specifies to try first those values that were used most often in previous assignments. This heuristic is implemented using a non deterministic `tryall` instruction with the order specified using the `nbOccur` function (`nbOccur(i,a)` denotes the number of occurrences of `i` in array `a` at a given step of the execution):

```
forall(t in Trans ordered by increasing <dsize(freq[t]),nbTrans[t.c]>)
   tryall(f in Freqs ordered by decreasing nbOccur(f,freq))
      freq[t] = f;
```

As mentioned, this search procedure is typical of many constraint satisfaction problems and consists of using a first heuristic to choose dynamically which variable to instantiate next (variable choice) and a second heuristic to choose which value to assign non deterministically to the selected variable (value choice). The `forall` instruction is of course deterministic, while the `tryall` instruction is non deterministic: potentially all possible values are chosen for the selected variable. These two instructions were introduced by OPL as part of the unification of mathematical modeling and constraint programming languages. Note that, on the instance depicted in Fig. 17.5.3, OPL returns a solution with 95 frequencies in about 3 seconds.

	Week 1	Week 2	Week 3	Week 4	Week 5	Week 6	Week 7
period 1	0 vs 1	0 vs 2	4 vs 7	3 vs 6	3 vs 7	1 vs 5	2 vs 4
period 2	2 vs 3	1 vs 7	0 vs 3	5 vs 7	1 vs 4	0 vs 6	5 vs 6
period 3	4 vs 5	3 vs 5	1 vs 6	0 vs 4	2 vs 6	2 vs 7	0 vs 7
period 4	6 vs 7	4 vs 6	2 vs 5	1 vs 2	0 vs 5	3 vs 4	1 vs 3

Figure 17.6.4. A Solution to the Sport-Scheduling Application with 8 Teams

17.6 Sport Scheduling

This section considers the sport-scheduling problem from ([172, 195]). This is a beautiful application which provides nice illustrations of the strengths of constraint programming. In particular, it prominently features the concept of "global constraints" which is critical to obtain solutions for large instances of this problem.

The Problem. The problem consists of scheduling games between n teams over $n-1$ weeks. In addition, each week is divided into $n/2$ periods. The goal is to schedule a game for each period of every week so that the following constraints are satisfied: (1) Every team plays against every other team; (2) A team plays exactly once a week; (3) A team plays at most twice in the same period over the course of the season. xA solution to this problem for 8 teams is shown in Fig. 17.6.4. In fact, the problem can be made more uniform by adding a "dummy" final week and requesting that all teams play exactly twice in each period. The rest of this section considers this equivalent problem for simplicity.
The sport-scheduling problem is an interesting application for constraint programming. On the one hand, it is a standard benchmark (submitted by Bob Daniel) to the well-known MIP library and it is claimed in [172] that state-of-the-art MIP solvers cannot find a solution for 14 teams. The `OPL` models presented here are computationally much more efficient. On the other hand, the sport-scheduling application demonstrates fundamental features of constraint programming including global and symbolic constraints. In particular, the model makes heavy use of arc-consistency [166], a fundamental constraint satisfaction techniques from artificial intelligence. This section presents an `OPL` model that solves the 14-teams problem in about 44 seconds. Appendix A show how to specialize it further to find a solution for 14 to 30 teams quickly. Both models are based on the constraint programs presented in [195].

Data and Variable Declarations. The simple model is depicted in Fig. 17.6.5. Its input is the number of teams `nbTeams`. Several ranges are defined from the input: the teams `Teams`, the weeks `Weeks`, and the extended weeks `EWeeks`, *i.e.*, the weeks plus the dummy week. The model also declares an

```
int nbTeams = ...;
range Teams 1..nbTeams;
range Weeks 1..nbTeams-1;
range EWeeks 1..nbTeams;
range Periods 1..nbTeams/2;
range Games 1..nbTeams*nbTeams;
enum Slots = { home, away };

int occur[t in Teams] = 2;
int values[t in Teams] = t;

var Teams team[Periods,EWeeks,Slots];
var Games game[Periods,Weeks];

struct Play { int f; int s; int g; };
{Play} Plays = { <i,j,(i-1)*nbTeams+j> | ordered i, j in Teams };
predicate link(int f,int s,int g) in Plays;

solve {
   forall(w in EWeeks)
      alldifferent( all(p in Periods & s in Slots) team[p,w,s]) onDomain;
   alldifferent(game) onDomain;
   forall(p in Periods)
      cardinality(occur,values,all(w in EWeeks & s in Slots) team[p,w,s])
         extendedPropagation;
   forall(p in Periods & w in Weeks)
      link(team[p,w,home],team[p,w,away],game[p,w]);
};

search {
   generate(game);
};
```

Figure 17.6.5. A Simple Model for the Sport-Scheduling Model

enumerated type `slot` to specify the team position in a game (`home` or `away`). The declarations

```
int occur[t in Teams] = 2;
int values[t in Teams] = t;
```

specifies two arrays that are initialized generically and are used to state constraints later on. The array `occur` can be viewed as a constant function always returning 2, while the array `values` can be tought of as the identify function over teams.

The main modeling idea in this model is to use two classes of variables: team variables that specify the team playing on a given week, period, and slot and the game variables specifying which game is played on a given week and period. The use of game variables makes it simple to state the constraint that every team must play against each other team. Games are uniquely identified by their two teams. More precisely, a game consisting of home team `h` and away team `a` is uniquely identified by the integer `(h-1)*nbTeams + a`. The instruction

```
var Teams team[Periods,EWeeks,Slots];
var Games game[Periods,Weeks];
```

declares the variables. These two sets of variables must be linked together to make sure that the game and team variables for a given period and a given week are consistent. The instructions

```
struct Play { int f; int s; int g; };
{Play} Plays = { <i,j,(i-1)*nbTeams+j> | ordered i, j in Teams };
```

specify the set of legal games `Plays` for this application. For 8 teams, this set consists of tuples of the form

```
 <1,2,2>
<1,3,3>
...
<7,8,56>
```

Note that this definition eliminates some symmetries in the problem statement since the home team is always smaller than the away team. The instruction

```
predicate link(int f,int s,int g) in Plays;
```

defines a symbolic constraint by specifying its set of tuples. In other words, `link(h,a,g)` holds if the tuple `<h,a,g>` is in the set `Plays` of legal games. This symbolic constraint is used in the constraint statement to enforce the relation between the game and the team variables. The ability to specify a constraint by listing its set of solutions originates from its roots in artificial intelligence and, more precisely, Constraint Satisfaction Problems (CSPs).

Constraints. The constraint declarations in the model follow almost directly the problem description. The constraint

```
alldifferent( all(p in Periods & s in Slots) team[p,w,s]) onDomain;
```

specifies that all the teams scheduled to play on week `w` must be different. It uses an aggregate operator `all` to collect the appropriate team variables by iterating over the periods and the slots and an annotation `onDomain` to enforce arc consistency. The instruction illustrates what is perhaps the most well-known global constraint of constraint programming: the `alldifferent` constraint.

This constraint, which embeds matching and connected components algorithms, captures an interesting combinatorial substructure which arises in many applications. The underlying arc-consistency algorithm removes, from the domain of the variables, all values that cannot appear in any solution. This domain reduction process is critical to obtain solutions to large instances of this problem. See [193] for a description on how to enforce arc consistency on this global constraint. The constraint

```
cardinality(occur,values,all(w in EWeeks & s in Slots) team[p,w,s])
    extendedPropagation
```

specifies that a team plays exactly twice over the course of the "extended" season. Its first argument specifies the number of occurrences of the values specified by the second argument in the set of variables specified by the third argument that collects all variables playing in period `p`. The annotation `extendedPropagation` specifies to enforce arc consistency on this constraint. Once again, this is a global constraint capturing a combinatorial substructure arising in many applications. Its implementation, which embeds a feasible flow algorithm, is described in [194]. Once again, it removes all infeasible values from the variable domains. The constraint

```
alldifferent(game) onDomain;
```

specifies that all games are different, *i.e.*, that all teams play against each other team. Finally, the constraint

```
link(team[p,w,home],team[p,w,away],game[p,w]);
```

is most interesting. It specifies that the game `game[p,w]` consists of the teams `team[p,w,home]` and `team[p,w,away]`. `OPL` enforces arc-consitency on this symbolic constraint.

The Search Procedure. The search procedure in this statement is extremely simple and consists of generating values for the games using the first-fail principle. Note also that generating values for the games automatically assigns values to the team by constraint propagation. As mentioned, this model finds a solution for 14 teams in about 44 seconds on a modern PC (400mhz).

17.7 Job-Shop Scheduling

The next application describes vertical extensions of `OPL`, another significant features of scheduling applications. `OPL` has a variety of domain-specific concepts for these applications that are translated into decision variables, global constraints, and search procedures (among other techniques). The scheduling concepts in `OPL` include the concepts of activities, unary, discrete, and state resources, reservoirs, and breaks as well as the global constraints linking them. Some of them are illustrated in this, and a subsequent section.

```
int nbMachines = ...;
range Machines 1..nbMachines;
int nbJobs = ...;
range Jobs 1..nbJobs;
int nbTasks = ...;
range Tasks 1..nbTasks;
Machines resource[Jobs,Tasks] = ...;
int+ duration[Jobs,Tasks] = ...;
int totalDuration = sum(j in Jobs, t in Tasks) duration[j,t];

ScheduleHorizon = totalDuration;
Activity task[j in Jobs, t in Tasks](duration[j,t]);
Activity makespan(0);
UnaryResource tool[Machines];

minimize
   makespan.end
subject to {
   forall(j in Jobs)
      task[j,nbTasks] precedes makespan;
   forall(j in Jobs & t in 1..nbTasks-1)
      task[j,t] precedes task[j,t+1];
   forall(j in Jobs & t in Tasks)
      task[j,t] requires tool[resource[j,t]];
};

search {
   LDSearch() {
      forall(r in Machines ordered by increasing localSlack(tool[r]))
         rank(tool[r]);
   }
}
```

Figure 17.7.6. A Job-Shop Scheduling Model (`jobshop.mod`)

The Problem. The problem is to schedule a number of jobs on a set of machines to minimize completion time, often called the *makespan*. Each job is a sequence of tasks and each task requires a machine.

Data and Variable Declarations. Fig. 17.7.6 describes a simple job-shop scheduling model. It first declares the number of maxchines, the number of jobs, and the number of tasks in the jobs. The main data of the problem, *i.e.*, the duration of all the tasks and the resources they require, are then given. The next set of instructions

```
ScheduleHorizon = totalDuration;
```

```
Activity task[j in Jobs, t in Tasks](duration[j,t]);
Activity makespan(0);
UnaryResource tool[Machines];
```

is most interesting. The first instruction describes the schedule horizon, *i.e.*, the date by which the schedule should be completed at the lastest. In this application, the schedule horizon is given as the summation of all durations, which is clearly an upper bound on the duration of the schedule. The next instruction declares the activities of the problem. Activities are first-class objects in OPL and can be viewed (in a first approximation) as consisting of variables representing the starting date, the duration, and the end date of a task, as well as the constraints linking them. The variables of an activity are accessed as fields of records. In our application, there is an activity associated with each task of each job. The instruction

```
UnaryResource tool[Machines];
```

declares an array of unary resources. Unary resources are, once again, first-class objects of OPL; they represent resources that can be used by at most one activity at anyone time. In other words, two activities using the same unary resource cannot overlap in time. Note that the makespan is modeled for simplicity as an activity of duration zero.

The Constraints. Consider now the problem constraints. The first set of constraints specifies that the activities associated with the problem tasks precede the makespan activity. The next two sets specify the precedence and resource constraints. The resource constraints specify which activities require which resource. Note the simplicity in expressing the problem constraints which is a consequence of the model-based methodology supported by vertical extensions.

The Search Procedure. Finally, the search procedure

```
search {
   LDSearch() {
      forall(r in Machines ordered by increasing localSlack(tool[r]))
         rank(tool[r]);
   }
}
```

illustrates a typical search procedure for job-shop scheduling and the use of limited discrepancy search (LDS) [98] as a search strategy. The search procedure

```
forall(r in Machines ordered by increasing localSlack(tool[r]))
   rank(u[r]);
```

consists of ranking the unary resources, *i.e.*, choosing in which order the activities execute on the resources. Once the resources are ranked, it is easy to find a solution. The procedure ranks first the resource with the smallest local slack (*i.e.*, the machine that seems to be the most difficult to schedule) and then considers the remaining resource using a similar heuristic. The instruction `LDSearch()` specifies that the search space specified by the search procedure defined above must be explored using limited discrepancy search. This strategy, which is effective for many scheduling problems, assumes the existence of a good heuristic. Its basic intuition is that the heuristic, when it fails, probably would have found a solution if it had made a small number of different decisions during the search. The choices where the search procedure does not follow the heuristic are called *discrepancies*. As a consequence, LDS systematically explores the search tree by increasing the number of allowed discrepancies. Initially, a small number of discrepancies is allowed. If the search is not successful or if an optimal solution is desired, the number of discrepancies is increased and the process is iterated until a solution is found or the whole search space has been explored. Note that, besides the default depth-first search and LDS, `OPL` also supports best-first search, interleaved depth-first search, and depth-bounded limited discrepancy search. It is interesting to mention that this simple model solves MT10 in about 40 seconds and MT20 in about 0.4 seconds.

17.8 Scene Allocation

This section considers a scene-allocation model that combines interesting feasibility constraints with an objective function. The problem and the models we describe raise several modeling issues and illustrate the benefits of a constraint and integer programming cooperation. The rest of this section specifies the problem, presents a simple constraint model, shows how to remove symmetries, and illustrates how constraint and integer programming may cooperate using a redundant modeling approach.

Problem Description. The scene-allocation problem was communicated to us by Irvin Lustig, who also provided an integer-programming model and a specific instance. It consists of deciding when to shoot scenes for a movie. Each scene involves a number of actors and at most 5 scenes a day can be filmed. All actors of a scene must, of course, be present on the day the scene is shot. The actors have fees representing the amount to be paid per day they spent in the studio. The goal of the application is to minimize the production costs.

A Simple Constraint-Programming Model. Fig. 17.8.7 describes a simple constraint program to solve the scene shooting problem, while Fig. 17.8.8 depicts the instance data. The model starts by defining the ranges for the scenes and the days, the set of actors and their daily fees, and the set of actors who appear in each of the scenes. The declaration

```
int maxScene = ...;
range Scene 1..maxScene;
int maxDay = ...;
range Day 1..maxDay;
enum Actor = ...;
int pay[Actor] = ...;
{Actor} appears[Scene] = ...;

{Scene} which[a in Actor] = { i | i in Scene :  a in appears[i] };
var Day shoot[Scene];

minimize
   sum(a in Actor, d in Day) pay[a] * (or(s in which[a]) shoot[s] = d)
subject to
   atmost(all(i in Day) 5,all(i in Day) i,shoot);

search {
   forall(s in Scene
   ordered by increasing <dsize(scene[s]),-sum(a in appears[s]) pay[a]>)
      tryall(d in Day:  isInDomain(shoot[s],d))
         shoot[s] = d;
};
```

Figure 17.8.7. The Scene Allocation Model

```
{Scene} which[a in Actor] = { i | i in Scene :  a in appears[i] };
```

computes the scenes in which an actor appears. The decision variables associate with every scene `s` its shooting day `shoot[s]`. The objective function computes the sum of all fees. The expression

```
or(s in which[a]) shoot[s] = d
```

denotes whether actor `a` is present on day `d`. The only constraint of the model

```
atmost(all(i in Day) 5,all(i in Day) i,shoot);
```

is a global cardinality constraint expressing that at most 5 scenes a day can be shot. The search procedure selects the scenes in sequence, using the *first-fail principle* and breaking ties by selecting first the most expensive scenes. It then assigns a shooting day to the selected scene.

This model solves the instance depicted in 17.8.8 in about 421 seconds and with about 650,000 nodes. The integer-programming model in OPL (The OPL implementation uses a state-of-the-art MIP system.) solves the problem in about 100 seconds. There are two reasons for the poor performance of the constraint-programming model:

```
maxScene = 19;
maxDay = 5;
Actor = {
   Patt, Casta, Scolaro, Murphy, Brown, Hacket, Anderson,
   McDougal, Mercer, Spring, Thompson
};
pay = #[
   Patt :  26481, Casta:  25043, Scolaro:  30310,
   Murphy:  4085, Brown:  7562, Hacket:  9381,
   Anderson:  8770, McDougal:  5788, Mercer:  7423,
   Spring:  3303, Thompson:  9593
]#;
appears = [
   { Hacket },
   { Patt, Hacket, Brown, Murphy },
   { McDougal, Scolaro, Mercer, Brown },
   { Casta, Mercer },
   { Mercer, Anderson, Patt, McDougal, Spring },
   { Thompson, McDougal, Anderson, Scolaro, Spring },
   { Casta, Patt },
   { Mercer, Murphy },
   { Casta, McDougal, Mercer, Scolaro, Thompson },
   { Casta, McDougal, Scolaro, Patt },
   { Patt },
   { Hacket, Thompson, McDougal, Murphy, Brown },
   { Hacket, Murphy, Casta, Patt },
   { Anderson, Scolaro },
   { Thompson, Murphy, McDougal, Patt },
   { Scolaro, McDougal, Casta, Mercer },
   { Scolaro, Patt, Brown },
   { Scolaro, McDougal, Hacket, Thompson },
   { Casta }
];
```

Figure 17.8.8. Instance Data for the Scene-Allocation Problem

1 the problem has many symmetries, increasing the size of the search space;

2 the lower bound computed by constraint programming is extremely poor in this model.

We show how to remedy both limitations in the rest of this section.

Removing Symmetries. As mentioned, the scene-allocation problem has many symmetries. In particular, the exact shooting day of a scene is not significant. What really matters is how the scenes are clustered, not the particular days to which they are assigned. However, it is not easy to state constraints removing these symmetries, mainly because of the interaction with the search procedure.

As a first approximation, let us ignore the search procedure. A step towards the solution then consists of observing what happens when the scenes are assigned a day in sequence. It is perfectly valid to assign the first scene on day 1, since the days are all equivalent at that stage. For the second scene, only two days must be considered: either the scene is shot on the same day as scene 1 (day 1) or it is scheduled on another day, say day 2. There is no need to consider days 3-5, since they are equivalent to day 2 for the purpose of this application. The same reasoning can be generalized to any scene. If days $1 \dots d$ have been assigned for the first $n-1$ scenes, then scene n needs only to consider days $1 \dots d+1$, *i.e.*, the days where some scenes are already allocated and one additional new day, if any, are left. It is thus possible to write the constraints

```
shoot[1] = 1;
forall(s in Scene: s > 1)
    shoot[s] <= 1 + max(p in 1..s-1) shoot[p];
```

The problem with these constraints is that they impose an implicit ordering of scenes that interferes with the dynamic ordering of the search procedure. Indeed, if $< s_1, \dots, s_n >$ is the dynamic ordering chosen by the search procedure, then the set of constraints should really enforce

```
shoot[s1] = 1
shoot[s2] <= 1 + shoot[s1]
shoot[s3] <= 1 + max(shoot[s1],shoot[s2])
...
```

Since this ordering is dynamic, it is not easy to enforce these constraints statically. Fortunately, it is simple to generate these constraints dynamically during search. Fig. 17.8.9 depicts a model that removes symmetries. The main novelty is the declaration of the search procedure whose formal parameter denotes the number of days allocated so far to the scenes. The procedure first selects, in a deterministic way, a scene that is not yet assigned (*i.e.*, `not bound(shoot[s])`) using the same heuristic as before. Once this scene is selected, the procedure tries to assign the scene a day from the set `1..day+1`,

```
int maxScene = ...;
range Scene 1..maxScene;
int maxDay = ...;
range Day 1..maxDay;
enum Actor = ...;
int pay[Actor] = ...;
{Actor} appears[Scene] = ...;

{Scene} which[a in Actor] = { i | i in Scene :  a in appears[i] };
var Day shoot[Scene];

SearchProcedure label(int day)
{
   select(s in Scene:  not bound(shoot[s])
   ordered by increasing <dsize(shoot[s]),-sum(a in appears[s]) pay[a]>)
      tryall(d in Day:  d <= day + 1 & isInDomain(shoot[s],d)) {
         shoot[s] = d;
         if d = day + 1 then
            label(d)
         else
            label(day)
         endif;
      };
};

minimize
   sum(a in Actor, d in Day) pay[a] * (or(s in which[a]) shoot[s] = d)
subject to
   atmost(all(i in Day) 5,all(i in Day) i,shoot);

search
   label(0);
```

Figure 17.8.9. The Scene-Allocation Model with Symmetry Removal

i.e., the days already allocated and one new day (if available). These are the only days that need to be considered since the other "new" days are "equivalent" to `day+1`. The `tryall` instruction contains a filter `d <= day + 1` for that purpose. The search procedure then is called recursively, possibly with a new value for the parameter if the new day was allocated to the scene. The model search procedure simply calls `label(0)` to generate a day to all the scenes, since no days are allocated initially.

It is interesting to note that this model now solves the instance in 6.5 seconds using around 11,000 nodes. However, as mentioned, it still computes a poor lower bound. The next section addresses this issue.

```
int maxScene = ...; range Scene 1..maxScene;
int maxDay = ...; range Day 1..maxDay;
enum Actor = ...;
int pay[Actor] = ...;
{Actor} appears[Scene] = ...;
{Scene} which[a in Actor] = { i | i in Scene :  a in appears[i] };

var Day shoot[Scene];
var int isPaid[Actor,Day] in 0..1;
var int isShot[Scene,Day] in 0..1;

SearchProcedure label(int day) {
   select(s in Scene:  not bound(shoot[s])
   ordered by increasing <dsize(shoot[s]),-sum(a in appears[s]) pay[a]>)
      tryall(d in Day:  d <= day + 1 & isInDomain(shoot[s],d)) {
         shoot[s] = d;
         if d = day + 1 then label(d) else label(day) endif;
      };
};

minimize with linear relaxation
   sum(a in Actor, d in Day) pay[a] * isPaid[a,d]
subject to {
   atmost(all(i in Day) 5,all(i in Day) i,shoot);

   forall(s in Scene, a in appears[s], d in Day)
      isShot[s,d] <= isPaid[a,d];
   forall(s in Scene)
      sum(d in Day) isShot[s,d] = 1;
   forall(d in Day)
      sum(s in Scene) isShot[s,d] <= 5;

   forall(s in Scene)
      isShot[s,shoot[s]] = 1;
};
search label(0);
```

Figure 17.8.10. The Scene-Allocation Model with Redundant Modeling and Cooperation

Redundant Modeling. To improve the lower bound of the previous models, Fig. 17.8.10 uses a redundant modeling approach that includes both the integer and constraint programming formulations. Consider the novelties of this model. The first addition

```
var int isPaid[Actor,Day] in 0..1;
var int isShot[Scene,Day] in 0..1;
```

is the declaration of the integer-programming variables that specify whether an actor is paid on a given day and whether a scene is filmed on a given day. The second novelty is the objective function

```
sum(a in Actor, d in Day) pay[a] * isPaid[a,d]
```

which comes from the integer-programming formulation. The last novelty is the constraint statement that now includes the existing constraints as well as the integer-programming constraints, and the constraints linking the two modelings. A constraint `isShot[s,shoot[s]] = 1` in fact connects the constraint variable `shoot[s]` to the integer-programming variables `isShot[s,1]`, ..., `isShot[s,5]`. Observe that these constraints feature the `element` constraint, *i.e.*, the ability to index an array of variables with a variable. The `element` constraint performs a variety of deductions. In particular, it makes sure that, when `shoot[s]` is given a value in the search procedure, the corresponding integer variables in array `isShot` are given the appropriate values. The ability to index arrays by expressions containing variables is called an `element` constraint for historic reasons. This functionality originated in the CHIP system which is a constraint logic programming language without arrays but with lists. CHIP featured a constraint `element(i,L,e)` that holds if `e` is element `i` of list `L`. The `element` constraint is generally considered as the first global constraint since it captures a disjunction of more primitive constraints. Indeed, if `L` is the list $[v_1,\ldots,v_n]$, the `element` constraint can be specified as

$$(i = 1 \,\&\, e = v_1) \;\vee\; \ldots \;\vee\; (i = n \,\&\, e = v_n).$$

The solver may enforce arc consistency or other forms of consistency on the `element` constraint. Expressions indexing arrays by variables are in fact transformed into `element` constraints when the expressions are parsed or constructed.

Note also the keyword `with linear relaxation` that specifies that `OPL` should use bound cooperation to solve this model. This keyword specifies that all linear constraints should be sent both to the constraint- and linear-programming solvers.

It is interesting to mention the performance of the redundant model for the scene-allocation problem. The instance is now solved in 5.9 seconds using only around 700 nodes. Observe the significant reduction in the number of nodes and the progress achieved compared to the original constraint- and integer-programming models.

17.9 The Trolley Application

This section illustrates the use of `OPL` for a more advanced scheduling application. It discusses abstractions such as discrete and state resources and transition times. It also shows how to use OPLSCRIPT to find good solutions quickly and demonstrates the use of code generation. To ease understanding, the application is presented in stepwise refinements starting with a simplified version of the problem and adding more sophisticated concepts incrementally. The basic model is presented in this section. The complete model is in Appendix B.

The Basic Model. Consider an application where a number of jobs must be performed in a shop equipped with a number of machines. Each job corresponds to the processing of an item that needs to be sequentially processed on a number of machines for some known duration. Each item is initially available in area A of the shop. It must be brought to the specified machines with a trolley. After it has been processed on all machines, it must be stored in area S of the shop. Moving an item from area x to area y consists of (1) loading the item on the trolley at area x; (2) moving the trolley from area x to area y and (3) unloading the item from the trolley at area y. The goal is to find a schedule minimizing the makespan. In this version of the problem, we ignore the time to move the trolley and we assume that the trolley has unlimited capacity. Subsequent sections will remove these limitations.

The specific instance considered here consists of 6 jobs, each of which requires processing on two specified machines. As a consequence, a job consists of 8 tasks

1. load the item on the trolley at area A;
2. unload the item from the trolley at the area of the first machine required by the job;
3. process the item on the first machine;
4. load the item on the trolley at the area of this machine;
5. unload the item from the trolley at the area of the second machine required by the job;
6. process the item on the second machine;
7. load the item on the trolley at the area of this machine;
8. unload the item from the trolley at Area S.

Figs. 17.9.11 and 17.9.12 depict an OPL model for this problem, while Fig. 17.9.13 describes the instance data.

The statement starts by defining the set of jobs, the set of tasks to be performed by the jobs, and the possible locations of the trolley. As can be seen in the instance data (see Fig. 17.9.13), the tasks correspond to the description given previously. The trolley has five possible locations, one for each available machine, one for the arrival area, and one for the storage area. The statement then defines the machines and the data for the jobs, *i.e.*, it specifies the two machines required for each job and the duration of the activities to be performed on these machines. The machines are identified by their locations for simplicity.

```
enum Jobs ...;
enum Tasks ...;
enum Location ...;
{Location} Machines = ...;
struct jobRecord {
   Location machine1; int durations1; Location machine2; int durations2;
};
jobRecord job[Jobs] = ...;
int loadDuration = ...;

Location location[Jobs,Tasks];
initialize {
   forall(j in Jobs) {
      location[j,loadA] = areaA;
      location[j,unload1] = job[j].machine1;
      location[j,process1] = job[j].machine1;
      location[j,load1] = job[j].machine1;
      location[j,unload2] = job[j].machine2;
      location[j,process2] = job[j].machine2;
      location[j,load2] = job[j].machine2;
      location[j,unloadS] = areaS;
   };
};

int duration[Jobs,Tasks];
initialize {
   forall(j in Jobs) {
      duration[j,loadA] = loadDuration;
      duration[j,unload1] = loadDuration;
      duration[j,process1] = job[j].durations1;
      duration[j,load1] = loadDuration;
      duration[j,unload2] = loadDuration;
      duration[j,process2] = job[j].durations2;
      duration[j,load2] = loadDuration;
      duration[j,unloadS] = loadDuration;
   }
};
```

Figure 17.9.11. The Trolley Problem: Part I

```
scheduleHorizon = 2000;
UnaryResource machine[Machines];
StateResource trolley(Location);
Activity act[i in Jobs,j in Tasks](duration[i,j]);
Activity makespan(0);

minimize
   makespan.end
subject to {
   forall(j in Jobs & ordered t1, t2 in Tasks)
      act[j,t1] precedes act[j,t2];
   forall(j in Jobs) {
      act[j,process1] requires machine[job[j].machine1];
      act[j,process2] requires machine[job[j].machine2];
   };
   forall(j in Jobs, t in Tasks :  t <> process1 & t <> process2)
      act[j,t] requiresState(location[j,t]) trolley;
   forall(j in Jobs)
      act[j,unloadS] precedes makespan;
};

search {
   setTimes(act);
};
```

Figure 17.9.12. The Trolley Problem: Part II

```
Jobs = {j1,j2,j3,j4,j5,j6};
Tasks = {loadA,unload1,process1,load1,unload2,process2,load2,unloadS};
Location = {m1,m2,m3,areaA,areaS};
Machines = { m1, m2, m3 };
job = [
   <m1,80,m2,60>, <m2,120,m3,80>, <m2,80,m1,60>,
   <m1,160,m3,100>, <m3,180,m2,80>,<m2,140,m3,60> ];
loadDuration = 20;
```

Figure 17.9.13. The Trolley Problem: the Instance Data

The statement also specifies the duration of a loading task, which concludes the description of the input data.

The remaining instructions in Fig. 17.9.11 specify derived data that are useful in stating the constraints. They specify the locations where each task of the application must take place and the durations of all tasks. The subsequent instructions, shown in Fig. 17.9.12, are particularly interesting. The instruction `UnaryResource machine[Machines]` declares the machines of this application. Note that the array `machine` is indexed by a set of values. The instruction `StateResource trolley(Location)` defines the trolley as a state resource whose states are the five possible locations of the trolley. A state resource is a resource that can only be in one state at any given time: Hence any two tasks requiring a different state cannot overlap in time. The instructions

```
Activity act[i in Jobs,j in Tasks](duration[i,j]);
Activity makespan(0);
```

define the decision variables for this problem. They associate an activity with each task of the application and an activity to model the makespan. Note how the subscripts `i` and `j` are used in the declaration to associate the proper duration with every task. These generic declarations are often useful to simplify problem description. The rest of the statement specifies the objective function and the constraints. The objective function consists of minimizing the end date of the makespan activity. The instruction

```
forall(j in Jobs & ordered t1, t2 in Tasks)
    act[j,t1] precedes act[j,t2];
```

specifies the precedence constraints inside a job. It also illustrates the rich aggregate operators in `OPL`. The instruction

```
forall(j in Jobs) {
    act[j,process1] requires machine[job[j].machine1];
    act[j,process2] requires machine[job[j].machine2];
};
```

specifies the unary resource constraints, *i.e.*, it specifies which task uses which machine. The instruction

```
forall(j in Jobs, t in Tasks :  t <> process1 & t <> process2)
    act[j,t] requiresState(location[j,t]) trolley;
```

specifies the state resource constraints for the trolley, *i.e.*, it specifies which tasks require the trolley to be at a specified location. The instruction

```
forall(j in Jobs)
    act[j,unloadS] precedes makespan;
```

makes sure that the makespan activity starts only when all the other tasks are completed. The search procedure is rather simple in this model and uses a procedure `setTimes(act)` that assigns a starting date to every task in the array `act` by exploiting dominance relationships.

```
Model m("trolley.mod","trolley.dat");
int fails := 1000;
m.setFailLimit(fails);
int solved := 0;
while m.nextSolution() do {
   solved := 1;
   cout << "solution with makespan:  " << m.objectiveValue() << endl;
   m.setFailLimit(m.getNumberOfFails() + fails);
}
m.setFailLimit(m.getNumberOfFails() + fails);
if solved then {
   m.restore();
   cout << "final solution with makespan:  " << m.objectiveValue() << endl;
}

cout << "Time:  " << m.getTime() << endl;
cout << "Fails:  " << m.getNumberOfFails() << endl;
cout << "Choices:  " << m.getNumberOfChoicePoints() << endl;
cout << "Variables:  " << m.getNumberOfVariables() << endl;
```

Figure 17.9.14. A Script for the Trolley Problem (`trolley.osc`)

Controlling Execution Time. We now discuss how to control the execution of this model using script, a language for composing and controlling OPL models. OPLSCRIPT() is particularly appropriate for applications that require solving several instances of the same model, a sequence of models, or a combination of both as in column-generation applications. See [236] for an overview of these functionalities. OPLSCRIPT can also be used for controlling OPL models in order to find good solutions quickly or to improve efficiency by exploiting more advanced techniques (*e.g.*, shuffling in job-shop scheduling). This section illustrates how OPLSCRIPT can be used to find a good solution quickly on the final trolley application.

The motivation here is that it is sometimes appropriate to limit the time devoted to the search of an optimal solution by restricting the number of failures, the number of choice points, or the execution time. Fig. 17.9.15 depicts a script for the trolley problem that limits the number of failures when searching for a better solution. The basic idea of the script is to allow for an initial credit of failures (say, i) and to search for a solution within these limits. When a solution is found with, say, f failures, the search is continued with a limit of $i + f$ failures, *i.e.*, the number of failures needed to reach the last solution is increased by the initial credit. Consider now the script in detail. The instruction `Model m("trolley.mod","trolley.dat")` defines a OPLSCRIPT model in terms of its model and data files. Models are first-class objects in OPLSCRIPT: They can be passed as parameters of procedures and stored in data structures and they also support a variety of methods. For instance, the method `nextSolution` on a model can be used to obtain the successive solutions of a model or, in optimization problems, to produce a sequence of solutions, each of which improves the best value of the objective function found so far. The instruction `m.setFailLimit(fails)` specifies that the next call to `nextSolution` can perform at most `fails` failures, *i.e.*, after `fails` failures, the execution aborts and `nextSolution()` returns 0. The main loop of the script produces a sequence of solutions, each of which having a smaller makespan. Note the `setFailLimit` method that retrieves the number of failures needed since the creation of model `m` and sets a new limit by adding `fails` to this number. The next call to `nextSolution` takes into account this new limit when searching for a better solution. Note also the instruction `m.restore()` to restore the last solution found by `nextSolution()`. This script displays a result of the form

```
solution with makespan:  2310
...
solution with makespan:  1620
final solution with makespan:  1620
Time:  7.0200
Fails:  3578
Choices:  3615
Variables:  204
```

Code Generation. Once a reasonable model has been successfully designed in OPL, it can be integrated in a larger application through C++ code generation. This is one of the novel aspects of the development environment of OPL. It makes it possible to develop models at a very high level of abstraction, while allowing a smooth integration of the model in a larger context. In addition, the generated code directly refer to the modeling objects of the OPL model, thus simplifying the integration work. We briefly illustrate these features on the trolley application. The basic idea behind code generation consists of producing a C++ class associated with each object in the model and a top-level class for the model. In other words, the generated code is specialized to the model and is strongly typed. Theses classes can then be used to access and modify the data and, of course, to solve the model and collect the results. Fig. 17.9.14 depicts C++ code to obtain the first solution to the trolley application and to display the main activities. Instruction `IloSolver_trolley solver;` defines an instance `solver` that encapsulates the functionality of the trolley model. The class definition is available in the `.h` that is not shown here. The instruction `IloArray_act act = solver.get_act();` is used to obtain the array of activities, while the instructions

```
IloEnum_Jobs Jobs = solver.get_Jobs();
```

```
int main(int argc,char* argv[])
{
   IloSolver_trolleyComplete solver;
   if (solver.nextSolution()) {
      IloArray_act act = solver.get_act();
      IloEnum_Jobs Jobs = solver.get_Jobs();
      IloEnum_Tasks Tasks = solver.get_Tasks();
      IloEnumIterator_Jobs iteJobs(Jobs);
      while (iteJobs.ok()) {
         IloEnumIterator_Tasks iteTasks(Tasks);
         while (iteTasks.ok()) {
            cout << "act[" << *iteJobs << "," << *iteTask << "]=";
            cout << act[*iteJobs][*iteTask] << endl;
            ++iteTasks;
         }
         ++iteJobs;
      }
      cout << endl;
   }
   solver.end();
   return 0;
}
```

Figure 17.9.15. C++ Code for the Trolley Problem

```
IloEnum_Tasks Tasks = solver.get_Tasks();
IloEnumIterator_Jobs iteJobs(Jobs);
```

obtain the enumerated types and define an iterator for the jobs. The remaining instructions iterate over the enumerated sets and display the activities.

17.10 Visualization

ILOG OPL STUDIO is the development environment for the OPL language. It provides a GUI and a set of commands that contain all the classical features of a development environment: projects management, editing commands, editor window, and run and debug commands and windows to name only. We do not review these standard functionalities, but we concentrate instead on some advanced features not found in existing modeling systems. Among those, the visualization of the search performed by the optimization engine is probably the most significant. Other visualizations are available, such as Gantt charts of activities and resource utilization over time for scheduling problems. Fig. 17.10.16 illustrates such a display.

Visualization of the Search Procedures. There exist different approaches to visualize search trees produced by search procedures; They range from dy-

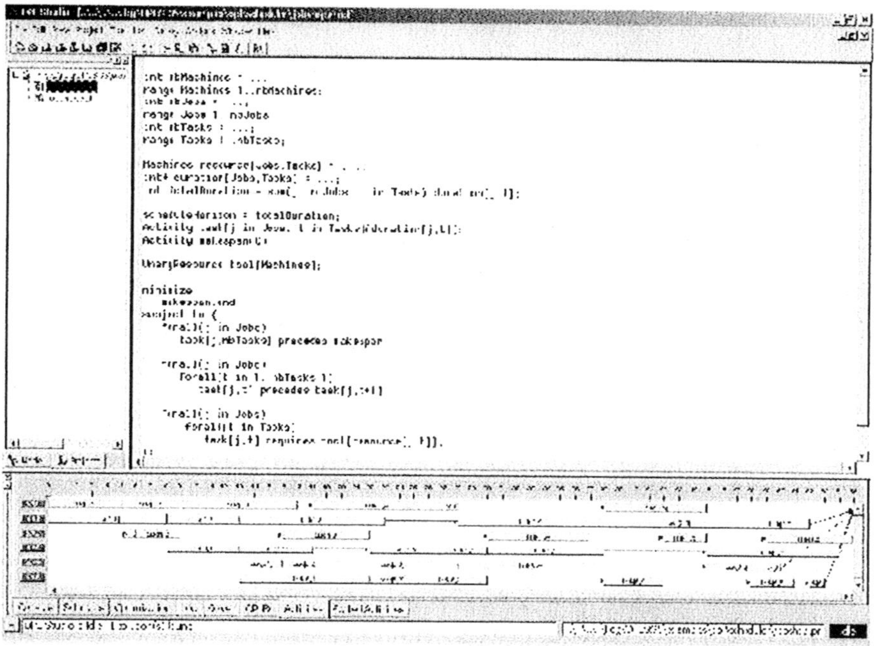

Figure 17.10.16. Gantt chart visualization for a scheduling problem

namic visualizations to post-mortem analyses ([214, 1]). `OPL Studio` supports dynamic visualizations, making it possible to monitor the exploration order as the model excutes. Since both performance evaluation and educational aspects are important in a tool such as `OPL Studio`, it is critical to visualize the current visited node, the traversal order of the search tree, as well as the state of the nodes. `OPL Studio` uses colors to indicate the state of the node and a yellow arrow to indicate the current node. During execution, the yellow arrow jumps up and down in the tree. The possible states of a node are: created but still unexplored (white), explored (blue), pruned without exploration (black), solution (green), and failure (red). When the yellow arrow points to the root node, the algorithm is performing the initial domain reduction, often called presolve in integer programming.

Tree Reduction. Because search trees can be very big, `OPL Studio` provides compact tree visualizations. For instance, it is possible to collapse subtrees in a single node, which can be expanded later if necessary. A collapsed subtree is represented by a triangle following the same color conventions as the node representation. See Fig. 17.10.17 for a visual illustration. Also, the `OPL` search constructs provide ways to compact search trees by abstracting the internal binary trees into the n-ary trees specified by the `forall` and `tryall` `OPL`

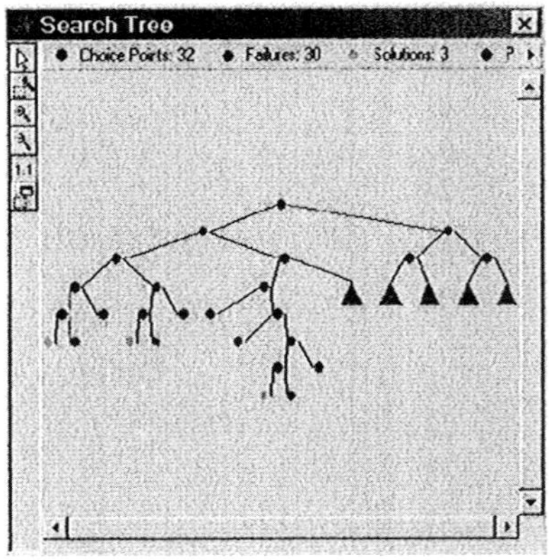

Figure 17.10.17. Search Tree Visualization with Five Collapsed Subtrees

keywords. Recall that `forall` is an iterated "AND" and `tryall` is an iterated "OR" which, together, specifies an AND-OR tree [241]. Contrary to [2], this compaction does not affect the underlying implementation which remains in terms of binary trees; It requires instrumentation of the search procedures to memorize the correct parent of each node.

The Choice Stack. Visualizing the shape of the tree is rarely sufficient for program debugging and performance tuning. One needs to know which choice occurred at a specific node and the list of choices made so far in the current branch. This information is captured in the choice stack which is displayed in its own panel in `OPL Studio`. Fig. 17.10.18 depicts the corresponding choice stack and search tree for a given problem. In this problem being illustrated, the choices are of the form *variable* = *value*. After double-clicking on a node in the search tree view, the corresponding frames are highlighted in the choice stack panel. In Fig. 17.10.18, the first node below the root is selected and the corresponding frame in the choice stack indicate that the corresponding choice is the assignment of value `London` to variable `Supplier[0]`. Note that the depth of a node in the tree is given between brackets.

Visualization of Propagation Events. In constraint programming models, constraint propagation is called before the search and at each node of the search tree. The propagation occurring before the search, at the root node, is called the *initial propagation*. The `OPL Studio` debugger provides a way of displaying the trace of the propagation events and the result of the propagation in terms of domain reduction. This information is collected in a panel called the "propagation spy". This is a special hierarchical sheet containing a tree hierarchy in the first column. Fig. 17.10.19 shows the initial propagation for a problem.

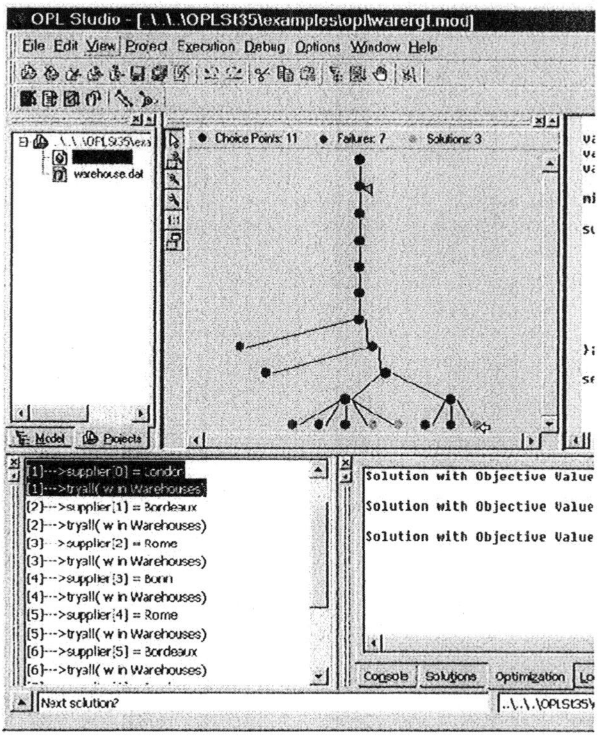

Figure 17.10.18. The Choice Stack and the Corresponding n-ary Search Tree

Each line is a propagation event, such as updating the upper bound ("Set Max"), assigning a value ("Set Value"), adding a constraint ("Post Constraint"), propagating a constraint ("Propagate Constraint"), and failure ("Constraint Fail"). Each column describes the impact on one variable and time passes vertically, from top to bottom. The textual description of the events is appended inside the first column. When an event line is selected, the columns are rearranged so that the variables of interest are displayed first. The cell at the junction of an event line and a variable column is colored according to the type of event and shows the result of the event in textual form. This considerably simplifies the visualization of the domain reduction history. Because the number of events can be very high, the propagation spy is not always active: it is activated on demand by the user. We describe the user interaction in more detail in the next paragraph.

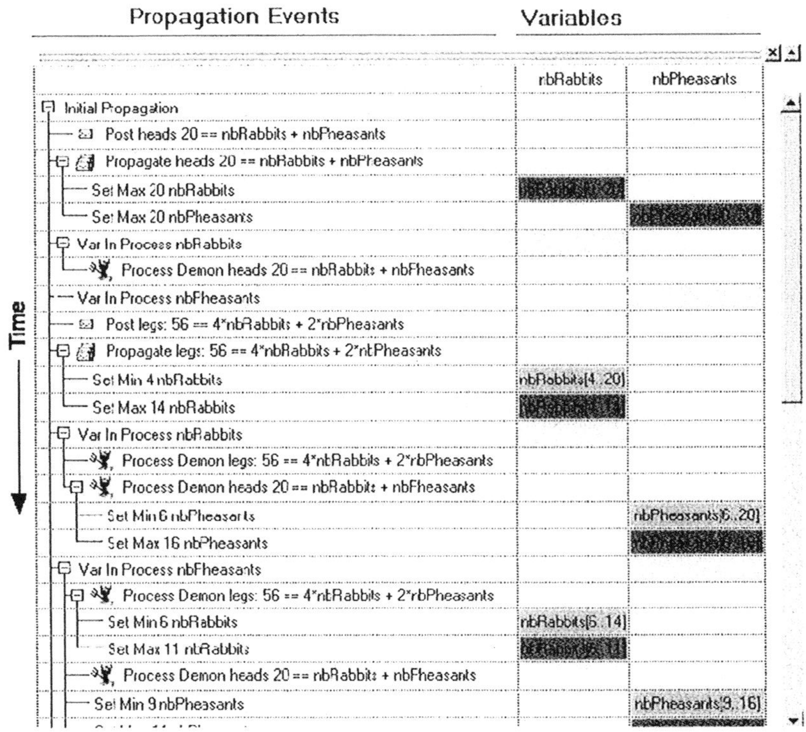

Figure 17.10.19. The Propagation Spy for the Pheasants and Rabbits Problem

User Interaction with the Debugger. In designing `OPL Studio`, special care was devoted to provide an interaction mechanism similar to those in debuggers for traditional languages (*e.g.*, Microsoft Visual Studio, Karmira BugSeeker™ for Java, or dbx on UNIX). In such debuggers, users can interrupt the execution, step into, out of, or over a function, and restart. In addition, the function call stack is a common abstraction of the execution flow and the current state of the variables can be inspected. In the context of constraint programming, these concepts are respectively replaced by the choice stack and the state of the domains. Fig. 17.10.20 presents the user interaction with the `OPL Studio` debugger and the mapping between debugging concepts in a traditional and constraint programming debuggers.

When users launch an execution, they see the tree search being drawn in the search tree viewer. If necessary, they can interrupt the execution. The debugger stops at the beginning of a node visit. Then, if users choose to step into the node, each step is a propagation event. So a line is added to the propagation spy at each "Step Into" command. Users can then step out, *i.e.*, stop tracing the propagation and continue to the next visited node. Because of the non-deterministic nature

Action	Traditional debugger	CP debugger
Step Into	Step into a function	Step into a node, tracing prop.
Step Out	Step out of a function	Step out of a node, stop tracing
Step Over	Step over a function	Step over a node, or prop. event
Skip Step	Step skipping predefined files	Trace prop. and stop at next node
Break	Break at next instruction	Break at next node or event
Breakpoint	Break at line	Break at node
Current	Current line in the code	Current visited node
Stack	Function Call Stack	Choice Stack

Figure 17.10.20. Concept Mapping between a Traditional and a CP Debugger

of the search, it is not possible to draw the tree in advance. A second run is necessary to focus on a segment of the tree. After a first run, users may identify interesting regions and place breakpoints at interesting nodes. At the second run, the tree nodes are "laundered", (*i.e.*, they become white) and then recolored. The debugger stops at breakpoints, *i.e.*, when the cooresponding search tree nodes are visited.

A Synthetic View: the Christmas Tree. The search tree viewer of `OPL Studio` provides a way of combining statistics on propagation information with the tree representation. The information added to the tree is the number of propagation events and the global domain reduction at each node. It is important to distinguish between these two pieces of information because many propagation events can be triggered with little impact on the domain reduction. The size of the nodes becomes proportional to the number of propagation events fired at each node (which is highly correlated with the time spent at each node). The meaning of the colors remains unchanged, except that the color is lighter or darker depending on the domain reduction obtained during the propagation at this node. The search tree now has big and small balls with different colors, which justifies calling it a "Christmas Tree". Fig. 17.10.21 depicts an example of such a Christmas tree. The statistical information connected with the initial propagation is concentrated at the tree root. It becomes obvious that all nodes are not equal. For instance, it is possible to detect, in failure nodes (red nodes), whether a failure is discovered early (small node) or late (large node), which is important in discovering the quality of the model in terms of pruning.

17.11 Conclusion

This chapter gave a brief overview of `OPL Studio`, a recent modeling system with a number of innovative features. The overview mostly focused on the novel features of `OPL Studio`. These includes the `OPL` modeling language which integrates mathematical and constraint programming language and supports constraint and integer programming cooperation, the domain-specific script-

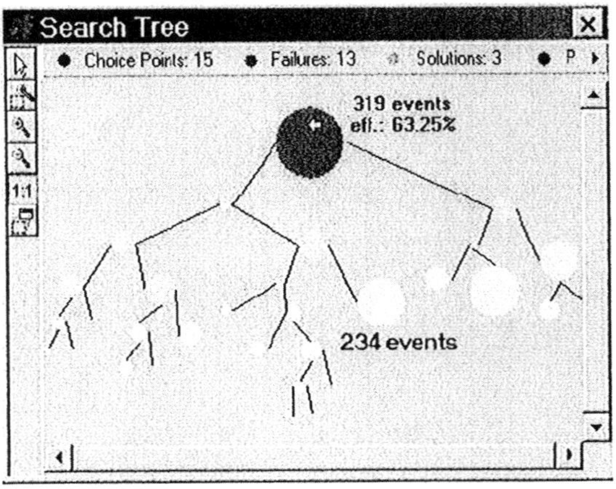

Figure 17.10.21. Christmas Tree

ing language OPLSCRIPT, code generation, as well as its advanced visualization facilities.

We believe that `OPL Studio` have opened novel avenues for modeling languages and that many more developments are likely to take place in the coming years. In particular, the success of hybrid algorithms in a variety of areas may have a profound impact of the next generation modeling languages. Similarly, advanced constraint programming techniques, such as symmetry breaking, will require even more flexibility in search procedures. An intriguing direction for future research is to explore whether modeling techniques can simplify the design and implementation of dynamic and stochastic problems which are receiving increasing attention in recent years. Finally, we believe that advanced modeling systems and visualizations will gain increasing acceptance to understand model behavior and to help the development of robust and effective models.

Appendix: Advanced Models

A A Round-Robin Model for Sport-Scheduling

The simple model has many symmetries that enlarge the search space considerably. In this section, we describe a model that uses a round-robin schedule to determine which games are played in a given week. As a consequence, once the round-robin schedule is selected, it is only necessary to determine the period of each game, not its schedule week. In addition, it turns out that a simple round-robin schedule makes it possible to find solutions for large numbers of teams. The model is depicted in Figs. 17.A.1 and 17.A.2.

The main novelty in the statement is the array `rr` that specifies the games for every week. Assuming that n denotes the number of teams, the basic idea is to fix the set of games of the first

```
int nbTeams = ...;
range Teams 1..nbTeams;
range Weeks 1..nbTeams-1; range EWeeks 1..nbTeams;
range Periods 1..nbTeams/2;
range Games 1..nbTeams*nbTeams;
enum Slots = { home, away };
int occur[t in Teams] = 2;
int values[t in Teams] = t;
var Teams team[Periods,EWeeks,Slots];
var Games game[Periods,Weeks];
struct Play { int f; int s; int g; };
{Play} Plays = { <i,j,(i-1)*nbTeams+j> | ordered i, j in Teams };
predicate link(int f,int s,int g) in Plays;

Play rr[Weeks,Periods];
initialize {
   rr[1,1].f = 1;
   rr[1,1].s = 2;
   forall(p in Periods :  p > 1) {
      rr[1,p].f = p+1;
      rr[1,p].s = nbTeams - (p-2);
   };
   forall(w in Weeks:  w > 1) {
      forall(p in Periods) {
         if rr[w-1,p].f <> 1 then
            if rr[w-1,p].f = nbTeams then rr[w,p].f = 2
            else rr[w,p].f = rr[w-1,p].f + 1 endif
         else
            rr[w,p].f = rr[w-1,p].f;
         endif;
         if rr[w-1,p].s = nbTeams then rr[w,p].s = 2
         else rr[w,p].s = rr[w-1,p].s + 1 endif;
      }
   };
   forall(w in Weeks, p in Periods)
      if rr[w,p].f < rr[w,p].s then
         rr[w,p].g = nbTeams*(rr[w,p].f-1) + rr[w,p].s
      else
         rr[w,p].g = nbTeams*(rr[w,p].s-1) + rr[w,p].f
      endif;
};
{int} domain[w in Weeks] = { rr[w,p].g | p in Periods };
```

Figure 17.A.1. A Round-Robin Model for the Sport-Scheduling Model (Part I)

```
solve {
   forall(p in Periods & w in Weeks)
      game[p,w] in domain[w];
   forall(w in EWeeks)
      alldifferent( all(p in Periods & s in Slots) team[p,w,s]) onDomain;
   alldifferent(game) onDomain;
   forall(p in Periods)
      cardinality(occur,values,all(w in EWeeks & s in Slots) team[p,w,s])
         extendedPropagation;
   forall(p in Periods & w in Weeks)
      link(team[p,w,home],team[p,w,away],game[p,w]);
};

search {
   forall(p in Periods) {
      generateSeq(game[p]);
      forall(po in Periods :  po > 1)
         generate(game[po,p]);
   };
};
```

Figure 17.A.2. A Round-Robin Model for the Sport-Scheduling Model (Part II)

nb. of teams	14	16	18	20	22	24	26	28	30
CPU Time (sec.)	3.91	4.97	1.00	6.41	10.36	11.81	45.66	36.2	42.38

Figure 17.A.3. Experimental Results for the Sport-Scheduling Model

week as

$$< 1, 2 > \cup \{< p+1, n-p+2 > \mid p > 1\}$$

where p is a period identifier. Games of the subsequent weeks are computed by transforming a tuple $< f, s >$ into a tuple $< f', s' >$ where

$$f' = \begin{cases} 1 & \text{if } f = 1 \\ 2 & \text{if } f = n \\ f+1 & \text{otherwise} \end{cases} \quad and \quad s' = \begin{cases} 2 & \text{if } s = n \\ s+1 & \text{otherwise} \end{cases}$$

This round-robin schedule is computed in the `initialize` instruction and the last instruction computes the game associated with the teams. The instruction

```
{int} domain[w in Weeks] = { rr[w,p].g | p in Periods };
```

defines the games played in a given week. This array is used in the constraint

```
game[p,w] in domain[w];
```

which forces the game variables of period p and of week `w` to take a game allocated to that week.

The model also contains a novel search procedure that consists of generating values for the games in the first period and in the first week, then in the second period and the second week, and so on. Table 17.A.3 depicts the experimental results for various numbers of teams. It is possible to improve the model further by exploiting even more symmetries: see [195] for complete details.

B The Complete Trolley Model

This section discusses how to extend the trolley model with transition times and capacity constraints.

Transition Times. Assume now that the time to move the trolley from an area to another must be taken account. This new requirement imposes transition times between successive activities. In OPL, transition times can be specified between any two activities requiring the same unary or state resource. Given two activities a and b, the transition time between a and b is the amount of time that must elapse between the end of a and the beginning of b when a precedes b. Transition times are modeled in two steps in OPL. First, a transition type is associated with each activity. Second, a transition matrix is associated with the appropriate state or unary resource. To determine the transition time between two successive activities a and b on a resource r, the transition matrix is indexed by the transition types of a and b.

In the trolley application, since the transition times depend on the trolley location, the key idea is that each activity may be associated with a transition type that represents the location where the activity is taking place. For instance, task `unload1` of job `j1` is associated with state `m1` if the first machine of `j1` is machine 1. The state resource can be associated with a transition matrix that, given two locations, return the time to move from one to the other. The model shown in the previous section can thus be enhanced easily by adding a declaration

```
int transition[Location,Location] = ...;
```

and by modifying the state resource and activity declarations to become

```
StateResource trolley(Location,transition);
UnaryResource machine[Machines];
Activity act[i in Jobs,j in Tasks](duration[i,j])
    transitionType location[i,j];
```

Using a transition matrix of the form

```
[
    [ 0, 50, 60, 50, 90 ],
    [ 50, 0, 60, 90, 50 ],
    [ 60, 60, 0, 80, 80 ],
    [ 50, 90, 80, 0,120 ],
    [ 90, 50, 80,120, 0 ]
];
```

would lead to an optimal solution of the form

```
act[j1,loadA]    = [  0 -- 20 -->  20]
act[j1,unload1]  = [ 70 -- 20 -->  90]
act[j1,process1] = [ 90 -- 80 --> 170]
act[j1,load1]    = [370 -- 20 --> 390]
act[j1,unload2]  = [530 -- 20 --> 550]
act[j1,process2] = [550 -- 60 --> 610]
act[j1,load2]    = [850 -- 20 --> 870]
```

```
act[j1,unloadS]   = [920 -- 20 --> 940]
...
act[j6,unloadS]   = [920 -- 20 --> 940]

makespan = [940 -- 0 --> 940]
```

Capacity Constraints. Consider now adding the requirement that the trolley has a limited capacity, *i.e.*, it can only carry so many items. To add this requirement in OPL, it is necessary to model the trolley by two resources: a state resource as before and a discrete resource that represents its capacity. Several activities can require the same discrete resource at a given time provided that their total demand does not exceed the capacity. In addition, it is necessary to model the tasks of moving from a location to another. As a consequence, each job is enhanced by three activities that represents the move from area A to the first machine, from the first machine to the second machine, and from the second machine to area S. Each of these trolley activities uses one capacity unit of the trolley. The declarations

```
int trolleyMaxCapacity = 3;
DiscreteResource trolleyCapacity(trolleyMaxCapacity);
enum TrolleyTasks {onTrolleyA1,onTrolley12,onTrolley2S};
Activity tact[Jobs,TrolleyTasks];
```

serve that purpose. It is now important to state that these activities require the trolley capacity and when these tasks must be scheduled. The constraint

```
forall(j in Jobs, t in TrolleyTasks)
    tact[j,t] requires trolleyCapacity;
```

specify the resource consumption, while the constraints

```
forall(j in Jobs) {
    tact[j,onTrolleyA1].start = act[j,loadA].start;
    tact[j,onTrolleyA1].end = act[j,unload1].end;
    tact[j,onTrolley12].start = act[j,load1].start;
    tact[j,onTrolley12].end = act[j,unload2].end;
    tact[j,onTrolley2S].start = act[j,load2].start;
    tact[j,onTrolley2S].end = act[j,unloadS].end;
};
```

specify the temporal relationships, *e.g.*, that the activity of moving from area A to the first machine in a job should start when the item is being loaded on the trolley and is completed when the item is unloaded. The trolley application is now completed and the final model is depicted in Figs. 17.B.4 and 17.B.5. This last model in fact is rather difficult to solve optimally despite its reasonable size.

```
enum Jobs ...;
enum Tasks ...;
enum Location ...;
{Location} Machines = ...;
struct jobRecord {
   Location machine1; int durations1; Location machine2; int durations2;
};
jobRecord job[Jobs] = ...;
int loadDuration = ...;
int transition[Location,Location] = ...;
int trolleyMaxCapacity = ...;
Location location[Jobs,Tasks];
initialize {
   forall(j in Jobs) {
      location[j,loadA] = areaA;
      location[j,unload1] = job[j].machine1;
      location[j,process1] = job[j].machine1;
      location[j,load1] = job[j].machine1;
      location[j,unload2] = job[j].machine2;
      location[j,process2] = job[j].machine2;
      location[j,load2] = job[j].machine2;
      location[j,unloadS] = areaS;
   };
};
int duration[Jobs,Tasks];
initialize {
   forall(j in Jobs) {
      duration[j,loadA] = loadDuration;
      duration[j,unload1] = loadDuration;
      duration[j,process1] = job[j].durations1;
      duration[j,load1] = loadDuration;
      duration[j,unload2] = loadDuration;
      duration[j,process2] = job[j].durations2;
      duration[j,load2] = loadDuration;
      duration[j,unloadS] = loadDuration;
   }
};
```

Figure 17.B.4. The Final Trolley Model: Part I

```
scheduleHorizon = 2000;
UnaryResource machine[Machines];
StateResource trolley(Location);
DiscreteResource trolleyCapacity(trolleyMaxCapacity);
Activity act[i in Jobs,j in Tasks](duration[i,j])
   transitionType location[i,j];
Activity tact[Jobs,TrolleyTasks];
Activity makespan(0);

minimize
   makespan.end
subject to {
   forall(j in Jobs & ordered t1, t2 in Tasks)
      act[j,t1] precedes act[j,t2];
   forall(j in Jobs) {
      act[j,process1] requires machine[job[j].machine1];
      act[j,process2] requires machine[job[j].machine2];
   };
   forall(j in Jobs, t in Tasks :  t <> process1 & t <> process2)
      act[j,t] requiresState(location[j,t]) trolley;
   forall(j in Jobs, t in TrolleyTasks)
      tact[j,t] requires trolleyCapacity;
   forall(j in Jobs) {
      tact[j,onTrolleyA1].start = act[j,loadA].start;
      tact[j,onTrolleyA1].end = act[j,unload1].end;
      tact[j,onTrolley12].start = act[j,load1].start;
      tact[j,onTrolley12].end = act[j,unload2].end;
      tact[j,onTrolley2S].start = act[j,load2].start;
      tact[j,onTrolley2S].end = act[j,unloadS].end;
   };
   forall(j in Jobs)
      act[j,unloadS] precedes makespan;
};

search {
   setTimes(act);
};
```

Figure 17.B.5. The Final Trolley Model: Part II

Chapter 18

PCOMP: A MODELING LANGUAGE FOR NONLINEAR PROGRAMS WITH AUTOMATIC DIFFERENTIATION

Klaus Schittkowski
Department of Mathematics
University of Bayreuth
D - 95440 Bayreuth
klaus.schittkowski@uni-bayreuth.de

Abstract PCOMP is a modeling language to facilitate the implementation of highly nonlinear equations arising in mathematical programming or in dynamical systems. Main applications are in engineering and natural sciences, where in addition to function values also first and second derivatives must be computed. In this paper, we briefly introduce the forward and reverse modes of automatic differentiation as implemented in PCOMP. Especially we show by means of a complex function that the relative effort for computing gradients by reverse accumulation is bounded. The structure of the PCOMP language and the usage of the software is outlined. To present a typical example, we show how PCOMP is attached to an interactive software system called EASY-FIT for estimating parameters in dynamical systems, where derivatives of the objective function, but also of initial values and right-hand sides of differential equations subject to different parameter sets must be computed.

Keywords: PCOMP, automatic differentiation, forward mode, backward mode, modeling language, Nonlinear Programming, data fitting, dynamical systems

18.1 Introduction

PCOMP is an implementation of a modeling language designed for situations, where highly nonlinear functions, often based on a series of complex analytical formulae, are to be implemented. Typical applications are more technically oriented mathematical models in engineering and natural sciences. The underlying numerical problem could be from nonlinear programming, optimal control, data fitting, roots of nonlinear equations, or differential equations, to give only a few examples. In particular we are interested in situations, where not only function values, but also derivative information must be provided.

The development of PCOMP goes back to a cooperation with MBB (now EADS) as part of a complex mechanical structural optimization system called MBB-LAGRANGE, see [146]. In this case, the whole FE system is too big to link additional codes to the system whenever user-provided nonlinear functions are to be processed. Whereas the main system functions are built in, e.g., bounds on stresses, displacements, frequencies, an additional option was attached to define arbitrary problem-dependent constraints or objective functions. Main requirements were the possibility to have access to internal design parameters and automatic differentiation by direct interpretation of the code.

Thus, the design goals of PCOMP are

1. modeling nonlinear functions or dynamical equations for technical and scientific applications in engineering and natural sciences,

2. FORTRAN-similar syntax,

3. direct interpretation of given code for function and derivative computations,

4. generation of efficient FORTRAN code for function and gradient evaluations,

5. flexible program organization in form of tool box without fixed attachment to domain-specific modeling system,

6. user-provided attachment of additional functions and link to external codes.

PCOMP is not designed to solve large scale linear or nonlinear programming problems or any large scale operations research problems. Especially, there is no matrix generator involved.

Automatic differentiation is an interesting and important tool for all numerical algorithms that require derivatives. The basic idea is to avoid not only numerical approximations, which are expensive with respect to CPU time and contain round-off errors, but also *hand-coded* differentiation. In Section 18.2, we briefly introduce the forward and backward accumulation methods and show the benefits by means of an example.

The main intention of our approach is to provide flexible and portable FORTRAN code for practical applications. The modeling language is described in the form of a formal grammar and is a subset of FORTRAN with a few extensions. Besides a parser that generates an intermediate code and that can be executed independently from the evaluation routines, there are other subroutines for the direct computation of function and gradient values, which can be called directly from a user program. On the other hand, it is possible to generate FORTRAN code for function and gradient evaluation that can be compiled and linked separately. The syntax and the program structure of PCOMP are outlined in Sections 18.3 and 18.4.

PCOMP is used for facilitating the input of nonlinear functions or equations, respectively, for a couple of applications, e.g.,

- simulation and control of a tubular reactor given in form of a distributed system, see [21], where also a high-order Runge-Kutta formula is automatically differentiated,
- optimal control of ordinary and one dimensional partial differential equations (PDECON), see [25], or [26],
- interactive nonlinear programming including multicriteria, least squares, min-max, and L_1 optimization (EASY-OPT), see [209],
- data fitting in dynamical systems (EASY-FIT), see [211].

The latter application is briefly introduced in Section 18.5 to illustrate a typical application of PCOMP. EASY-FIT is an interactive software system to identify parameters in explicit model functions, steady-state systems, Laplace transformations, systems of ordinary differential equations, differential algebraic equations, and systems of one-dimensional time-dependent partial differential equations with or without algebraic equations. Proceeding from given experimental data, *i.e.*, observation times and measurements, the minimum least squares distance of measured data from a fitting criterion is computed that depends on the solution of the dynamical system. The software system is implemented in form of a Microsoft Access database running under MS-Windows. The underlying numerical algorithms are coded in FORTRAN and are executable independently from the interface. A practical example is outlined to show how the PCOMP language is used and how remaining problem data must be defined.

18.2 Automatic Differentiation

Let $f(x)$ be a nonlinear differentiable function defined for all $x \in \mathbb{R}^n$. By automatic differentiation, we understand the numerical computation of a derivative value $\nabla f(x)$ of f at a given point x without truncation errors and without hand-coded formulae.

Numerical differentiation requires at least n additional function evaluations for one gradient calculation and induces truncation errors. Although very easy to implement, the numerical errors are often not tolerable, for example, when the derivatives are used within another numerical approximation scheme. A typical example is the differentiation of solutions of unstable differential equations in a parameter estimation problem.

Automatic differentiation overcomes the drawbacks mentioned, and is a very useful tool in practical applications that require derivatives. Meanwhile, there exists a large variety of different computer codes for automatic differentiation, see [134] for a review. They differ in the underlying design strategy, domain of application, mathematical method, implementation, and numerical performance.

There are several different approaches to apply automatic differentiation. First, there is the possibility to proceed from a given code in a language like C or FORTRAN, and to generate code for derivatives in the same language. A frequently used routine is ADIFOR developed by [22]. Another possibility is to proceed from a language that allows operator overloading like C++, PASCAL SC, or FORTRAN 90, and to compute higher order derivatives by this technique, see ADOL-C written by [92] or ADOL-F of [220].

Finally, there are codes that allow function input in a special modeling language and evaluation of derivative values during runtime by automatic differentiation, especially useful in case of interactive modeling systems. A possible implementation, PCOMP, is discussed in this chapter, see [55] or [157] for further details. A review of further literature and a more extensive discussion of symbolic and automatic differentiation is found in [89]. A summary of related papers is published by [91].

First, we have to investigate the question, how a nonlinear function is evaluated to prepare the application of an automatic differentiation technique. The idea is to break a given expression into elementary operations that can be evaluated either by internal compiler operations or by external function calls. For a given function f, the existence of a sequence $\{f_i\}$ of $m-n$ elementary functions is assumed, where each individual function f_i is scalar and defined on $\mathbb{R}^{n_i}$, $1 \leq n_i \leq m-1$ for $i = n+1, \ldots, m$. We define now the situation more formally by a pseudo-program.

Definition 18.1 *Let f be a scalar function defined on $\mathbb{R}^n$ and $x = (x_1, \ldots, x_n)^T \in \mathbb{R}^n$. A sequence of scalar functions f_i defined on $\mathbb{R}^{n_i}$, $i = n+1, \ldots, m$, is called a sequence of elementary functions for f, $m \geq n$, if there exists an index set J_i with $J_i \subset \{1, \ldots, i-1\}$, $|J_i| = n_i$ for each function f_i, $i = n+1, \ldots, m$, such that a function value $f(x)$ is evaluated according to the following program:*

$$\begin{aligned} &\textit{For} \quad i = n+1, \ldots, m \ \textit{let} \\ &\qquad x_i = f_i(x_k, k \in J_i) \ . \\ &\textit{Let} \quad f(x) = x_m \ . \end{aligned} \tag{18.2.1}$$

The proposed procedure for evaluating function values is implemented in every compiler or interpreter of a higher programming language, if we neglect code optimization. In computer science terminology, we say that a postfix expression is built in the form of a stack, which is then evaluated successively.

Thus, elementary function values can be obtained easily, and the corresponding technique is found in any introductory computer science textbook.

Note that for every function f, there exists at least one trivial sequence of elementary functions with $m = n + 1$ and $f_{n+1}(x) = f(x)$. For practical use, however, we assume that the functions f_i are basic machine operations and intrinsic or external functions, where the relative evaluation effort is limited by a constant independent of n. Under this condition, suitable bounds for the work ratio can be proved. The algorithm can be implemented efficiently using stack operations, which reduce the storage requirements as much as possible. There is no need to store all intermediate variables $x_{n+1}, \ldots, x_m$.

When investigating the above program for evaluating a function value $f(x)$, we see immediately that the gradient $\nabla f(x)$ is computed simultaneously in a very straightforward way. If we know how derivatives of elementary functions are obtained, we only have to include another program line for gradient updates by exploiting the chain rule. Therefore, we denote the resulting approach *forward mode*, also called forward accumulation, see [55]. The complexity of the algorithm is bounded by a constant times n, the number of variables, *i.e.*, the numerical work is of the same order of magnitude as for numerical differentiation.

To improve the efficiency of the gradient evaluation, another method can be considered, the so-called backward or reverse mode. We have to store all intermediate variable values x_i, $i = n + 1, \ldots, m$, before starting the reverse loop. Again, the chain rule is exploited to compute intermediate derivatives, see [55] for more details or any of the literature mentioned above. The reverse mode is also called reverse accumulation. An interesting result is the observation that the relative computational work, the quotient of the computational effort to evaluate $f(x)$ and $\nabla f(x)$ versus the computational effort to evaluate $f(x)$ alone, is bounded by the maximum work ratio of all elementary functions $f_i(x)$ included in the evaluation list of $f(x)$. See [89] for detailed assumptions needed to prove this statement.

The drawback of the reverse mode, however, is the necessity to store all intermediate data x_i, $i = n+1, \ldots, m$, which could become unacceptable depending on the storage organization of an actual implementation and the complexity of f. Thus, `PCOMP` uses both algorithms. For interpreting the functions and evaluating gradients at run time, the forward accumulation mode is preferred to avoid difficulties with limited memory. On the other hand, the alternate possibility to generate FORTRAN codes for function and gradient evaluation uses reverse accumulation.

For more information about the mathematical background of differentiation, see [135], [143], or [192].

Example 18.1 *A sophisticated test case is the evaluation of the Helmholtz equation, also studied by [89], given in the form*

$$f(x) = RT \sum_{i=1}^{n} x_i \log \left(\frac{x_i}{1 - b^T x} \right) - \frac{x^T A x}{\sqrt{8} b^T x} \log \left(\frac{1 + (1 + \sqrt{2}) b^T x}{1 + (1 - \sqrt{2}) b^T x} \right) . \tag{18.2.2}$$

For a numerical test, let A be the Hilbert-matrix with coefficients $a_{i,j} = 1/(i + j - 1)$, $i, j = 1, \ldots, n$, where $a_{i,j}$ denotes an element of A, and we set $b_i = 0.00001$ for $i = 1, \ldots, n$, where b_i is the i-th element of the vector b. Parameter values are $x_i = 2$ for $i = 1, \ldots, n$. Application of PCOMP leads to the results of Table 18.1. While the reverse version of automatic differentiation has a bounded work ratio W_{rev}, the work ratios for the forward mode W_{fwd} and for numerical approximation by forward differences W_{num} are proportional to n. However, the forward method exploits known intermediate values computed for function evaluation, and leads to a much more efficient gradient calculation.

Table 18.1. Work Ratios for Forward, Reverse, and Numerical Differentiation.

n	W_{rev}	W_{fwd}	W_{num}
25	3.03	4.04	26.01
50	3.13	6.63	51.01
100	3.76	11.88	100.99
200	3.33	22.47	200.89
400	3.70	45.77	400.88
800	3.49	81.59	802.56

18.3 The PCOMP Language

PCOMP is a special FORTRAN-similar modeling language allowing automatic differentiation, see [55] and [56] for details. Data, variables, and functions defining the mathematical model must be written on a text file and are parsed and precompiled internally before starting the optimization cycle. Starting from the generated intermediate code, function and gradient values are evaluated during run time. Particular advantage is that gradients, as far as needed, are calculated automatically without any numerical approximation errors, see also Section 18.2.

Declaration and executable statements must satisfy the usual FORTRAN input format and have to begin at column 7 or later. Comments beginning with `C` in the first column may be included in a program text wherever needed. Statements are continued on subsequent lines by adding a continuation mark in column 6. Either capital or small letters are allowed for identifiers of the user and key words of the language. Variables and functions must be declared separately only if they are used for automatic differentiation. PCOMP possesses special constructs to identify program blocks:

* PARAMETER
  Declaration of constant integer parameters to be used throughout the program, particularly for dimensioning index sets.

* SET OF INDICES
  Definition of index sets that can be used to declare data, variables and functions or to define sum or prod statements, in form of ranges, individual sets, or computed sets subject to numerical or formal parameters.

* INDEX
  Definition of an index variable, which can be used in a FUNCTION program block. An index is not taken into account during the automatic differentiation process.

* REAL CONSTANT
  Definition of real constants, either without index or with one- or two-dimensional index. An index may be a variable or a constant number within an index set. Arithmetic expressions can be included.

* INTEGER CONSTANT
  Definition of integer constants, either without index or with one- or two-dimensional index.

* TABLE <identifier>
  Assignment of constant real numbers to one- or two-dimensional array elements. In subsequent lines, one has to specify one or two indices followed by one real value per line in a free format (starting at column 7 or later).

* VARIABLE
  Declaration of variables, also indexed, with respect to which automatic differentiation is to be performed.

* CONINT <identifier>
  Declaration of a piecewise constant interpolation function by a sequence of data points, one pair per line.

* LININT <identifier>
  Declaration of a piecewise linear interpolation function.

* SPLINE <identifier>
  Declaration of a spline interpolation function.

* `MACRO <identifier>`
  Definition of a macro, an arbitrary set of PCOMP statements that define an auxiliary function to be inserted into subsequent declaration blocks. Macros are identified by a name to be used in any right-hand side of an assignment statement.

* `FUNCTION <identifier>`
  Declaration of functions, also indexed, for which function and derivative values are to be evaluated. The subsequent statements must assign a numerical value to the function identifier.

* `END`
  End of the program.

Following a syntax similar to FORTRAN, assignment statements are used to pass a numerical value to an identifier, which may be either the name of the function that is to be defined or of an auxiliary variable that is used in subsequent expressions. Allowed intrinsic functions are

```
ABS, SIN, COS, TAN, ASIN, ACOS, ATAN, SINH, COSH
TANH, ASINH, ACOSH, ATANH, EXP, LOG, LOG10, SQRT
```

Alternatively, the corresponding double precision FORTRAN names possessing an initial `D` can be used as well.

Sums and products over predetermined index sets are formulated by `SUM` and `PROD` expressions, where the corresponding index and the index set must be specified, for example in the form

```
*     FUNCTION f
      f = 100*PROD(x(i)**a(i), i IN inda)
```

In the above example, `x(i)` is a variable vector defined by an index set and `a(i)` a data array.

To control the execution of a program, the conditional statements

```
   IF ⟨condition⟩ THEN
     ⟨statements⟩
ENDIF
```

or

```
IF ⟨condition⟩ THEN
   ⟨statements⟩
ELSE
   ⟨statements⟩
ENDIF
```

can be inserted into a program. Conditions are defined as in FORTRAN by the comparative operators `.EQ.`, `.NE.`, `.LE.`,

`.LT.`, `.GE.`, `.GT.`, which can be combined using brackets and the logical operators `.AND.`, `.OR.` and `.NOT.`.

The `GOTO`– and the `CONTINUE`–statements are further possibilities to control the execution of a program. The syntax for these statements is

`GOTO` ⟨label⟩

and

⟨label⟩ `CONTINUE`

where `label` has to be a number between 0 and 9999. The `<label>` part of the `CONTINUE`-statement must be located between columns 2 and 5 of an input line. Together with an index, the `GOTO`-statement can be used to simulate `DO`-loops.

Whenever indices are used within arithmetic expressions, it is allowed to insert polynomial expressions of indices from a given set. However, functions must be treated in a particular way. Since the design goal is to generate short, efficient codes, indexed function names can be used only in exactly the same way as defined. In other words, if a set of functions is declared by

```
* FUNCTION  f(i),  i IN index
```

then only an access to `f(i)` is allowed, not to `f(1)` or `f(j)`, for example. `PCOMP` does not extend indexed functions to a sequence of single expressions similar to the treatment of `SUM` and `PROD` statements.

Example 18.2 *We consider again the Helmholtz energy function (18.2.2). The corresponding* PCOMP *implementation as used in the numerical tests of Section 18.2, is listed:*

```
*     PARAMETER
      n = 100
C
*     SET OF INDICES
      index = 1..n
C
*     REAL CONSTANT
      R = 8.314
      T = 273
      c1 = 1 + SQRT(2)
      c2 = 1 - SQRT(2)
      c3 = SQRT(8)
      A(I,J) = 1/(i+j-1), i IN index, j IN index
      B(i) = 0.00001, i IN index
C
*     VARIABLE
      x(i), i IN index
C
*     FUNCTION f
      Bx = SUM(B(i)*x(i), i IN index)
      xAx = SUM(x(i)*SUM(A(i,j)*x(j), j IN index), i IN index)
      f = R*T*SUM(x(I)*LOG(x(i)/(1 - Bx)),i IN index)
     /             - xAx*LOG((1 + c1*Bx)/(1 + c2*Bx))/(c3*Bx)
```

```
C
*     END
```

18.4 Program Organization

The PCOMP system consists of three sets of subroutines that can be executed completely independently from each other. There are also some auxiliary routines, in particular an error routine called SYMERR to make error messages readable, and a routine with the name SYMPRP to read intermediate code generated by the parser. All routines are implemented in FORTRAN 77 and tested on the following systems: VAX/VMS, HP-UX, MS-DOS, MS-Windows (Watfor, Microsoft, Lahey, Compaq, Absoft). In this section, we give only a short overview of the way PCOMP works, for more details we refer to [55] and [56].

1. Parser: The source code is analyzed and compiled into an intermediate code, which can then be processed by the other routines. The subroutine to be executed has the name SYMINP. The syntax of PCOMP is described in the form of a formal grammar. The parser was generated in C by the *yacc* compiler-compiler of UNIX and then transformed to FORTRAN *by hand.* If there is any error during the processing of the input file, SYMINP is interrupted and the error code can be retrieved from the integer parameter IERR. This error code and the corresponding line number can be passed to SYMERR to generate an error message on standard output. If no errors have been detected, the generated intermediate code and additional data are stored in a double precision and an integer working array. These two working arrays must be passed to all subsequently called subroutines for function and derivative evaluation or for code generation, respectively.

2. Direct interpretation of intermediate code: The intermediate code generated by SYMINP is passed to SYMFUN in the form of a real and an integer working array. Given any variable vector, this subroutine computes the corresponding function values by interpreting the intermediate code. The subroutines SYMGRA and SYMHES are very similar to SYMFUN. The only distinction is that they compute the gradients or gradients and Hessians of the symbolically defined functions in addition by the forward mode of automatic differentiation. An additional index array indicates for which variables the first or second derivatives are to be evaluated. Moreover, it is possible to calculate function and derivative values only for a certain subset of all functions declared.

3. Generation of FORTRAN code: Proceeding from the intermediate code generated by the parser SYMINP, the subroutine SYMFOR generates two FORTRAN subroutines for function and gradient evaluation in a given output file. These routines have to be compiled and linked separately. The calling sequences of the generated subroutines XFUN and XGRA are similar to that of SYMFUN

and SYMGRA, except that no working arrays are required. In this case, the reverse mode of automatic differentiation is applied.

The basic idea of PCOMP is to factorize a given function $f(x)$ with respect to a set of library functions, for which derivatives can easily be computed, for instance elementary arithmetic and intrinsic functions. In some applications, it is desirable to extend the list of library functions and to allow for user-provided symbols in the source code. Thus, PCOMP can easily be extended to accept additional functions by inserting information about structure, type, and symbolic name into the parser and by defining subroutines for function, gradient, and Hessian evaluation called EXTFUN, EXTGRA, and EXTHES. For more details and an example see [55].

18.5 Case Study: Interactive Data Fitting with EASY-FIT

PCOMP became part of an interactive software system called EASY-FIT that performs parameter estimation by a least squares fit, see [210] or [211]. The mathematical model has to belong to one of the following categories:

explicit model functions (175)
steady state systems (38)
Laplace transformations of differential equations (8)
ordinary differential equations (463)
differential algebraic equations (34)
one-dimensional partial differential equations (247)
one-dimensional partial differential algebraic equations (35)

Proceeding from given experimental data, *i.e.*, observation times and measurements, the minimum least squares distance of measured data from a fitting criterion is computed that depends on the solution of the dynamical system. EASY-FIT is implemented in form a database that contains model information, experimental data, and results, whereas the numerical routines are written in FORTRAN. They are executed by the shell procedure of Visual Basic and can also be used independently from the GUI. A context sensitive help option is included for additional technical and organizational information, for example about the input of data and optimization tolerances. Most applications are from pharmacy, biochemistry, chemical engineering, physics, and mechanical engineering. At present, EASY-FIT contains 1,000 *real life* and academic data fitting test examples, distributed among the different model variants as shown in the brackets above.

The main goal of EASY-FIT is to identify parameters in dynamical systems, *i.e.*, parameters that are *hidden* in additional time-dependent state equations that must be solved implicitly. The basic mathematical structure is a least squares formulation of our data fitting problem, *i.e.*, minimization of a sum of squares of nonlinear functions of the form

$$
\begin{array}{lll}
 & \min \sum_{r=1}^{l} f_r(x)^2 & \\
 & g_j(x) = 0 \,, & j = 1, ..., m_e \,, \\
p \in \mathbb{R}^n : & g_j(x) \geq 0 \,, & j = m_e + 1, ..., m \,, \\
 & x_l \leq x \leq x_u & .
\end{array}
\qquad (18.5.3)
$$

Here we assume that the parameter vector x is n-dimensional and that all nonlinear functions are continuously differentiable with respect to x. Upper and lower bounds are treated independently from the remaining constraints.

To get a data fitting problem, we suppose that one vector-valued model function $h(x, t)$ is available, the so-called *fitting criterion function*, depending on the parameter vector x to be identified, furthermore on an additional variable t called time. To simplify the notation, we omit other independent model parameters that could be required, for example for the concentration of a substrate.

Now we proceed from n_e sets of experimental measurements in the form (t_i, y_i^k), $i = 1, \ldots, n_t$, $k = 1, \ldots, n_e$, where n_t time values and $n_t n_e$ corresponding experimental measurement values are defined. Together with an n_e dimensional model function $h(x, t)$, we get the above least squares formulation by defining

$$
f_r(x) = w_i^k (h_k(x, t_i) - y_i^k) \,, \qquad (18.5.4)
$$

where r runs from 1 to $l = n_t n_e$ in any order. Moreover we assume that there are suitable weighting factors $w_i^k \geq 0$ given by the user that are to reflect the individual influence of a measurement for the whole experiment.

The underlying idea is to minimize the distance between the model function at certain time and concentration points and the corresponding measurement values. This distance is called the residual of the problem. In the ideal case, the residuals are zero indicating a perfect fit of model function by measurements.

If we assume that $h(x, t)$ does not depend on the solution of additional state equations of a dynamical system, we call it an explicit model function. Otherwise, $h(x, t)$ may depend in addition on the solution vector of an auxiliary problem, e.g., an ordinary differential equation, that is implicitly defined.

To outline the basic idea and the usage of the modeling language PCOMP, let us consider only one specific case of a dynamical model, a system of ordinary differential equations. The fitting criterion is given in the form

$$
h(x, t) = \overline{h}(x, y(x, t), t) \,, \qquad (18.5.5)
$$

where $y(x, t)$ denotes the solution of a system of ordinary differential equations

$$
\dot{y} = F(x, y, t) \,, \quad y(0) = y^0(x) \,. \qquad (18.5.6)
$$

Without loss of generality we assume that the initial time is zero. The initial values $y^0(x)$ may depend on the system parameters to be estimated.

There are many practical situations, where a model changes during integration over the time, and where initial values are to be adopted. A typical example is a chemical reactor model with non-continuous, time-dependent input functions. Thus, break or switching points τ_i with $0 < \tau_1 < \ldots < \tau_{n_b}$ are allowed, where the initial values for each sub-interval are given by any functions $y_j^i(x, y)$ depending on the parameters to be estimated and the solution of the previous interval at the break point τ_i. Internally, the integration of the differential equation is restarted at a break point. It is possible that break points become variables to be adapted during the optimization process, if they are not known in advance.

Constraints of the form (18.5.3) are permitted, where the restriction functions

$$g_j(x) = \overline{g}_j(x, y(p, t_j), t_j) \quad , \quad j = 1, ..., m. \tag{18.5.7}$$

may depend on the solution of the dynamical system at predetermined time values, t_j. To compute derivatives of the fitting function $h(x, t) = \overline{h}(x, y(x, t), t)$ w.r.t. x, we have to get the partial derivatives of $\overline{h}(x, y, t)$ w.r.t. x and w.r.t. y and the partial derivative of the solution of the differential equation, $y(x, t)$, w.r.t. x at fixed t values to be able to apply the chain rule. The computation of $\nabla_x h(x, t)$ should be as exact as possible to provide the underlying least squares algorithm, for example a Gauss-Newton method, with highly accurate gradients.

More complex is the differentiation of the solution of a differential equation, see [211] for a review. Among the various alternatives, there are the simultaneous solution of sensitivity equations or, more or less equivalent, the internal numerical differentiation, where the integration scheme, say an explicit higher order Runge-Kutta method, is differentiated [27]. In both cases, we need the derivatives of the right-hand side of (18.5.6), *i.e.*, of $F(x, y, t)$ w.r.t. x and y, and the differentiation of the initial values $y^0(x)$ w.r.t. the parameters. It is essential to have possibility to compute derivatives for certain subsets of given functions in different situations w.r.t. different subsets of variables.

Another reason for applying automatic differentiation, is the integration of (18.5.6) by implicit methods in case of stiff equations. In this case, the Jacobian of $F(x, y, t)$ w.r.t. the state variables y is required, again with high precision because of the inherent instability in the dynamical system.

Example 18.3 *To give an example, we consider a particular application, the cooling of a crystallization process by the Miller and Parsival formulation. The system of five differential equations depends on some constants and spline interpolation of a few hundred noisy tabular data. The fitting criterion is a highly nonlinear function of some of the state variables* m, μ_a^0, μ_a^1, μ_a^2, *and* μ_a^3. *Parameters to be estimated, are* k_g *and* k_b *proceeding from 270 experimental data for the first and 96 data for the second criterion.*

```
C
*       REAL CONSTANT
```

```
      mh2o = 1659.8
      rhos = 2.11
      kv = 1
      ka = 6
      ls = 0.2
      g = 1.32
      b = 1.78
C
C-----------------------------------------------------
C
*     SPLINE Te
       0.0         32.0400
       0.35000     32.0190
       0.60000     31.9953
       ...
       80.3600     28.9227
       80.6100     28.9189
       80.8600     28.9252
       81.1100     28.9297
C
C-----------------------------------------------------
C
C   - Independent variables in the following order:
C     1. parameters to be estimated (x)
C     2. variables identifying solution of ordinary
C        differential equations (y)
C     3. concentration variable, if exists (c)
C     4. time variable (t)
C
*     VARIABLE
      kg, kb, m, mue0a, mue1a, mue2a, mue3a,  t
C
C-----------------------------------------------------
C
C   - Functions defining right-hand side of system of
C     ordinary differential equations:
C
*     FUNCTION dm_t
      cs = 0.1286 + 5.88E-3*Te(t) + 1.721E-4*Te(t)**2
      S = (m/mh2o - cs)/cs
      IF (S.GT.0) THEN
        xG = kg*S**g
      ELSE
        xG = 0
      ENDIF
      dm_t = -3*kv*rhos*xG*mue2a
C
*     FUNCTION dmue0a_t
      IF (S.GT.0) THEN
        xB = EXP(kb)*mue3a*S**b
      ELSE
        xB = 0
      ENDIF
      dmue0a_t = xB
C
*     FUNCTION dmue1a_t
      IF (S.GT.0) THEN
```

```
          xG = kg*S**g
        ELSE
          xG = 0
        ENDIF
        dmue1a_t = xG*mue0a
C
*       FUNCTION dmue2a_t
        IF (S.GT.0) THEN
          xG = kg*S**g
        ELSE
          xG = 0
        ENDIF
        dmue2a_t = 2*xG*mue1a
C
*       FUNCTION dmue3a_t
        IF (S.GT.0) THEN
          xG = kg*S**g
        ELSE
          xG = 0
        ENDIF
        dmue3a_t = 3*xG*mue2a
C
C-------------------------------------------------------------
C
C   - Initial values for solving systems of ordinary
C     differential equations (t=0):
C
*       FUNCTION m0
        m0 = 0.4922*mh2o
C
*       FUNCTION mue0a0
        mue0a0 = 0.811*mh2o
C
*       FUNCTION mue1a0
        mue1a0 = 1.59E-2*mh2o
C
*       FUNCTION mue2a0
        mue2a0 = 3.11E-4*mh2o
C
*       FUNCTION mue3a0
        mue3a0 = 6.1E-6*mh2o
C
C-------------------------------------------------------------
C
C   - Fitting criteria:
C
*       FUNCTION FIT1
        FIT1 = m/mh2o
C
*       FUNCTION FIT2
        X = m/mh2o
        rho28 = 0.4214*X + 1.0236
        rho32 = 0.5433*X + 0.9667
        rhof  = (rho32 - rho28)/4.0*(Te(t) - 28) + rho28
        Vges  = (m + mh2o)/rhof + kv*mue3a
        convert = mh2o/Vges
        FIT2 = EXP(-ka/2*ls*convert*mue2a/mh2o)
```

```
C
C--------------------------------------------------
C
*       END
C
```

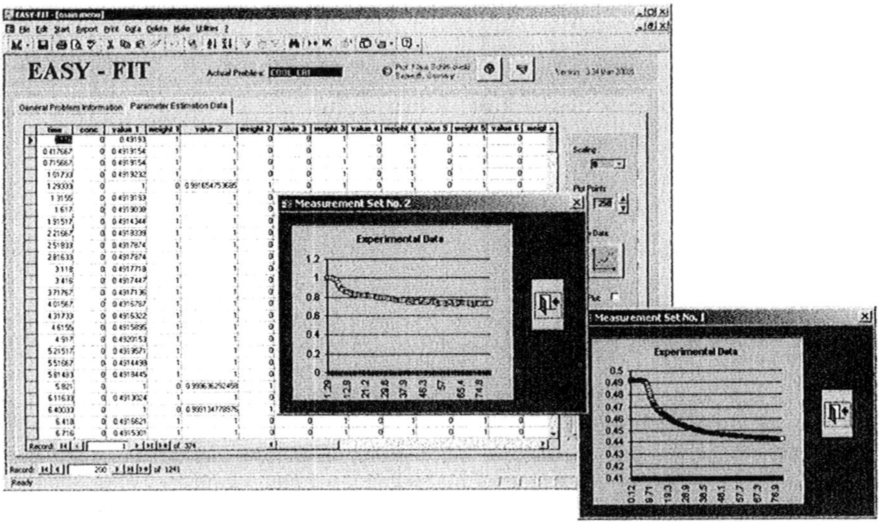

Figure 18.5.1. Experimental Data

The interactive implementation of the model and the identification of unknown parameters is summarized by five steps:

Step 1: *Create a new problem in the database, insert some information strings and in particular experimental data, see Fig. 18.5.1.*

Step 2: *Choose type of dynamical model (ODE, PDE, ...), define the model structure and set discretization parameters, see Fig. 18.5.2.*

Step 3: *Implement model equations and check correct syntax, see Fig. 18.5.3 or the listing given below.*

Step 4: *Define parameters to be estimated, select least squares solver, set termination tolerances, and start data fitting run, see Fig. 18.5.4.*

Step 5: *Check report, especially parameter values and residuals, see Fig. 18.5.5.*

Starting from $k_g = 0.76$ *and* $k_b = 14.204$, *the least squares code DFNLP of [208] computes the optimal parameters* $k_g = 0.76$ *and* $k_b = 14.204$ *after 12 iterations subject to a termination accuracy of* 10^{-7}. *The total residual is reduced from 1.14 to 0.00573. If necessary, model equations, data, or tolerances are refined and a data fitting run is repeated, until parameters are identified up to desired precision.*

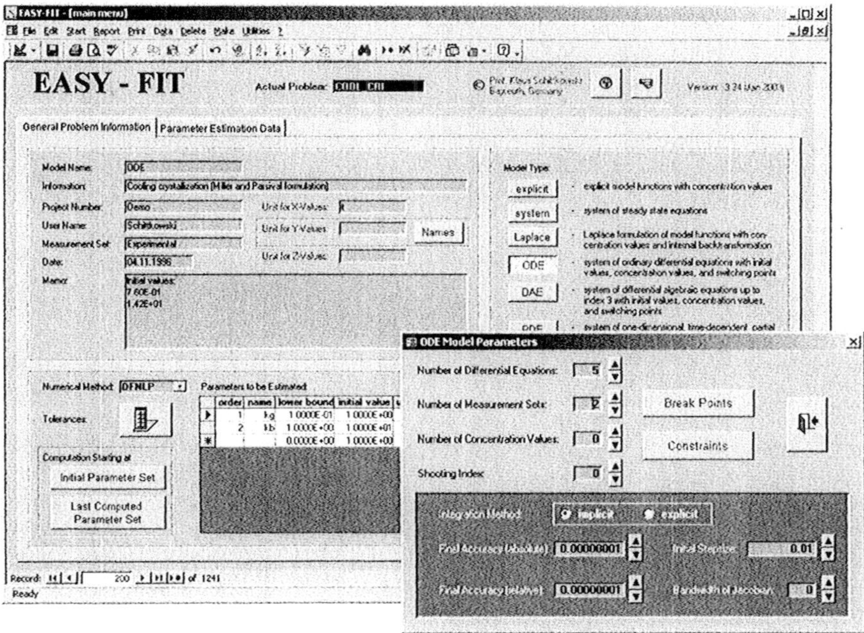

Figure 18.5.2. Model Structure and Discretization

18.6 Summary

We introduced a modeling language called PCOMP which is designed for applications in engineering and science, where highly nonlinear equations are used to describe the behavior of a real-life system. Of particular interest is the question how derivatives of these functions w.r.t. to some variables can be computed. PCOMP consists of a collection of FORTRAN subroutines which can be adapted to more specific domains of applications. A typical example is its usage within the interactive software system EASY-FIT for estimating parameters in dynamical systems, where derivatives of fitting functions, but also of solutions of differential equations w.r.t. to some parameters need to be computed as accurately as possible. Proceeding from intermediate code generated by the parser, the PCOMP routines can be implemented for direct interpretation of this code or for generating FORTRAN code for function and derivative calculations. The automatic differentiation routines are based on forward and reverse accumulation.

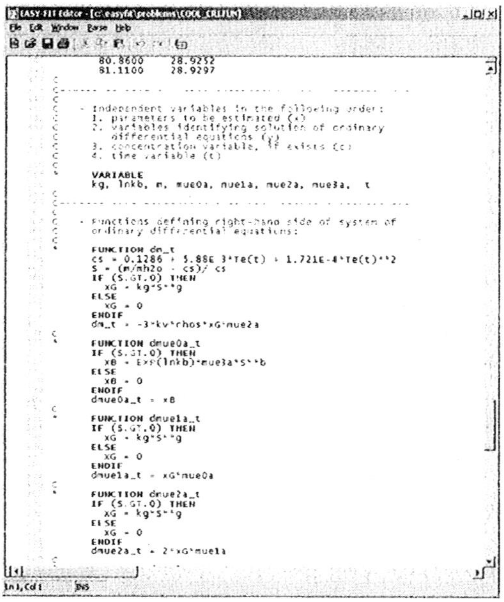

Figure 18.5.3. Dynamical Equations

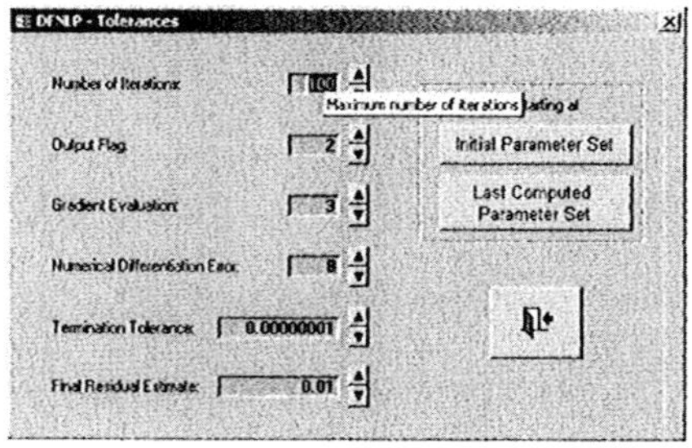

Figure 18.5.4. Prepare Data Fitting

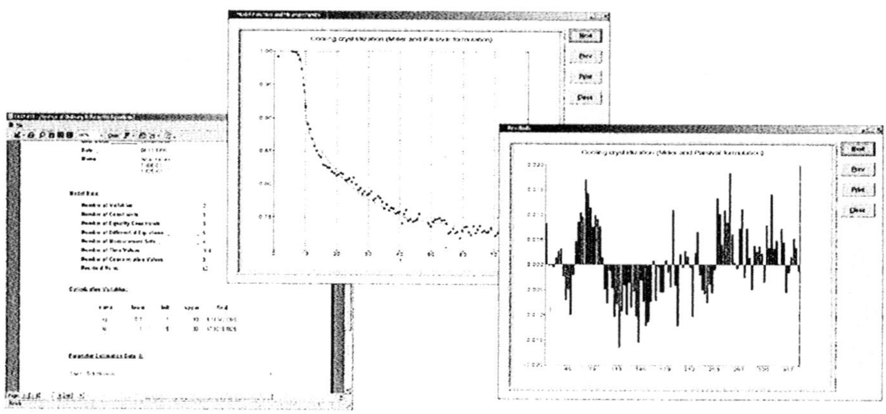

Figure 18.5.5. Interpretation of Results

Chapter 19

THE TOMLAB OPTIMIZATION ENVIRONMENT

Kenneth Holmström
Mälardalen University, Department of Mathematics and Physics
Västerås, Sweden
hkh@tomlab.BIZ

Marcus M. Edvall
Tomlab Optimization Inc.
Arcata, CA, USA
medvall@tomlab.BIZ

Abstract The TOMLAB Optimization Environment is a powerful optimization tool in MATLAB, which incooperates many results from the last 40 years of research in the field. More than 70 different algorithms for linear, discrete, global and nonlinear optimization are implemented in TOMLAB, and a large number of C and Fortran solvers are also fully integrated. The environment is call-compatible with MathWorks' Optimization Toolbox, and supports problems formulated in AMPL. This chapter discusses the design and contents of TOMLAB, and examplifies its usage on a practical optimization problem. The objective is to present the overall design and describe how to efficiently model a problem in TOMLAB using the standard structures and assign statements. More information about TOMLAB is available at URL: `http://tomlab.BIZ`.

Keywords: TOMLAB, MATLAB, Mathematical Software, Optimization.

19.1 Introduction

One major feature of algebraic modeling languages, such as AMPL [76] and GAMS [23, 31], is that they separate the model building from the solution process. Once the model is formulated, several different solvers can be used to solve the problem, and the default settings for the solvers are usually sufficient to obtain the desired solution. MATLAB [177] is a very strong modeling language, but it is missing a uniform approach to the optimization solution process. The TOMLAB optimization environment rectifies this situation and improves the solver interaction, with the aid of a GUI, and enables advanced use of callback algorithms when solving mixed-integer problems. TOMLAB by itself is not a complete modeling language, but combined with MATLAB it can be viewed as an alternative to one of the algebraic modeling languages.

The last two decades have seen an increase in the use of MATLAB as an advanced tool for mathematical model building and analysis. The language makes it easy to formulate any type of optimization problem, with added complexity from the included mathematical libraries. A number of language extensions in the form of toolboxes give the users special solutions in many fields; an example is the FEMLAB toolbox [47] for multi-physics modeling. As part of the advanced modeling and simulation, optimization problems often need to be solved. The TOMLAB Optimization Environment introduces support for advanced large-scale optimization, which has been very limited in MATLAB so far.

This chapter presents some of the design ideas of TOMLAB and how a uniform format for optimization is obtained. The chapter is organized as follows. Section 19.2 describes the use of MATLAB, Section 19.3 outlines the development of TOMLAB, and in Section 19.4 the design of TOMLAB is discussed, with a special emphasis on the modeling aspects. An example of a constrained nonlinear problems finally illustrates the usage of TOMLAB in Section 19.5.

19.2 MATLAB as a Modeling Language

MATLAB was originally created as a preprocessor to LINPACK [59] and EISPACK [222] [80], the standard FORTRAN subroutine libraries in numerical linear algebra. The latest version of MATLAB efficiently integrates the numerical linear algebra package LAPACK [3].

MATLAB of today is an advanced and powerful tool, which integrates mathematics with graphics (see Fig. 8.3.3), animation and tools for building GUI's. The current MATLAB version has many advanced language features that TOMLAB utilizes in its design like structure arrays, object classes and multi-dimensional cell arrays. High-level matrix operations on both dense and sparse matrices are available, which makes it suitable for modeling of optimization problems. Pre- and post-processing becomes easy and complicated objective functions are readily formulated, such as parameter estimation problems in partial differential equations or design parameter optimization in simulation studies. Another main features that TOMLAB uses, is the ability to interface programs written in FORTRAN and C.

MATLAB lacks some of intuitive problem formulations enabled by algebraic languages but by combining the MATLAB language features with TOMLAB it has become less of an issue. It is difficult for a specialized modeling language to follow the general development and enhancements that constantly are expanding a general-purpose language like MATLAB. The widespread use of MATLAB (500,000+ users) and multi-platform support is also to TOMLAB's advantage [1], as it minimizes training requirements, reduces development time of optimization applications and gives novices direct access to leading optimization packages.

19.3 The TOMLAB Development

The lack of good optimization tools in MATLAB motivated the early development of some optimization solvers already in 1990. As the number of solvers and tools increased they were collected in two toolboxes; NLPLIB TB for nonlinear programming and parameter estimation [112]; and OPERA TB for linear and discrete optimization [113]. The integrated concept of TOMLAB emerged in 1997 when the first TOMLAB version 1.0 was released [107, 106]. It handled small and medium size dense problems and was free for academic use. Around 1,000 users downloaded the software during the year it was introduced.

The next version, TOMLAB v2.0 [109, 110], also handled large sparse problems and the user could convert the code to C++ and compile it for faster execution, by using the now obsolete MIDEVA system in MATLAB. The TOMLAB system was completely revised in version 3.0. In addition to some 95,000 lines of code in about 450 files, TOMLAB also had around twenty FORTRAN solvers interfaced [111]. TOMLAB v4.0 improves the use of the environment by introducing even more problem types and extends the list of solver packages [2]. The current version also includes some unique features, like Semi-definite Programming with bilinear matrix inequalities (see Section 4.2.2), costly (CPU-intensive) global black-box optimization (see Section 1.2.1) and special algorithms for exponential model fitting and approximation [114]. TOMLAB as of now more than well meets the requirements by the practitioners, as set forth in Section 5.2. Users of v4.0 report solutions of *very* large-scale problems, where the available memory is the limit.

TOMLAB is used for several different purposes. In the Applied Optimization Group at Mälardalen University, it has been used to develop algorithms and solve a number of applied optimization projects [108, 17, 102]. At the Stanford Systems Optimization Laboratory it is used as a subproblem solver in the

[1] TOMLAB is readily available on Windows, Linux, SGI, HP, MAC, DEC and SUN systems.

[2] The following solvers packages are available TOMLAB: TOMLAB BASE, MINOS, NPSOL, SNOPT, SOL, CGO, Xpress-Optimizer, CPLEX, MINLP, PENSDP, PENBMI, KNITRO, OQNLP, CONOPT, NLPQL, and shortly Xpress-MISLP, PDEFIT, MODFIT, DECIS, PORTF and PENNLP

development of new numerical algorithms. Feedback from hundreds of users shows that TOMLAB is used for a large variety of applied optimization problems.

- Halliburton Energy Services uses TOMLAB for application development.
- US Naval Research's program DIDO [200] integrates TOMLAB .
- TOMLAB has been embedded in portfolio optimization software.
- Lumileds uses TOMLAB for production planning optimization.

19.4 The Design of TOMLAB

The scope of TOMLAB is large and broad, thereby increasing the need for a well-designed system. It is natural to use the power of the MATLAB language to make the system flexible as well as easy to use and maintain. The concept of structure and cell arrays are used and the ability to execute MATLAB code defined as string expressions. Dense and sparse matrices with multiple indices may also be defined in MATLAB. A multiple index matrix can be converted to its two-dimensional counterpart in MATLAB by concatenating the different sub-matrices. The user would simply iterate through the higher indices until the matrix has two dimensions. This is a very useful feature when modeling practical large-scale problems, as natural sets of variables and constraints are often found, and as index sets are commonly used to get a more structured problem description.

Vectors of structure arrays are suitable for large sets of problems. TOMLAB is designed for nonlinear multi-stage optimization with any level of recursion.

19.4.1 Structure Input and Output

Normally, when solving an optimization problem in MATLAB, a direct call to a solver is made with a long list of solver-dependent parameters in the call. TOMLAB solves the problem in two steps. First, the problem is defined and assigned to a MATLAB structure. Second, the solver is called with at least one argument. Solvers that were not originally developed for the TOMLAB environment need a long list of parameters. This is handled by the driver routine *tomRun*, which can call any available solver, thereby hiding the details of the call from the user. The solver output is collected in a standardized result structure and returned to the user.

19.4.2 Description of the Input Problem Structure

The Input Problem Structure, here referred to as *Prob*, is one of the most central aspects of working with TOMLAB. It contains numerous fields and sub-structures that define the problem and influence the behavior and performance of the solvers. Table 19.1 shows a small selection of the fields available in *Prob*. TOMLAB uses several fields in the *Prob* for the parameters controlling the execution of the optimization solvers. The user can easily specify sub-solvers and much more using these pre-defined fields.

Table 19.1. Model information stored in the problem structure *Prob*. Fields defining sub-structures are defined in Table 19.2.

Field	**Description**
A	Matrix with linear constraints, one constraint per row.
b_L / b_U	Lower/Upper bounds on the linear constraints.
c_L / c_U	Lower/Upper bounds on the general constraints.
ConsDiff / NumDiff	Determines the numerical approximation method to be used for the constraint derivatives and objective function, respectively.
ConsPattern	Non-zero pattern in the constraint gradient matrix.
HessPattern	Matrix with non-zero pattern in the Hessian matrix.
JacPattern	Matrix with non-zero pattern in the Jacobian matrix.
LargeScale	Flag if the problem is large scale.
x_0	Starting point.
x_L / x_U	Lower/Upper bounds on the variables x.

Table 19.2. Fields defining sub-structures in the problem structure *Prob*.

Field	**Description**
QP	Structure with special fields for linear and quadratic problems.
LS	Structure with special fields for least squares problems.
MIP	Structure with special fields for mixed-integer programming.
ExpFit	Structure with special fields for exponential fitting problems.
PartSep	Structure with special fields for partially separable functions.
USER	Structure with user defined names of the m-files computing the objective, gradient, Hessian etcetera. The *Prob.USER* fields are listed below.
f	Name of m-file computing the objective function $f(x)$.
g	Name of m-file computing the gradient vector $g(x)$. If *Prob.USER.g* is empty then numerical derivatives will be used.
H	Name of m-file computing the Hessian matrix $H(x)$.
c	Name of m-file computing the vector of constraint functions $c(x)$.
dc	Name of m-file computing the matrix of constraint normals $\partial c(x)/dx$.
d2c	Name of m-file computing the 2nd part of 2nd derivative matrix of the Lagrangian function, $\sum_i \lambda_i \partial^2 c(x)/dx^2$.
r	Name of m-file computing the residual vector $r(x)$.
J	Name of m-file computing the Jacobian matrix $J(x)$.
d2r	Name of m-file computing the 2nd part of the Hessian for nonlinear least squares problem, *i.e.*, $\sum_{i=1}^{m} r_i(x) \frac{\partial^2 r_i(x)}{\partial x_j \partial x_k}$.

19.4.3 Defining an Optimization Problem

The different types of optimization problems that can be modeled in TOMLAB are listed in Table 19.3. The problem formulations in TOMLAB follows a standardized structure. As an example, the *constrained nonlinear optimization* problem (**con**) is defined as

Table 19.3. Problems and assignments in TOMLAB.

probSet	probType	Description of problem set
uc	1	Unconstrained problems.
qp	2	Quadratic programming problems.
con	3	Constrained problems.
ls	4	Nonlinear least squares problems.
lls	5	Linear least squares problems.
cls	6	Linear constrained nonlinear least squares problems.
mip	7	Mixed-integer programming problems.
lp	8	Linear programming problems.
glb	9	Box-bounded global optimization problems.
glc	10	Global MINLP problems.
miqp	11	Constrained mixed-integer quadratic problems.
minlp	12	Constrained mixed-integer nonlinear problems.
lmi	13	Semi-definite programming with Linear Matrix Inequalities.
bmi	14	Semi-definite programming with Bilinear Matrix Inequalities.
exp	15	Exponential fitting problems.
nts	16	Nonlinear time series problems.
◇Assign		Assign statement, where ◇ is one of the *probSets*

$$\min_x f(x) \quad , \quad s.t. \begin{array}{ccccc} x_L & \leq & x & \leq & x_U, \\ b_L & \leq & Ax & \leq & b_U, \\ c_L & \leq & c(x) & \leq & c_U \end{array} \tag{19.4.1}$$

where $x, x_L, x_U \in \mathbb{R}^n$, $f(x) \in \mathbb{R}$, $A \in \mathbb{R}^{m_1 \times n}$, $b_L, b_U \in \mathbb{R}^{m_1}$ and $c_L, c(x), c_U \in \mathbb{R}^{m_2}$. Equality constraints are defined by setting the lower bound equal to the upper bound, *i.e.*, for linear constraint i: $b_L(i) = b_U(i)$. The problem is easily re-defined to a mixed-integer problem by setting the variables $x \in I$, where I is the index subset of $1, ..., n,$.

All problem types have a similar structure in TOMLAB. Once the applicable variables have been defined according to Section 19.4.2, the problem can be assigned to the *Prob* structure using an applicable assign statement. An LP problem is assigned to the *Prob* by executing the following command.

```
Prob = lpAssign(c, A, b_L, b_U, x_L, x_U, x_0);
```

TOMLAB has several different test problems sets available for distribution, including AMPL problems, Hoch-Schittkowski constrained and unconstrained problems [104], CUTE (ordinary and large) constrained and unconstrained problems [28] as *dll*-files, as well as More, Garbow, Hillstrom nonlinear least squares problems [178].

19.4.4 Solving Optimization Problems

The user can either use the driver routine *tomRun* or call the solver directly. The problem structure *Prob* is a required input for both options. When using *tomRun* a solver name also needs to be specified.

Depending on the type of problem, the user needs to supply *low-level* routines that calculate the objective function, constraint functions and derivatives if possible, see Table 19.2. TOMLAB *gateway* routines ($\diamond\_f$, $\diamond\_g$, $\diamond\_H$, $\diamond\_c$, $\diamond\_dc$, $\diamond\_d2c$, $\diamond\_r$, $\diamond\_J$, $\diamond\_d2r$, where $\diamond =$ nlp) ensure that each solver obtains the values in the correct format and simplifies the coding process. For example, when working on a least squares problem, it is natural to code the function that computes the vector of residual functions $r_i(x_1, x_2, \ldots)$. A dedicated least squares solver probably operates on the residual while a general nonlinear solver needs a scalar function, in this case $f(x) = \frac{1}{2}r^T(x)r(x)$. Such issues are automatically handled in TOMLAB. Fig. 19.4.1 illustrates the chain of calls when computing the objective function value in *ucSolve* (one of the TOMLAB solvers) for a nonlinear least squares problem defined in *mgh_prob*, *mgh_r* and *mgh_J*. The gateway

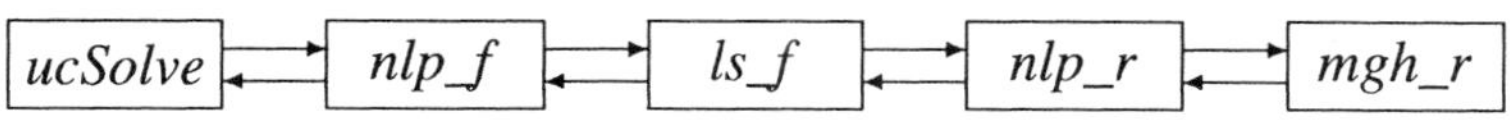

Figure 19.4.1. The chain of calls when computing the objective function value in *ucSolve* for a nonlinear least squares problem defined in *mgh_prob*, *mgh_r* and *mgh_J*.

routines also extract the search directions and line search steps, count iterations, handle separable functions, keep track of the differentiation method and handle special algorithms like separable nonlinear least squares.

Only the low level routines relevant for a certain type of optimization problem need to be coded. If user routines for the gradient, Jacobian and constraint gradient are unavailable, one of the five numerical differentiation methods in TOMLAB will be used instead. It is recommended that the routines are explicitly coded as derivative estimation is costly and reduces numerical accuracy. Automatic differentiation may also be used as TOMLAB is fully integrated with the MAD and ADMAT toolboxes in MATLAB.

To conclude, the system design is flexible and easy to expand in many different ways.

19.5 A Nonlinear Programming Example

This example shows how a nonlinear problem is modeled and solved in TOMLAB using the solver MINOS. The following equations define the problem.

$$
\begin{array}{llcccl}
\min_x & f(x) = exp(x_1)(4x_1^2 + 2x_2^2 + 4x_1x_2 + 2x_2 + 1) & & & & \\
 & -10 & \leq & x_1 & \leq & 10 \\
 & --10 & \leq & x_2 & \leq & 10 \\
s/t & 0 & \leq & x_1 + x_2 & \leq & 0 \\
 & 1.5 & \leq & -x_1x_2 + x_1 + x_2 & & \\
 & -10 & \leq & x_1x_2 & &
\end{array}
\qquad (19.5.2)
$$

The user needs to extract some information about the problem, such as lower bounds, upper bounds, linear constraints and possibly identify problem patterns. The MATLAB code below illustrates how TOMLAB solves the problem.

```
Name='Nonlinear problem, 1 linear eq.+ 2 nonlinear ineq.';
% This problem probably have several minima.
A       = [1 1];      % One linear constraint
b_L     = 0;          % Lower bound on linear constraint
b_U     = 0;          % b_L == b_U implies equality
c_L     = [1.5;-10]   % Two nonlinear inequality constraints
c_U     = [];         % Empty means Inf (default)
x_0     = [-5;5];     % Initial values on x
x_L     = [-10;-10];  % Lower bounds on x
x_U     = [10;10];    % Upper bounds on x
fLowBnd = 0;          % Lower bound on the function value

HessPattern = [];  % All elements in Hessian are nonzero
ConsPattern = [];  % All elements in the constraint
                   % Jacobian are nonzero.
pSepFunc    = [];  % The function f is not defined as
                   % separable
% Generate the problem structure
Prob = conAssign('con1_f', 'con1_g', 'con1_H', ...
            HessPattern, x_L, x_U, Name, x_0, ...
            pSepFunc, fLowBnd, A, b_L, b_U, 'con1_c', ...
             'con1_dc', [], ConsPattern, c_L, c_U);

Result  = tomRun('minos', Prob, [], 1);
                   % Final argument is the print level.
```

The user also has to supply the *con1_◇* functions referenced in *conAssign* as M-files.

III

THE FUTURE OF MODELING SYSTEMS

Chapter 20

THE FUTURE OF MODELING LANGUAGES AND MODELING SYSTEMS

Josef Kallrath

BASF-AG, GVC/S-B009 (Scientific Computing) & Dept. of Astronomy, University of Florida
D-67056 Ludwigshafen, Germany *Gainesville, FL 32611*
josef.kallrath@t-online.de kallrath@astro.ufl.edu

Abstract This chapter gives some ideas about the future of modeling system and the modeling companies. A list of features which might find a way into future releases of the modeling languages is listed. The chapter focuses on how economic constraints might influence the development of the modeling systems.

Keywords: Founders, future technical features, future of modeling companies

What does the future hold for the users of modeling languages? The author makes an attempt to answer this question and expresses his personal view on the future of modeling languages. One clear trend seems to be that modeling languages are evolving more and more into modeling systems [see page xviii for a definition] which support all aspects of solving real-world problems. While some keep the declarative language part separate from their API [in this group we see `OPL Studio` with *OPL Script*, `LINGO` with *LINGO-API*, and `MPL` with *OptiMax 2000*], other modeling languages [in this group we see, for instance, `AMPL` and `GAMS`] are moving away from pure declarative languages and add procedural features such as scripts to code algorithmic methods, recursive or multi-level structures of models. Adding procedural features gives, of course, more flexibility, but on the other hand it is more error prone and susceptible to new releases of the modeling language, which might not support these scripts any more. An example of this is multi-criteria optimization. Dash's `mp-model` allowed to define several objective functions and a Lexicographic Goal Programming approach was supported by Dash's solver Xpress-Optimizer. While the solver is still able to solve such problems, the successor, `Mosel`, does not support the formulation of multi-criteria problems, but leaves it to the user to program this feature. The migration of applications developed using the original language to the new one costs time and money. Of course, there are benefits such as better performance and new features. Ideally, it should be left to the user when to migrate. Taking out features makes life easier for the software providers but not for the user. This trend is problematic because companies have a tendency to avoid in-house programming — and, nowadays, that includes coding a model in a modeling language. There is a chance that what is not contained in a language or is supported in an easy way, will not be used. The opposite might also be true. As was stated by Robert Fourer during the Bad Honnef symposium "... features built in might help the solver community to improve their algorithms and do more testing"; this is also true for the real-world problems modeled by such systems. If optimization under uncertainty is supported then it will be used. If not, it is unlikely that companies using modeling systems in the current uncertain economic climate are innovative enough to come up with deep technical request and thus are able to have a positive technical impact on modeling languages.

The developers of modeling languages put more effort into easing the integration of optimization models into larger applications. One approach is code generation, as exemplified by `OPL Studio`, which allows the generation of a C++ representation of the model written in the modeling language. Another approach is to allow a call of a modeling language (*e.g.*, via COM) from a general-purpose language such as C or VB. This is the idea behind the latest products from Dash Optimization, ILOG Inc., Lindo Systems, and Maximal Software. Yet another way to deal with the application development issue comes from Paragon Decision Technology — `AIMMS` comes with a built-in GUI development environment.

As was discussed and became obvious during the Bad Honnef symposium most of the companies offering modeling languages and systems are still run by and are under full control of their original founders. This has the positive consequence that the field will be rich for the near future at least and will reflect that all these strong individuals like Johannes Bisschop, Bob Daniel, Robert Fourer, David Gay, Bjarni Kristjansson and Alexander Meeraus keep the modeling languages and systems close to their original ideas and visions. These individuals come from different backgrounds which also explains the different focus in their software.

However, all these companies have now existed for many years and have grown in size. This of course puts more economic pressure on them, and thus over the years we can expect to see some homogenization due to market forces. This effect is already observed now. Many features are present in most of the languages; the difference in syntax reflects the founder's view on how a language should look and be designed. Most of them operate in a Windows environment; here again we see the market force and the dominance of Microsoft acting. The grey hair of most of the founders is increasing strongly. Thus, we have to expect that in five or ten years these companies will be run by another generation. This will probably be a great loss and the convergence process will accelerate. It is natural that vendors want to make money; they try to design products that will appeal to a relevant or broad cross-section of the market while staying within resource (*e.g.*, capital, brainpower) limitations. Technically, the Operations Research market has a fairly low barrier to entry, and software vendors are always on the look-out for productivity-enhancing products. It has been said in many places in this book that modeling languages are a pretty good solution to most prototyping and development needs. New ideas still have a chance to be commercialized. `LPL` is an example of such a newcomer. Dash Optimization has taken the risk to replace their well established language `mp-model` by Yves Colombani's rather different approach manifested in `Mosel`.

In addition to what has been said it needs to be considered that there is another market force acting these days. This has to do with the fact that the group with the largest economic weight and impact is no longer the mathematical optimization community but rather hordes of people with a pure IT background but no skills in mathematics. These people are used to Microsoft standard and less to the more scientific standard systems, UNIX or LINUX. Their requirements cause the modeling companies to change their focus and also their development activities. The trend here is similar to other areas. While in 1960's and 1970's most of the CPU time was used on number crunching, nowadays the CPU time is mostly used in the GUI of Microsoft Windows operating systems; the achievement of work is measured in terms of mouse-clicks per second. The world-wide-web was invented by scientists for the scientific community. Nowadays, the scientific community is a minority. So, as in those two areas, the focus of modeling systems changes from technically minded users to non-technical mass users. Technically-minded people are bringing up new ideas or new requests and suggestions improving the language, but the modeling systems compete against these mass demands, and hence they need to hear arguments that this request is only important to a tiny percentage of all users and thus the effort to bring it into the system or to maintain it is not justified. This argument can be understood in a commercial sense, but it changes the relationship between the technically minded customers and the modeling language providers. The mathematical audience will get less back, their dissatisfaction with the existing modeling systems will rise, and thus there might be a tendency that they will contribute less to existing systems and develop their own ones. This will probably lead to a decrease of the technical progress rate.

Despite these tendencies, as was communicated by the modeling language providers during the symposium there are some improvements that the technical audience, can expect in the near future:

- *graphs* and combinatorial problems can be represented more easily and compactly,

- language and syntax elements supporting *optimization under uncertainties*,
- modeling languages will have joint elements from *Mathematical Programming* and *Constraint Programming.*

But let us nevertheless have some dreams. Maybe they will become reality one day and we will find modeling languages and systems supporting:

- *graphical representation* of optimization models,
- automatic generation of the *documentation* of models in LaTeX and MS Word,
- matrix support including syntax elements which allows us to specify that a matrix has to be positive definite, its determinant is non-zero, its condition number is larger than a lower bound — this will be relevant in Cone Programming,
- *interval data* can be passed to global optimization solvers.

Let us conclude this chapter on the future of modeling languages with some thoughts about the future of the companies forming the field. The Bad Honnef symposium formed a positive opinion that there are many new ideas around waiting for implementation, that the founders are still active and the modeling language companies provide a broad spectrum of different approaches. This pluralism is of value in itself. The market seems to support such a broad spectrum of different approaches. At least it is safe to say, this has been the case in the last decades and is still true today. Will it be so it in the future? Will some languages or modeling language providers disappear? The field has not seen mergers. Will some modeling companies combine their languages? Some of them cover such complementary aspects that one might expect there is a good chance that they could combine both their features and customers. While users benefit strongly from the modeling languages available, the end of this book is the right place to focus also on how the modeling language providers can be supported. The answer is by using their software, communicating with them and exchanging ideas, recommending use of their software to university students, to colleagues at work or to other practitioners, and perhaps also running special sessions on modeling languages at conferences.

Acknowledgments

I want to thank Susanne Heipcke (Dash Optimization) and Bjarni Kristjansson (Maximal Software) for their constructive comments on this chapter.

References

[1] A. Aggoun and H. Simonis. Search Tree Visualization. In *Analysis and Visualization Tools for Constraint Programming*, pages 191–208, Pierre Deransart, Manuel V. Hermenegildo, and Jan Maluszynski, editors, LNCS-1870, 2000.

[2] C. Aillaud and P. Deransart. Towards a Language for CLP Choice-tree Visualisation. In *Analysis and Visualization Tools for Constraint Programming*, pages 209–236, Pierre Deransart, Manuel V. Hermenegildo, and Jan Maluszynski, editors, LNCS-1870, 2000.

[3] E. Anderson, Z. Bai, C. Bischof, S. Blackford, J. Demmel, J. Dongarra, J. D. Croz, A. Greenbaum, S. Hammerling, A. McKenney, and D. Sorensen. *LAPACK User's Guide.* SIAM, Philadelphia, PA, 3rd edition, 1999.

[4] I. Androulakis, C. Maranas, and C. Floudas. αBB: A Global Optimization Method for General Constrained Nonconvex Problems. *J. Global Optim.*, 7:337–363, 1995.

[5] R. W. Ashford and R. C. Daniel. LP-MODEL: XPRESS-LP's Model Builder. *IMA Journal of Mathematics in Management*, 1:163–176, 1987.

[6] R. W. Ashford and R. C. Daniel. *XPRESS-MP Reference Manual.* Dash Associates, Blisworth House, Northants NN73BX, 1995.

[7] P. Barton and C. Pantelides. Modeling of Combined Discrete/Continuous Processes. *AIChE J.*, 40:966–979, 1994.

[8] E. M. L. Beale. Integer programming. In D. Jacobs, editor, *The State of the Art in Numerical Analysis*, pages 409–448. Academic Press, 1977.

[9] E. M. L. Beale and J. A. Tomlin. Special Facilities in a General Mathematical System for Non-Convex Problems Using Ordered Sets of Variables. In J. Lawrence, editor, *Proceedings of the Fifth International Conference on Operational Research*, pages 447–454. Tavistock Publications, London, 1970.

[10] J. E. Beasley. OR-Library: Distributing Test Problems by Electronic Mail. *Journal of the Operational Research Society*, 41(11):1069–1072, 1990.

[11] R. Bellman. *Dynamic Programming*. Princeton University Press, Princeton, NJ, 1957.

[12] A. Ben-Tal and A. Nemirovski. Robust Solutions of Linear Programming Problems Contaminated with Uncertain Data. *Mathematical Programming*, 88:411–424, 2000.

[13] M. Benichou, J.-M. Gauthier, G. Hentges, and G. Ribiere. The Efficient Solution of Large-Scale Linear Programming Problems. *Mathematical Programming*, 13:280–322, 1977.

[14] J. L. Bentley. Little Languages. *Communication of the ACM*, 29(8):711–721, 1987.

[15] J. L. Bentley, L. W. Jelinski, and B. W. Kernighan. CHEM — A Language for Phototypesetting Chemical Structure Diagrams. *Computers and Chemistry*, 11(4):281–297, 1987.

[16] J. L. Bentley and B. W. Kernighan. GRAP — A Language for Typesetting Graphs. *Communications of the ACM*, 29(8):782–792, 1986.

[17] J. M. Berg and K. Holmström. On Parameter Estimation Using Level Sets. *SIAM Journal on Control and Optimization*, 37(5):1372–1393, 1999.

[18] L. T. Biegler. Solution of Dynamic Optimization Problems by Successive Quadratic Programming and Orthogonal Collocation. *Comput. Chem. Eng.*, 8:243–248, 1984.

[19] L. T. Biegler, J. J. Damiano, and G. E. Blau. Nonlinear Parameter Estimation: A Case Study Comparison. *AIChE J.*, 32:29–45, 1986.

[20] J. R. Birge and F. Louveaux. *Introduction to Stochastic Programming*. Springer, 1997.

[21] J. Birk, M. Liepelt, K. Schittkowski, and F. Vogel. Computation of Optimal Feed Rates and Operation Intervals for Tubular Reactors. *Journal of Process Control*, 9:325–336, 1999.

[22] C. Bischof, A. Carle, G. Corliss, A. Griewank, and P. Hovland. ADIFOR: Generating Derivative Codes from Fortran Programs. *Scientific Programming*, 1:11–29, 1992.

[23] J. Bisschop and A. Meeraus. On the Development of a General Algebraic Modeling System in a Strategic Planning Environment. *Mathematical Programming Study*, 20:1–29, 1982.

[24] J. Bisschop and M. Roelofs. *AIMMS: The Language Reference*. Paragon Decision Technology B.V., Haarlem, The Netherlands, 1999.

[25] M. Blatt and K. Schittkowski. PDECON: A Fortran Code for Solving Optimal Control Problems based on Ordinary, Algebraic and Partial Differential Equations. Technical report, Department of Mathematics, University of Bayreuth, D-95440 Bayreuth, Germany, 1997.

[26] M. Blatt and K. Schittkowski. Optimal Control of One-dimensional Partial Differential Algebraic Equations with Applications. *Annals of Operations Research*, 98:45–64, 2000.

[27] H. Bock. Recent Advantages in Parameter Identification Techniques for ODE. In *International Workshop on Numerical Treatment of Inverse Problems in Differential and Integral Equations*, pages 95–121, Boston, Basel, Berlin, 1983. Birkhäuser.

[28] I. Bongartz, A. R. Conn, N. I. M. Gould, and P. L. Toint. CUTE: Constrained and Unconstrained Testing Environment. *ACM Transactions on Mathematical Software*, 21(1):123–160, 1995.

[29] B. Borchers and J. E. Mitchell. An Improved Branch and Bound Algorithm for Mixed Integer Nonlinear Programs. Technical Report RPI Math Report No. 200, Renssellaer Polytechnic Institute, 1991.

[30] C. Bracchi, C. Gefflot, and F. Paulin. Combining Propagation Information and Search Tree Visualization using ILOG OPL Studio. In *Proceedings of the Eleventh Workshop on Logic Programming Environments (WLPE'01)*, Paphos, Cyprus, December 2001.

[31] A. Brooke, D. Kendrick, and A. Meeraus. *GAMS: A User's Guide*. The Scientific Press, Redwoord City, CA, 1988.

[32] A. Brooke, D. Kendrick, and A. Meeraus. *GAMS - A User's Guide (Release 2.25)*. Boyd & Fraser Publishing Company, Danvers, Massachusetts, 1992.

[33] A. E. Bryson and Y.-C. Ho. *Applied Optimal Control*. Hemisphere Publishing Corporation, Washington, 1988.

[34] M. R. Bussieck, A. S. Drud, and A. Meeraus. MINLPLib - A Collection of Test Models for Mixed-Integer Nonlinear Programming. *Informs J. Comput.*, 15(1):114–119, 2003.

[35] M. D. Canon, C. D. Cullum, and E. Polak. *Theory of Optimal Control and Mathematical Programming*. McGraw Hill, New York, 1970.

[36] K. S. Carlton Lemke, Harvey Salkin. Set Covering by Single Branch Enumeration with Linear Programming Sub Problems. *Operations Research*, 19:998–1022, 1971.

[37] C. C. Carœ and R. Schultz. Dual Decomposition in Stochastic Integer Programming. *Operations Research Letters*, 24:37–45, 1999.

[38] CDC. *APEX-III Reference Manual Version 1.2*. Control Data Corporation, Minneapolis, MI, 1980.

[39] J. W. Chinneck. An Effective Polynomial-Time Heuristic for the Minimum-Cardinality IIS Set-Covering Problem. *Annals of Mathematics and Artificial Intelligence*, 17:127–144, 1996.

[40] J. W. Chinneck. Analyzing Mathematical Programs using MProbe. *Annals of Operations Research*, 104:33–48, 2001.

[41] M. Chlond, R. C. Daniel, and S. Heipcke. Fiveleapers a-leaping. INFORMS Transactions on Education (to appear), 2003.

[42] V. Chvátal. *Linear Programming*. W. H. Freeman and Company, New York, USA, 1983.

[43] T. A. Ciriani, Y. Colombani, and S. Heipcke. Optimisation Modelling Innovations. In S. I. di Matematica Appilcata, editor, *Proceedings of "VI Congresso Nazionale SIMAI", Chia Laguna (Sardinia, Italy), 27-31 May, 2002*, page 248, SIMAI, Viale del Policlinico 137, 00161 Roma (Italy), 2002. SIMAI.

[44] T. A. Ciriani, Y. Colombani, and S. Heipcke. Embedding Optimisation Algorithms with Mosel. *4OR*, 1(2):155–168, 2003.

[45] T. A. Ciriani, S. Gliozzi, E. L. Johnson, and R. Tadei, editors. *Operational Research in Industry*. Macmillan, Houndmills, Basingstoke, UK, 1999.

[46] The COCONUT Project Home Page. URL = `http://www.mat.univie.ac.at/coconut`.

[47] COMSOL AB, Tegnergatan 23, SE-111 40 Stockholm, Sweden. *FEMLAB 2.3 User's Guide and Introduction*, 2003.

[48] A. R. Conn, N. I. M. Gould, and P. L. Toint. *LANCELOT. A Fortran Package for Large-Scale Nonlinear Optimization*. Springer, Berlin, 1992.

[49] J. E. Cuthrell and L. T. Biegler. On the Optimization of Differential-Algebraic Process Systems. *AIChE J.*, 33:1257–1270, 1987.

[50] S. Dallwig, A. Neumaier, and H. Schichl. GLOPT - A Program for Constrained Global Optimization. In I. M. Bomze, T. Csendes, R. Horst, and P. Pardalos, editors, *Developments in Global Optimization*, pages 19–36. Kluwer Academic Publishers, Dordrecht, The Netherlands, 1997.

[51] G. B. Dantzig. Programming in a Linear Structure. Technical report, Comptroller, USAF, Washington, DC, 1948.

[52] G. B. Dantzig. *Linear Programming and Extension*. Princeton University Press, Princeton, New Jersey, 1963.

[53] S. Dempe. *Foundations of Bilevel Programming*. Kluwer Academic Publishers, Dordrecht, The Netherlands, 2002.

[54] L. C. W. Dixon and M. C. Bartholomew-Biggs. Adjoint-Control Transformations for Solving Practical Optimal Control Problems. *Opt. Cont. Appl. & Meth.*, 2:365–381, 1981.

[55] M. Dobmann, M. Liepelt, and K. Schittkowski. Algorithm 746: PCOMP: A Fortran Code for Automatic Differentiation. *ACM Transactions on Mathematical Software*, 21:233–266, 1995.

[56] M. Dobmann, M. Liepelt, K. Schittkowski, and C. Trassl. PCOMP: A Fortran Code for Automatic Differentiation – Language Description and User's Guide (Version 5.3). Technical report, Department of Mathematics, University of Bayreuth, D-95440 Bayreuth, Germany, 1995.

[57] E. D. Dolan, R. Fourer, J.-P. Goux, and T. S. Munson. Kestrel: An Interface from Modeling Systems to the NEOS Server. Technical report, Argonne National Laboratory, 2002. URL = `http://www-neos.mcs.anl.gov/neos/ftp/kestrel2.pdf`.

[58] E. D. Dolan and T. S. Munson. The Kestrel Interface to the NEOS Server. Technical report, Argonne National Laboratory, 2002. URL = `http://www-neos.mcs.anl.gov/neos/ftp/kestrel.pdf`.

[59] J. J. Dongarra, C. B. Moler, J. R. Bunch, and G. W. Stewart. LINPACK User's Guide. *SIAM*, 1979.

[60] A. S. Drud. CONOPT - A Large-Scale GRG Code. *ORSA Journal of Computing*, 6(2):207–218, 1994.

[61] M. A. Duran and I. E. Grossmann. An Outer-Approximation Algorithm for a Class of Mixed-Integer Nonlinear Programms. *Mathematical Programming*, 36:307–339, 1986.

[62] E. F. D. Ellison and G. Mitra. UIMP: User Interface for Mathematical Programming. *ACM Transactions on Mathematical Software*, 8(3):229–255, 1982.

[63] H. Elmqvist, S. Mattsson, and M. Otter. Modelica: The New Object-oriented Modeling language, 1998. URL = http://citeseer.nj.nec.com/elmqvist98modelica.html.

[64] M. Ferris, R. Fourer, and D. M. Gay. Expressing Complementarity Problems in an Algebraic Modeling Language and Communicating Them to Solvers. *SIAM J. Optimization*, 9(4):991–1009, 1999.

[65] M. C. Ferris and D. M. Shepard. Optimization of Gamma Knife Radiosurgery. In D. Z. Du, P. Pardalos, and J. Wang, editors, *Discrete Mathematical Problems with Medical Applications*, volume 55 of *DIMACS Series in Discrete Mathematics and Theoretical Computer Science*, pages 22–77. American Mathematical Society, 2000.

[66] R. Fletcher and S. Leyffer. Solving Mixed Integer Nonlinear Programs by Outer Approximation. *Math. Prog.*, 66(3):327, 1994.

[67] C. A. Floudas. *Nonlinear and Mixed Integer Optimization: Fundamentals and Applications*. Oxford University Press, 1995.

[68] C. A. Floudas. *Deterministic Global Optimization: Theory, Methods and Applications*. Kluwer Academic Publishers, Dordrecht, The Netherlands, 2000.

[69] C. A. Floudas, A. Aggarwal, and A. R. Ciric. Global Optimal Search for Nonconvex NLP and MINLP Problems. *Comput. Chem. Eng.*, 13(10):1117, 1989.

[70] J. Forrest and J. Tomlin. Important Note: "OSL-MIP"; Seminal Contributions by Martin Beale and Colleagues. Working paper 2003, IBM Research; Yorktown Heights, 2003.

[71] R. Fourer. Modeling Languages Versus Matrix Generators for Linear Programming. *ACM Trans. Math. Software*, 9(2):143–183, 1983.

[72] R. Fourer and D. M. Gay. Extending an Algebraic Modeling Language to Support Constraint Programming. *INFORMS J. Computing*, 14(4):322–344, 2002.

[73] R. Fourer, D. M. Gay, and B. W. Kernighan. AMPL: A Mathematical Programming Language. Technical Report Computing Science Technical Report No. 133, AT&T Bell Laboratories, Murray Hill, NJ, USA, Jan. 1987. revised June 1989.

[74] R. Fourer, D. M. Gay, and B. W. Kernighan. A Modeling Language for Mathematical Programming. *Management Science*, 36(5):519–554, 1990.

[75] R. Fourer, D. M. Gay, and B. W. Kernighan. *AMPL: A Modeling Language for Mathematical Programming.* Duxbury Press, Brooks/Cole Publishing Company, Monterey, CA, 1993.

[76] R. Fourer, D. M. Gay, and B. W. Kernighan. *AMPL: A Modeling Language for Mathematical Programming.* Duxbury Press/Brooks/Cole Publishing Co., 2nd edition, 2003.

[77] R. Fourer and J.-P. Goux. Optimization as an Internet Resource. *Interfaces*, 31(2):130–150, 2001.

[78] E. Fragnière and J. Gondzio. Optimization Modeling Languages, 1999.

[79] GAMS Development Corporation. GAMS - The Solver Manuals, 2003.

[80] B. S. Garbow, J. M. Boyle, J. J. Dongara, and C. B. Moler. Matrix Eigensystem Routines-EISPACK Guide Extension. In *Lecture Notes in Computer Science*. Springer Verlag, New York, 1977.

[81] D. M. Gay. Automatic Differentiation of Nonlinear AMPL Models. In A. Griewank and G. Corliss, editors, *Automatic Differentiation of Algorithms: Theory, Implementation, and Application*, pages 61–73. SIAM, 1991.

[82] D. M. Gay. More AD of Nonlinear AMPL Models: Computing Hessian Information and Exploiting Partial Separability. In M. Berz, C. Bischof, G. Corliss, and A. Griewank, editors, *Computational Differentiation: Techniques, Applications, and Tools*. SIAM, 1996.

[83] A. M. Geoffrion. Generalized Benders Decomposition. *Journal of Optimization Theory and Applications*, 10:237–260, 1972.

[84] H. Gericke. *Mathematik in Antike, Orient und Abendland*. Fourier Verlag, Wiesbaden, sixth edition, 2003.

[85] P. E. Gill, W. Murray, and M. H. Wright. *Practical Optimization.* Academic Press, London, 1981.

[86] F. Glover and M. Laguna. *Tabu Search.* Kluwer Academic Publisher, Dordrecht, The Netherlands, 1997.

[87] C. J. Goh and K. L. Teo. Control Parameterization: a Unified Approach to Optimal Control Problems with General Constraints. *Automatica*, 24(1):3–18, 1988.

[88] H. J. Greenberg. A Functional Description of ANALYZE: a Computer-Assisted Analysis System for Linear Programming Models. *ACM Transactions On Mathematical Software*, 9:18–56, 1983.

[89] A. Griewank. On Automatic Differentiation. In M. Iri and K. Tanabe, editors, *Mathematical Programming: Recent Developments and Applications*, pages 83–107. Kluwer Academic Publishers, Boston, 1989.

[90] A. Griewank. *Evaluating Derivatives: Principles and Techniques of Algorithmic*. SIAM, 2000.

[91] A. Griewank and G. Corliss, editors. *Automatic Differentiation of Algorithms: Theory, Implementation and Application*, Philadelphia, 1991. SIAM.

[92] A. Griewank, D. W. Juedes, and J. Srinivasan. ADOL-C: A Package for the Automatic Differentiation of Algorithms Written in C/C++. Preprint MCS-P180-1190, Mathematics and Computer Science Division, Argonne National Laboratory, Argonne, Illinois, 1991.

[93] M. Guignard, S. Kim, and K. Spielberg. Multi-Commodity Nonlinear Distribution Planning. *Methods of Operations Research, Verlag Athenäum*, 58:191–202, 1989.

[94] M. Guignard and K. Spielberg. Logical Reduction Methods in Zero-One Programming (Minimal Preferred Variables). *Operations Research*, 29:49–74, 1981.

[95] M. Guignard and K. Spielberg. Double Contraction and Probing, Short Starts, and Branch-and-Bound Implications for Mixed (0,1) Programming. Working Paper 2003; presented at Pucon Conf. 2002, University of Pennsylvania; OPIM Dept., 2003.

[96] O. K. Gupta. *Branch and Bound Experiments in Nonlinear Integer Programming*. PhD thesis, Purdue University, 1980.

[97] O. K. Gupta and V. Ravindran. Branch and Bound Experiments in Convex Nonlinear Integer Programming. *Management Science*, 31:1533–1546, 1985.

[98] W. D. Harvey and M. L. Ginsberg. Limited Discrepancy Search. In *Proceedings of the 14th International Joint Conference on Artificial Intelligence*, Montreal, Canada, August 1995.

[99] C. A. Haverly. Practical Modelling. In T. A. Ciriani, S. Gliozzi, E. L. Johnson, and R. Tadei, editors, *Operational Research in Industry*, pages 93–101. Macmillan, Houndmills, Basingstoke, UK, 1999.

[100] Haverly Systems. PDS MaGen User Information Manual, 1977.

[101] S. Heipcke. *Applications of Optimization with Xpress-MP*. Dash Optimization, Blisworth, UK, 2002.

[102] T. Hellström and K. Holmström. Parameter Tuning in Trading Algorithms using ASTA. In Y. S. Abu-Mostafa, B. LeBaron, A. W. Lo, and A. S. Weigend, editors, *Computational Finance 1999*, Cambridge, MA, 1999. MIT Press.

[103] R. Hettich and K. O. Kortanek. Semi-infinite programming: Theory, methods, and applications. *SIAM Review*, 35(3):380–429, 1993.

[104] W. Hock and K. Schittkowski. *Test Examples for Nonlinear Programming Codes*. Lecture Notes in Economics and Mathematical Systems 187. Springer Verlag, Berlin, Heidelberg, New York, 1981.

[105] K. Holmberg. On the Convergence of the Cross Decomposition. *Math. Prog.*, 47:269, 1990.

[106] K. Holmström. TOMLAB - A General Purpose, Open Matlab Environment for Research and Teaching in Optimization. Technical Report IMa-TOM-1997-03, Department of Mathematics and Physics, Mälardalen University, Sweden, 1997. Presented at the 16th International Symposium on Mathematical Programming, Lausanne, Switzerland, August 24-29, 1997.

[107] K. Holmström. TOMLAB - An Environment for Solving Optimization Problems in Matlab. In M. Olsson, editor, *Proceedings for the Nordic Matlab Conference '97, October 27-28*, Stockholm, Sweden, 1997. Computer Solutions Europe AB.

[108] K. Holmström. Constrained Separable NLLS Algorithms for Chemical Equilibrium Analysis. In A. Løkketangen, editor, *Proceedings from the 5th Meeting of the Nordic Section of the Mathematical Programming Society*, ISBN 82-90347-76-6, Molde, 1998. Division of Operations Research, Molde University.

[109] K. Holmström. New Optimization Algorithms and Software. *Theory of Stochastic Processes*, 5(21)(1-2):55–63, 1999.

[110] K. Holmström. The TOMLAB v2.0 Optimization Environment. In E. Dotzauer, M. Björkman, and K. Holmström, editors, *Sixth Meeting of the Nordic Section of the Mathematical Programming Society. Proceedings*, Opuscula 49, ISSN 1400-5468, Västerås, 1999. Mälardalen University, Sweden.

[111] K. Holmström. Practical Optimization with the Tomlab Environment. In T. A. Hauge, B. Lie, R. Ergon, M. D. Diez, G.-O. Kaasa, A. Dale, B. Glemmestad, and A. Mjaavatten, editors, *Proceedings of the 42nd SIMS Conference*, pages 89–108, Porsgrunn, Norway, 2001. Telemark University College, Faculty of Technology.

[112] K. Holmström and M. Björkman. The TOMLAB NLPLIB Toolbox for Nonlinear Programming. *Advanced Modeling and Optimization*, 1(1):70–86, 1999.

[113] K. Holmström, M. Björkman, and E. Dotzauer. The TOMLAB OPERA Toolbox for Linear and Discrete Optimization. *Advanced Modeling and Optimization*, 1(2):1–8, 1999.

[114] K. Holmström and A. Göran. User's Guide for Tomlab v4.0. Technical Report IMa-TOM-2003-01, Department of Mathematics and Physics, Mälardalen University, Sweden, 2003. Revised August 1, 2003.

[115] R. Horst and P. M. Pardalos, editors. *Handbook of Global Optimization*. Kluwer Academic Publishers, Dordrecht, The Netherlands, 1995.

[116] R. Horst, P. M. Pardalos, and N. V. Thoai. *Introduction to Global Optimization*. Kluwer Academic Publishers, Dordrecht, The Netherlands, 1996.

[117] T. Hürlimann. *A Structured Language for Modeling Linear Programs*. PhD Dissertation, University of Fribourg, CH, Bern, CH, 1987.

[118] T. Hürlimann. Units in LPL. Working Paper No. 182, April, Fribourg, Institute for Automation and Operations Research, University of Fribourg, Fribourg, Switzerland, 1991.

[119] T. Hürlimann. Computer-Based Mathematical Modeling. Habilitation Thesis, 1997.

[120] T. Hürlimann. *Mathematical Modeling and Optimization, An Essay for the Design of Computer-Based Modeling Tools*, volume 31 of *Applied Optimization*. Kluwer Academic Publishers, Dordrecht, The Netherlands, 1999.

[121] T. Hürlimann. LPL : A Mathematical Modeling Language, An Introduction. Departement of Informatics, Fribourg, 2002.

[122] T. Hürlimann. Modeling Languages: A New Paradigm of Programming. *Annals of Operations Research*, 104 (On Modeling Languages and Applications), 2003. to appear.

[123] T. Hürlimann. *Reference Manual for the LPL Modelling Language*. Virtual Optima, www.virtual-optima.com/download/manual.pdf, 2003.

[124] T. Hürlimann. Virtual-optima's homepage: www.virtual-optima.com, 2003.

[125] IBM. *Mathematical Programming System Extended (MPSX) Control Language User's Manual*. IBM, Sammet, 1978.

[126] J. P. Ignizio. *Goal Programming and Extensions*. Heath, Lexington, Massachusetts, USA, 1976.

[127] Ilog OPL Studio 3.5. Reference Manual. Ilog SA, Gentilly, France, 2001.

[128] D. H. Jacobson and D. Q. Mayne. *Differential Dynamic Programming*. American Elsevier, New York, 1970.

[129] J. Jaffar and M. J. Maher. Constraint Logic Programming: A Survey. *Journal of Logic Programming*, 19/20:503–581, 1994.

[130] C. Jansson and O. Knüppel. *A Global Minimization Method: The Multidimensional Case*. Hamburg-Harburg, Techn. Report 92.1 edition, August 1992. URL = http://www.ti3.tu-harburg.de/Software.html.

[131] E. L. Johnson and B. Gopala-Krishnan. Airline Crew Scheduling: the State-of-the-Art. Working Paper 2003, Georgia Institute of Technology; Operations Research, 2003.

[132] E. L. Johnson and K. Spielberg. Logical Processing for IP and MIP. Working Paper, 2003, Georgia Institute of Technology; Operations Research, 2003.

[133] C. Jordan and A. Drexl. A Comparison of Constraint and Mixed-Integer Programming Solvers for Batch Sequencing With Sequence-Dependent Setups. *INFORMS J. Computing*, 7(2):160–165, 1995.

[134] D. W. Juedes. A Taxonomy of Automatic Differentiation Tools. In A. Griewank and G. Corliss, editors, *Proceedings of the Workshop on Automatic Differentiation of Algorithms: Theory, Implementation and Applications*, pages 315–330. SIAM, Philadelphia, 1991.

[135] H. Kagiwada, R. Kalaba, N. Rosakhoo, and K. Spingarn. *Numerical Derivatives and Nonlinear Analysis*. Plenum Press, New York and London, 1986.

[136] P. Kall and S. W. Wallace. *Stochastic Programming*. John Wiley and Sons, Chichester, 1994.

[137] J. Kallrath. Combined Strategic and Operational Planning - An MILP Success Story in Chemical Industry. *OR Spectrum*, 24(3):315–341, 2002.

[138] J. Kallrath. *Gemischt-Ganzzahlige Optimierung: Modellierung in der Praxis*. Vieweg, Wiesbaden, Germany, 2002.

[139] J. Kallrath. Exact Computation of Global Minima of a Nonconvex Portfolio Optimization Problem. In C. A. Floudas and P. M. Pardalos, editors, *Frontiers in Global Optimization*. Kluwer Academic Publishers, 2003. to appear.

[140] J. Kallrath and J. M. Wilson. *Business Optimisation Using Mathematical Programming*. Macmillan, Houndmills, Basingstoke, UK, 1997.

[141] N. Karmarkar. A New Polynomial-time Algorithm for Linear Programming. *Combinatorica*, 4:373–395, 1984.

[142] R. B. Kearfott. *Rigorous Global Search: Continuous Problems*. Kluwer Academic Publishers, Dordrecht, The Netherlands, 1996.

[143] G. Kedem. Automatic differentiation of computer programs. *ACM Transactions on Mathematical Software*, 6:150–165, 1980.

[144] KETRON. MPSIII Mathematical Programming System. URL = http://ketronms.com/products.html.

[145] W. K. Klein-Haneveld and M. H. van der Vlerk. Stochastic Integer Programming: General Models and Algorithms. *Annals of Operational Research*, 85:39–57, 1999.

[146] G. Kneppe. MBB-LAGRANGE: Structural Optimization System for Space and Aircraft Structures. Technical report, EADS Deutschland GmbH, D-81663 Munich, Germany, 1990.

[147] D. E. Knuth. Literate Programming. *Computer Journal*, 27(2):97–111, 1984.

[148] D. E. Knuth. *Literate Programming*. CSLI (Center for the Study of Language and Information), Lecture Notes Number 27, Cambridge University Press, Cambridge, MA, 1992.

[149] G. R. Kocis and I. E. Grossmann. Relaxation Strategy for the Structural Optimization of Process Flow Sheets. *Ind. Eng. Chem. Res.*, 26(9):1869, 1987.

[150] L. Ladany, J. Lee, and R. Lougee-Heimer. Implementation of an exact algotihm for a cutting- stock problem using components of COIN-OR. IBM Research Report RC22720, IBM Research, Yorktown Heights, 2003.

[151] L. S. Lasdon and A. D. Waren. Generalized Reduced Gradient Method for Linearly and Nonlinearly Constrained Programming. In H. J. Greenberg, editor, *Design and Implementation of Optimization Software*, pages 363–397. Sijthoff and Noordhoff, The Netherlands, 1978.

[152] L. S. Lasdon, A. D. Waren, A. Jain, and M. Ratner. Design and Testing of a Generalized Reduced Gradient Code for Nonlinear Programming. *ACM Trans. Math. Software*, 4:34–50, 1978.

[153] E. Lee, C. Mavroidis, and J. P. Merlet. Five Precision Points Synthesis of Spatial RRR Manipulators using Interval Analysis. In *Proceedings of DETC'02, 27th Biennial Mechanisms and Robotics Conference Montreal, Canada*, 2002.

[154] D. Leineweber, H. G. Bock, J. P. Schlöder, J. V. Gallitzendörfer, A. Schäfer, and P. Jansohn. A Bounday Value Problem Approach to the Optimization of Chemical Processes Described by DAE Models. In *Proceedings of the 15th IMACS World Congress on Scientific Computation, Modelling and Applied Mathematics*, Berlin, 1997.

[155] C. Lemke and K. Spielberg. Direct Search Algorithms for Zero-One and Mixed Integer Programming. *Operations Research*, 15:892–914, 1967.

[156] F. L. Lewis and V. L. Syrmos. *Optimal Control.* John Wiley & Sons, Inc., 1995.

[157] M. Liepelt and K. Schittkowski. Algorithm 746: New features of PCOMP, a Fortran Code for Automatic Differentiation. *ACM Transactions on Mathematical Software*, 26:352–362, 2000.

[158] J. S. Logsdon and L. T. Biegler. Accurate Solution of Differential-Algebraic Optimization Problems. *Ind. Eng. Chem. Res.*, 28:1628–1639, 1989.

[159] K. Louden. *Programming Languages — Principles and Practice.* PWS-KENT Publ. Comp., 1993.

[160] R. Lougee-Heimer. The Common Optimization INterface for Operations Research. *IBM Journal of Research & Development*, 47(1):57–66, 2003.

[161] I. J. Lustig and J. F. Puget. Program Does Not Equal Program: Constraint Programming and its Relationship to Mathematical Programming. *Interfaces*, 31(6):29–53, 2001.

[162] R. Luus. Application of Dynamic Programming to High-Dimensional Non-Linear Optimal Control Problems. *Int. J. Control*, 52:239–250, 1990.

[163] R. Luus. Optimal Control by Dynamic Programming Using Systematic Reduction in Grid Size. *Int. J. Control*, 5:995–1013, 1990.

[164] R. Luus. Piecewise Linear Continuous Optimal Control by Iterative Dynamic Programming. *Ind. Eng. Chem. Res.*, 32:859–865, 1993.

[165] R. Luus and O. Rosen. Application of Dynamic Programming to Final State Constrained Optimal Control Problems. *Ind. Eng. Chem. Res.*, 30:1525–1530, 1991.

[166] A. Mackworth. Consistency in Networks of Relations. *Artificial Intelligence*, 8(1):99–118, 1977.

[167] T. Mairs, G. Wakefield, E. Johnson, and K. Spielberg. On a Production Allocation and Distribution Problem. *Management Science*, 24:1622–1630, 1978.

[168] A. S. Manne. *Scheduling of Petroleum Refinery Operations*. Harvard University Press, Cambridge, Massachusetts, 1956.

[169] K. Marriott and P. J. Stuckey. *Programming with Constraints: An Introduction*. MIT Press, 1998.

[170] H. Mawengkang and B. A. Murtagh. Solving Nonlinear Integer Programs with Large-Scale Optimization Software. *Annals of Operations Research*, 5(6):425–437, 1985.

[171] Maximal Software. MPL for Windows Reference Manual. URL = http://www.maximal-usa.com, 2001.

[172] K. McAloon, C. Tretkoff, and G. Wetzel. Sport League Scheduling. In *Proceedings of the 3th Ilog International Users Meeting*, Paris, France, 1997.

[173] L. Michel and P. Van Hentenryck. Modeler++: A Modeling Layer for Constraint Programming Libraries. In *CP-AI-OR'2001*, Wye College (Imperial College), Ashford, Kent UK, April 2001.

[174] A. Miele. Recent Advances in Gradient Algorithms for Optimal Control Problems. *J. Opt. Theory Applic.*, 17:361–430, 1975.

[175] A. Miele, J. N. Damoulakis, J. R. Cloutier, and J. L. Tietze. Sequential Gradient-Restoration Algorithm for Optimal Control Problems with Nondifferential Constraints. *J. Opt. Theory Applic.*, 13:218–255, 1974.

[176] A. Miele, R. E. Pritchard, and J. N. Damoulakis. Sequential Gradient-Restoration Algorithm for Optimal Control Problems. *J. Opt. Theory Applic.*, 5:235–282, 1970.

[177] C. B. Moler. MATLAB — An Interactive Matrix Laboratory. Technical Report 369, Department of Mathematics and Statistics, University of New Mexico, 1980.

[178] J. J. Moré, B. S. Garbow, and K. E. Hillstrom. Testing Unconstrained Optimization Software. *ACM Trans. Math. Software*, 7:17–41, 1981.

[179] K. R. Morrison and R. W. H. Sargent. Optimization of Multistage Processes Described by Differential-Algebraic Equations. *Lect. Notes Math.*, 1230:86–102, 1986.

[180] B. A. Murtagh and M. A. Saunders. Large-scale Linearly Constrained Optimization. *Mathematical Programming*, 14:41–72, 1978.

[181] G. L. Nemhauser and L. A. Wolsey. *Integer and Combinatorial Optimization*. John Wiley and Sons, New York, 1988.

[182] A. Neumaier. NOP - A Compact Input Format for Nonlinear Optimization Problems. In I. M. Bomze, T. Csendes, R. Horst, and P. Pardalos, editors, *Developments in Global Optimization*, pages 1–18. Kluwer Academic Publishers, Dordrecht, The Netherlands, 1997.

[183] G. M. Ostrovsky, M. G. Ostrovsky, and G. W. Mikhailow. Discrete Optimization of Chemical Processes. *Comput. Chem. Eng.*, 14(1):111, 1990.

[184] M. Padberg. *Linear Optimization and Extensions*. Springer, Berlin - Heidelberg, 1996.

[185] J. Pasquier, P. Hättenschwiler, Hürlimann, and B. Sudan. A Convenient Technique for Constructing your own MPSX Generator Using dBASE II. *Angewandte Informatik*, 28(7):295–300, 1986.

[186] G. E. Paules, IV and C. A. Floudas. APROS: Algorithmic Development Methodology for Discrete-Continuous Optimization Problems. *Oper. Res.*, 37(6):902–915, 1989.

[187] P. Piela, T. Epperly, K. Westerberg, and A. Westerberg. ASCEND: An Object-Oriented Computer Environment for Modeling and Analysis: The Modeling Language. *Computers and Chemical Engineering*, 15:1:53–72, 1991.

[188] J. D. Pintér. LGO: An Integrated Model Development and Solver System for Continuous Global Optimization. *INFORMS Computing Society Newsletter*, 20(1), 1999.

[189] G. P. Pollard and R. W. H. Sargent. Off Line Computation of Optimum Controls for a Plate Distillation Column. *Automatica*, 6:59–76, 1970.

[190] The BoFit system by ProComm. http://www.procom.de/english/bofit.

[191] J. F. Puget and I. Lustig. Constraint Programming and Mathematical Programming. *Knowledge Engineering Review*, 16(1), 2001.

[192] L. Rall. *Automatic Differentiation - Techniques and Applications*. Lecture Notes in Computer Science, Vol. 120. Springer, Berlin, Heidelberg, New York, 2000.

[193] J.-C. Régin. A Filtering Algorithm for Constraints of Difference in CSPs. In *AAAI-94, Proceedings of the Twelth National Conference on Artificial Intelligence*, pages 362–367, Seattle, Washington, 1994.

[194] J.-C. Régin. Generalized Arc Consistency for Global Cardinality Constraints. In *AAAI-96, Proceedings of the Thirteenth National Conference on Artificial Intelligence*, pages 209–215, Portland, Oregon, 1996.

[195] J.-C. Régin. Sport League Scheduling. In *INFORMS*, Montreal, Canada, 1998.

[196] J. G. Renfro, A. M. Morshedi, and O. A. Asbjornsen. Simultaneous Optimization and Solution of Systems Described by Differential/Algebraic Equations. *Comput. Chem. Eng.*, 11:503–517, 1987.

[197] J. Rivera. Modeling with Extend. In *Winter Simulation Conference*, pages 257–262, 1998.

[198] C. Romero. *Handbook of Critical Issues in Goal Programming*. Pergamon Press, Oxford, 1991.

[199] H. Rommelfanger. *Fuzzy Decision Support-Systeme - Entscheiden bei Unschärfe*. Springer, Heidelberg, 2nd edition, 1993.

[200] I. M. Ross and F. Fahroo. *User's Manual For DIDO - A MATLAB Application Package for Dynamic Optimization.* Naval Postgraduate School, Monterey, CA, 2003.

[201] J. Rubin. Airline Crew Scheduling-the Non-Mathematical Problem. REPORT 320-3006, IBM New York Scientific Center, 1971.

[202] H. S. Ryoo and N. V. Sahinidis. A Branch-and-Reduce Approach to Global Optimization. *Journal of Global Optimization*, 8(2):107–139, 1996.

[203] N. Sahinidis. BARON: A General Purpose Global Optimization Software Package. *J. Global Optim.*, 8:201–205, 1996.

[204] R. W. H. Sargent and G. R. Sullivan. The Development of an Efficient Optimal Control Package. *Proc. IFIP Con. Optim. Tech.*, 8th:part 2, 1977.

[205] R. W. H. Sargent and G. R. Sullivan. Development of Feed Change-over Policies for Refinery Distillation Units. *Ind. Eng. Chem. Proc. Des. Dev.*, 18:113–124, 1979.

[206] H. Schichl and A. Neumaier. Interval Analysis on Directed Acyclic Graphs for Global Optimization, 2003. Preprint.

[207] H. Schichl, A. Neumaier, and S. Dallwig. The NOP-2 Modeling Language. *Ann. Oper. Research*, 104:281–312, 2001.

[208] K. Schittkowski. Solving Nonlinear Least Squares Problems by a General Purpose SQP-Method. In K.-H. Hoffmann, J.-B. Hiriart-Urruty, C. Lemarechal, and J. Zowe, editors, *Trends in Mathematical Optimization*, pages 295–309. Birkhäuser, Boston, Basel, Berlin, 1988.

[209] K. Schittkowski. EASY-OPT: An Interactive Optimization System with Automatic Differentiation - User's Guide. Technical report, Department of Mathematics, University of Bayreuth, D-95440 Bayreuth, Germany, 1999.

[210] K. Schittkowski. EASY-FIT: A Software System for Data Fitting in Dynamic Systems. *Structural and Multidisciplinary Optimization*, 23:153–169, 2001.

[211] K. Schittkowski. *Numerical Data Fitting in Dynamical Systems - A Practical Introduction with Applications and Software.* Kluwer Academic Publishers, Dordrecht, Boston, London, 2002.

[212] M. J. Schniederjans. *Goal Programming: Methodology and Applications.* Kluwer Academic Publishers, Boston, MA, 1995.

[213] L. Schrage. *Optimization Modeling with LINGO.* LINDO Systems, Inc., Chicago, Il, 1999.

[214] C. Schulte. Oz Explorer: A Visual Constraint Programming Tool. In *Proceedings of the 14th International Conference on Logic Programming (ICLP-91)*, pages 286–300, (Leuven, Belgium), June 1997. The MIT Press.

[215] R. Schultz. On Structure and Stability in Stochastic Programs with Random Technology Matrix and Complete Integer Recourse. *Mathematical Programming*, 70:73–89, 1995.

[216] R. Schultz. Stochastic Programming with Integer Variables. *Mathematical Programming Ser. B*, 97:285–309, 2003.

[217] C. A. Schweiger and C. A. Floudas. Interaction of Design and Control: Optimization with Dynamic Models. In W. W. Hager and P. M. Pardalos, editors, *Optimal Control: Theory, Algorithms, and Applications*, pages 388–435. Kluwer Academic Publishers, 1997.

[218] C. A. Schweiger, A. Rojnuckarin, and C. A. Floudas. *MINOPT: A Software Package for Mixed-Integer Nonlinear Optimization*. Dept. of Chemical Engineering, Princeton University, Princeton, NJ 08544, 1996.

[219] A. Serfaty. Optimal Debt Restructuring and Debt Management. In J. Rosend and R. D. Glisker, editors, *The Handbook of Fixed Income Technology*. The Summit Group Press, NY, NY, 1999.

[220] D. Shiriaev, A. Griewank, and J. Utke. A user guide to ADOL-F: Automatic Differentiation of Fortran codes. Technical report, Institute of Scientific Computing, Technical University Dresden, D-01069 Dresden, Germany, 1997.

[221] L. Slate and K. Spielberg. The Extended Control Language of MPSX/370 (and possible applications). *IBM Systems Journal*, 17:64–81, 1978.

[222] B. T. Smith, J. M. Boyle, J. J. Dongarra, B. S. Garbow, Y. Ikebe, V. C. Klema, and C. B. Moler. *Matrix Eigensystem Routines - EISPACK Guide Lecture Notes in Computer Science*. Springer-Verlag, New York, 2nd edition, 1976.

[223] P. Spelluci. *Numerische Verfahren der nichtlinearen Optimierung*. Birkhäuser, Basel, 1993.

[224] K. Spielberg. Algorithms for the Simple Plant Location Problem with some Side Conditions. *Operations Research*, 17:85–111, 1969.

[225] K. Spielberg. The Extended Control Language of MPSX/370, Simple and Advanced Applications. *Bulletin SIGMAP; NBS*, 29, 1980.

[226] K. Spielberg and U. Suhl. An Experimental Enumeration Software System for Integer Programming. IBM Research Report RC8219, IBM Research, Yorktown Heights, 1980.

[227] R. F. Stengel. *Stochastic Optimal Control*. John Wiley and Sons, New York, 1986.

[228] G. Still. Generalized semi-infinite programming: Theory and methods. *European Journal of Operational Research*, 119:301–313, 1999.

[229] W. Stöllinger. Untersuchung der Resourceneffizienz von Holz mit besonderem Augenmerk auf dessen Lebensdauer. Master's thesis, Technische Universität Wien, 2003. to appear.

[230] M. Tawarmalani and N. V. Sahinidis. *Convexification and Global Optimization in Continuous and Mixed-Integer Nonlinear Programming: Theory, Algorithms, Software, and Applications*, volume 65 of *Nonconvex Optimization And Its Applications*. Kluwer Academic Publishers, Dordrecht, The Netherlands, 2002.

[231] J. Thames. PROblem Solution Engineering. Numerical Problems Including Differentiation and Integration. *Research/Development*, 26(5), 1975.

[232] C. Timpe. Solving Planning and Scheduling Problems with Combined Integer and Constraint Programming. *OR Spectrum*, 24(4):431–448, 2002.

[233] J. A. Tomlin and J. S. Welch. Mathematical Programming System. Ibm research report rj 7400 (69202), IBM, Yorktown Heights, 1990.

[234] T. H. Tsang, D. M. Himmelblau, and T. F. Edgar. Optimal Control via Collocation and Nonlinear Programming. *Int. J. Control*, 21:763–768, 1975.

[235] P. Van Hentenryck. *The OPL Optimization Programming Language*. MIT Press, Cambridge, MA, 1999.

[236] P. Van Hentenryck and L. Michel. *New Trends in Constraints*, chapter OPL Script: Composing and Controlling Models. Lecture Note in Artificial Intelligence (LNAI 1865). Springer Verlag, 2000.

[237] P. Van Hentenryck and L. Michel. *OPL Script: Composing and Controlling Models.* Lecture Note in Artificial Intelligence, Springer, New York, NY, 2000.

[238] P. Van Hentenryck, L. Michel, and Y. Deville. *Numerica - A Modeling Language for Global Optimization.* MIT Press, Cambridge, MA, 1997.

[239] P. Van Hentenryck, L. Michel, P. Laborie, W. Nuijten, and J. Rogerie. Combinatorial Optimization in OPL Studio. In *Proceedings of the 9th Portuguese Conference on Artificial Intelligence International Conference (EPIA'99)*, Evora, Portugal, September 1999. (Invited Paper).

[240] P. Van Hentenryck, L. Michel, L. Perron, and J. C. Régin. Constraint Programming in OPL. In *Proceedings of the International Conference on the Principles and Practice of Declarative Programming (PPDP'99)*, Paris, France, September 1999. (Invited Paper).

[241] P. Van Hentenryck, L. Perron, and J.-F. Puget. Search and Strategies in OPL. *ACM Transactions on Computational Logic*, 1(2):1–36, October 2000.

[242] T. J. VanRoy and L. A. Wolsey. Solving Mixed Integer Programs by Automatic Reformulation. *Operations Research*, 35(1):45–57, 1987.

[243] S. Vasantharajan and L. T. Biegler. Simultaneous Strategies for Optimization of Differential-Algebraic Systems with Enforcement of Error Criteria. *Comput. Chem. Eng.*, 14:1083–1100, 1990.

[244] V. S. Vassiliadis, R. W. H. Sargent, and C. C. Pantelides. Solution of a Class of Multistage Dynamic Optimization Problems. 1. Problems without Path Constraints. *Ind. Eng. Chem. Res.*, 33:2111–2122, 1994.

[245] V. S. Vassiliadis, R. W. H. Sargent, and C. C. Pantelides. Solution of a Class of Multistage Dynamic Optimization Problems. 2. Problems with Path Constraints. *Ind. Eng. Chem. Res.*, 33:2123–2133, 1994.

[246] A. I. F. Vaz, E. M. G. P. Fernandes, and M. P. S. F. Gomes. SIPAMPL v2.0: Semi-Infinite Programming with AMPL. Technical Report ALG/EF/4-2002, Universidade do Minho, Braga, Portugal, December 2002. http://www.norg.uminho.pt/aivaz/.

[247] J. Viswanathan and I. E. Grossmann. A Combined Penalty Function and Outer-Approximation Method for MINLP Optimization. *Comp. Chem. Eng.*, 14(7):769–782, 1990.

[248] J. Vlassenbroeck and R. A. van Dooren. Chebyshev Technique for Solving Nonlinear Optimal Control Problems. *IEEE Trans. Automat. Contr.*, 33:333–340, 1988.

[249] H. P. Williams. *Model Building in Mathematical Programming*. John Wiley and Sons, Chichester, 3rd edition, 1993.

[250] L. A. Wolsey. *Integer Programming*. Wiley, New York, US, 1998.

[251] World Bank. The Agricultural Impact of the Kalabagh Dam. Technical Report 6884PAK, The World Bank, Washington, DC, 1986.

[252] H. J. Zimmermann. *Fuzzy Set Theory and its Applications*. Kluwer Academic Publishers, Boston, MA, 2nd edition, 1987.

[253] H. J. Zimmermann. *Fuzzy Sets, Decision Making, and Expert Systems*. Kluwer Academic Publishers, Boston, MA, 1987.

Index

Abbreviations, 24
Adjoint problem, 198
Adjoint variables, 198
Agent, **96**–97
AIMMS, xviii–xix, xxvii, 46, 54–55, 60, **71**–73, 75, 77–78, 81–82, 84–96, 99–104, 176, 241, 380
Airline crew scheduling, 271
Algebraic expression, 49
Algebraic Modeling Language, 24, 45–**46**, 46, 51
Algorithm, 27, 39
Algorithmic knowledge, 173
Algorithmic language, 48
All-different constraint, 57
AML see *Algebraic Modeling Language*, 24
AMPL Optimization LLC, 241
AMPL, 15, xviii–xix, 46, 49–50, 72, **105**–107, 109, 111–112, 114–116, 118–121, 123–130, 132–133, 135, 142–144, 187, 211, 241–242, 280, 369, 375, 379
ANALYZE, 55
AOA, 103
APEX, 152
API, 104, 145, 243, 281, 379
Application development, 220
Archimedian approach, 10
Architecture (modular), 212
ARKI Consulting, 261
Array
  dynamic, 216
  static, 216
ASCEND, 51
Assignment statements, 358
Automatic differentiation, 46, 51, 59, 188, 282, 351, **353**, 355–356, 363
  ADIFOR, 354
  code for derivatives, 354
  elementary functions, 354
  FORTRAN, 356
  forward mode, 355–356, 360
  Helmholtz energy function, 356, 359
  operator overloading, 354
  PCOMP, 356
  reverse mode, 355–356, 361
  TOMLAB, 375
  work ratio, 355–356
Automatic model documentation, 174, **177**, 180, 182
Automatic updates, 105
Backward mode, 351
BARON, 56, 144
Behavior of LP models, 293
Bilinear matrix inequalities, 371
Binary model file, 221
Binary variable, 57
Cardinality constraint, 57
CHIP, 212
COIN, 261, 278
Collocation, 203
Column generation, 116
COM, 102, 104, 380
Combinatorial optimization, 6, **123**, 307
Commutative operation, 236
Comparative operators, 359
Compatibility, 68
Complementarity problems, 6, **120**, 159
Complementarity, 105
Component library, 239
Conditional statements, 358
Connecting with solvers, 133
CONOPT, **261**, 103, 261, 371
Constraint Logic Programming, 47, 51, 60
Constraint Programming, 8, 24, 212, 307
Constraints, 3–4, 48, 57, 123
  disaggregated, 273
  discrete, 57
  types of, 57
  continuous, 57
  global, 57
Continuous constraint, 57
Continuous variable, 57
Control parameterization, 202
Control variables, 200, 202
Conventions, 24
CONVERT, 142
CP see *Constraint Programming*, 24
CPLEX, 103, 242, 250, 260–261, 268, 278, 297, 306, 371
Dash Optimization, **21**, 54, 212, 242, 380

Data access, 217
Data file, 217
Data fitting, 351, 353
Data input, 217
Data structures, 215
Data, 4, 30, 58
Database access, 105, 128
Database connection, 212, 239, 241, 258
Database connectivity, 104, 243
Decision support system, 64
Decision variable, 57
Declarative knowledge, 173
Declarative language, 45, 48
Deployment, 67, 71, 104, 239, 242, **245**, 262–263
Derived data, 46
Derived model, 61
Deterministic optimization, 13
Differential and algebraic equations, 185, 187, 195
Differential equations, 362
Directed rounding, 56
Disaggregation, 274
Discrete constraint, 57
Discrete variable, 57
Distributed model development, 68
Documentation, 8, 64–65, **178**, 182–183, 297, 382
Double contraction, 267
Dynamic library, 230
Dynamic optimization, 185
Dynamic shared object, 230
Dynamical systems, 351, 353, 361
EASY-FIT, 353, 361
ECLiPSe, 52
Embedding, 220, 242
End-user interface, 103
Enterprise Resource Planning, 24, 239
ERP see *Enterprise Resource Planning*, 24
Exclusion region, 57
Expected value, 15
EXTEND, 51
Flat model, 49
Flow control, 218
FortMP, 261
Forward accumulation, 355
Forward declaration, 220
Forward mode, 351, 355
Founders, 379–380
FrontLine Systems, 261
FrontLine, 261
Function identifier, 358
Function, 220
Functional language, 47
Future of modeling companies, 379
Future technical features, 379
Fuzzy set, 14
GAMS Applications
  coal and electricity market, 145
  radiosurgery / GAMMA Knife, 147
  scheduling at US Military Academy, 148
  water resources, 153
GAMS Development Corporation, 137, 241, 268
GAMS, xviii–xix, 45–46, 72, **137**–139, 141–147, 149–153, 187, 241–242, 280, 297, 369, 379
GDX, 145
Generalized Benders Decomposition, 190
Genetic Algorithms, 67
Global constraint, 57
Global optimization, 45, 56, 59, 159, 262, 279, 281
  black-box, 371
GLPK, 261
Gnu software, 261
Goals, 10
GPROMS, 51
Graphical environment, 212, 338
Graphical tools, 103
Graphical User Interface, 24, **60**, 338
GUI see *Graphical User Interface*, 24
Heuristics, 4, 149, 182, 211, 213, 222, 238
Hierarchical indexed sets, 174
Hierarchical modeling, 61
Histogram constraint, 57
History of modeling, 25, 37
Horn-rule, 47
HSLP, 306
ILOG Inc., 54, 242, 380
ILOG, 23, 242
Imperative language, 46
Independent infeasible sets, 65
Index sets, 49, 357
Index variable, 357
Index-based formulation, 49
Indexing, 244
Infeasibility tracing, 65
Infinite dimensional, 202
Integer Programming, 24, 159, 211
Integer variable, 57
Integration, 195
Internet optimization services, 129
Internet optimization services, 239
Interval arithmetic, 14, 56, 59
Interval data, 382
Intrinsic functions, 358
IP see *Integer Programming*, 24
Iteration, 105
Ketron Management, 261
Knowledge
  algorithmic, 173
  analytical, 279
  declarative, 173
  fuzzy, 14
  incomplete, 14
  partial, 14
  vague, 14
Language
  Constraint Logic Programming, 47

declarative, 48, 379
functional programming, 47
procedural, 379
construct, 218
declarative, 282
Logic Programming, 47
MATLAB, 370
Lexicographic Goal Programming, 10, 379
LGO, **262**, 144, 261–262
Library functions, 361
Lindo Systems, 242, 261, 380
LINDO, xxviii, 159, 242, 261
Linear Programming, 24, 48, 159, 211, 239
LINGO, xix–21, xxiii, xxviii, 46, 144, **159**–164, 166–171, 242, 379
Logic Programming language, 47
Logical
constraint, 57
modeling, 173
operators, 57
processing, 267–268, 270–271, 273
statements, 47
LOGMIP, 141
Loop, 219
LP see *Linear Programming*, 24
LPL, xix–xx, 22, 55, 65, **173**–174, 176–178, 180–183, 381
LPSolve, 261
LSGRG2, **262**, 261–262
Macro, 358
MaGen/OMNI, 152
Manage versions of models, 68
MapInfo, 145
Master problem, 191–192
MathCAD, xviii
Mathematica, xviii, 60
Mathematical model, 25, 29, 37, 48
Mathematical optimization, 3
Mathematical Programming, 105, 159, 211, 239, 307, 313
software, 211
MathML, 62
MATLAB, xviii, xx, 147, 288–289, 369–372, 375–376
ADMAT toolbox, 375
FEMLAB toolbox, 370
MAD toolbox, 375
Matrix generators, 18–19, 64, 352
FORTRAN, 20, 139, 152
Matrix operations, 13
Maximal Software, 54, **239**–241, 262, 380
MCP, 24, 141
Memory management, 66, 242, 244
Metaheuristics, 8, 65, 67, 223
MILP see *Mixed Integer Linear Programming*, 24
MINLP see *Mixed Integer Nonlinear Progr.*, 24
MINLP with differential constraints, 197
MINOPT, xviii–xx, 46, 144, **185**–195, 197–198, 202–205, 207, 209
MINOS, 103, 305–306, 371, 376
MIP see *Mixed Integer Programming*, 24
MIQP see Mixed Integer Quadratic Progr., 24
Mixed Complementarity Problem, 24
Mixed Integer Linear Programming, 24
Mixed Integer Nonlinear Programming, 24, 71, 185–186, 190
Mixed Integer Programming, 24, 267
Mixed Integer Quadratic Programming, 24
Mixed-integer optimal control problem, 200
Model documentation, 182
Model, 3, 25, 28, 37
Model-data separation, 105
Model-programming language, 55
Modelica, 51
Modeling diagram, 31, 35, 40
Modeling environment, 54
Modeling goal, 28
Modeling language, 45, 105, 159, **175**, 211, 351
algebraic, 46, 64
environment, 65
the future, 379
AIMMS, xviii–xix, xxvii, 46, 54–55, 60, 71–73, 75, 77–78, 81–82, 84–96, 99–104, 176, 241, 380
AMPL, 15, xviii–xix, 46, 49–50, 72, 106–107, 109, 111–112, 114–116, 118–121, 123–130, 132–133, 135, 142–144, 187, 211, 241–242, 280, 369, 375, 379
GAMS, xviii–xix, 21, 45–46, 72, 137–139, 141–147, 149–153, 187, 241–242, 280, 297, 369, 379
LINGO, xix–21, xxiii, xxviii, 46, 144, 160–164, 166–171, 242, 379
LPL, xix–xx, 22, 55, 65, 174, 176–178, 180–183, 381
MINOPT, xviii–xx, 46, 144, 185–195, 197–198, 202–205, 207, 209
Modelica, 51
Mosel, xviii–21, 23, 54–55, 176, 211–223, 226–227, 230–238, 242, 379, 381
mp-model, xviii, 21, 46, 213, 223, 242, 379, 381
MPL, xviii, xx, 23, 46, 54, 60, 212, 240–241, 245–253, 255, 258–264, 379
NOP-2, xix–xx, 23, xxvii, 46, 280–284, 286–288, 290–291
Numerica, 46, 59
OMNI, xix–xx, xxiv, 293–300, 303, 305–306
OPL Studio, xviii–xx, 23, 54, 212, 242, 379–380
PCOMP, xviii–xx, 351–353, 355, 357, 359–361, 367
SIF, 281
TOMLAB, xix–xx, 369–372, 374–376
Modeling software, 45, 63
Modeling statement, 212

Modeling systems, **xviii**, 45, 239, 379
  Easy Modeler, 277
  GIANO, 277
Modeling tool, 55
Modeling, 25, 31, 37, 45, 173, 239, 293, 307
  object oriented, 61
Modularity, xx
Modules see *Mosel,modules*, 222
Mosek, 103
Mosel, xviii–21, 23, 54–55, 176, **211**–223, 226–227, 230–238, 242, 379, 381
  language, 213
  libraries, 220
  module context, 234
  modules, 212, 222
  Native Interface, 213
  Solver Module, 212
  subroutine definition, 231
  types, 215
  User Module, 213, 230
Mp-model, xviii, 21, 46, 213, 223, 242, 379, 381
MPEC, 141
MPL, xviii, xx, 23, 46, 54, 60, 212, **239**–241, 245–253, 255, 258–264, 379
MPROBE, 55
MPS, 281
MPSGE, 141
MPSIII, 152
MPSX, xix–xx, 152, **267**–269, 271–272, 275–278, 295, 305
Multi-agent system, 71
Multi-criteria problems, 9
Multiple platforms, 243
Multiple solvers, 211
Negation, 237
NEOS, 129–130, 132, 142
NLP see *Nonlinear Programming*, 24
NLP with differential constraints, 195
Non-algebraic modeling language, 51
Nonlinear Programming, 24, 159, 211, 351, 353
Nonlinearity, 105
NOP, 46, 58, 280
NOP-2, xix–xx, 23, xxvii, 46, 58, **279**–284, 286–288, 290–291
Numerica, 46, 58–59
Numerical differentiation TOMLAB, 375
Object oriented, 47
Object-oriented modeling language, 51
Objective function, 3–4, 282
ODBC, **46**, 24, 51, 60, 94, 104, 129, 212, 243, 258
OML, 261
OMNI, xix–xx, xxiv, 152, **293**–300, 303, 305–306
Open design, 243
Operations Research, 24, 175
Operator, 219
  commutative, 236
  deduction, 236
OPL Studio, xviii–xx, 23, 54, 212, 242, 287, **307**, 379–380
Optimal control, 185, 200
Optimal Methods, 262
OptiMax 2000, 239, 241, 260, 262, 264
Optimization problem, 30, 48
Optimization, 3, 105, 159, 211, 239
  mathematical, 64
  multi-criteria, 67, 379
  combinatorial, 123
  deterministic, 13
  global, 14, 65, 262, 374, 382
  multi-stage stochastic, 14
  portfolio, 372
  robust, 13, 67
  stochastic integer, 65
  stochastic, 15, 67, 141
  under uncertainty, 13, 63, 65, 380, 382
OptiRisk Systems, 241, 261
OSL, 152, 261, 267–268, 274, 277–278
  user interface, 277–278
Outer Approximation, 71, 192
  Equality Relaxation (Augmented Penalty), 192
Overloading, **220**, 354
  operator, 234, 236, 354
Paragon Decision Technology, 22, 54, 241, 380
Parameter, 49, 357
Pareto optimal, 9
Parser, 188, 280, 352, 360–361, 367
Partial integer variable, 57
PATH, 103
PCOMP, xviii–xx, **351**–**356**, 356–361, 367
  code interpretation, 360
  EASY-FIT, 361
  external functions, 361
  integer constants, 357
  parser, 360
  program organization, 360
  real constants, 357
  table, 357
PCx, 261
Piecewise constant interpolation, 357
Piecewise linear interpolation, 357
Pinter Consultancy Service, 262
Practitioners, 63, 371, 382
Preconditioning of models, 267
Preprocessing, 33
Presolve, 33, 49
Primal problem, 190
Priorities, 10, 40, 133, 153, 271
Problem instance, 49, 59
Procedural language, 46
Procedural statement, 50
Procedure, 220
Product over index set, 358
Programming language, 211, 239
Programming

Cone, 382
Constraint, 382
Disjunctive, 141
Mathematical, 382
Semi-Definite, 371
Semi-Infinite, 11–12
PROSE, 139
QP see *Quadratic Programming*, 24
Quadratic Programming, 24, 224
Real-world object, 25, 28
Real-world problems, 8, 27, 30–31, 63, 73, 84
Refinery planning & scheduling, 293
Relation, 30
Report generators, 18–19, 243
Reverse accumulation, 355
Robustness, 66, 242, **245**, 260
Round-off, 58
Safety Programming, 12
Scalability, 66, 242, 244, 252
Scientific community, 381
SCM see *Supply-Chain Management*, 24
Scope of application, 28
Selection statement, 218
Semi-continuous variable, 57
Semi-definite Programming, 371
Sequential Linear Programming, 212
Set covering, 271
Set manipulation, 105
Set operation, 219
Sets, 105
constant, 216
dynamic, 216
SIF, 281
Simulated Annealing, 67
Simulation, 3–4, 6, 15
SNOPT, 103, 371
Soft constraint, 58
Solution algorithm, 225
Solver view, 60
Solvers, 46, 371
AOA, 103
BARON, 56, 144
COIN, 261, 278
CONOPT, 103, 261, 371
CPLEX, 103, 242, 250, 260–261, 268, 278, 297, 306, 371
FortMP, 261
FrontLine, 261
GLPK, 261
HSLP, 306
LGO, 144, 261–262
LINDO, xxviii, 159, 242, 261
LPSolve, 261
LSGRG2, 261–262
MATLAB, xviii, xx, 369–372, 375–376
MINOS, 103, 305–306, 371, 376
Mosek, 103
MPSX, xix–xx, 267–269, 271–272, 275–278, 295, 305
OML, 261
OSL, 261, 267–268, 274, 277–278
PATH, 103
PCx, 261
SNOPT, 103, 371
XA, 103, 261
Xpress-Optimizer, 103, 212, 215, 242, 260–261, 268, 306, 371, 379
Solving statement, 212
Sparsity, 39, 59, 73, 102, 105, 128, 137, 139, 150, 152, 166, 232, 241–242, 244, 246, **251**, 288–289, 309, 316, 370–372
Special Ordered Sets, 57, 65
Spline interpolation, 357
Spreadsheet modeling system, 54
Spreadsheets, 159
SQL, 212, 223, 258–259
Stochastic Programming, **15**, **126**, 261
Structure, 32
Subroutine libraries
EISPACK, 370
LAPACK, 370
LINPACK, 370
Subroutine, 220
Substraction, 237
Subtour elimination, 225
Sum over index set, 358
Supply-Chain Management, 24, 239
Tabu Search, 67
Targets, 10
TOMLAB, xix–xx, 369–372, 374–376
Toy problem, 46
Traveling Salesman Problem, 225
Type conversion, 236
Type definition, 233
Uncertain data, 58
Unit consistency checking, 103
Unit-based scaling, 71, 103
Units of measurement, 103, 174
Variables, 3–4, 26, 30, 49, 57, 357
binary, 65
integer, 65
partial-integer, 65
random, 15
semi-continuous, 65
stochastic, 15
Vector minimization, 9
VEDA, 147
Version management, 68
Visual model, 27
World-wide-web, 381
XA, 103, 261
XML, 62, 102, 104, 183
Xpress-IVE, 212, 215
Xpress-Optimizer, 103, 212, 215, 242, 260–261, 268, 306, 371, 379